I0036771

The Biology and Treatment of Myeloid Leukaemias

Special Issue Editors

Geoffrey Brown
Ewa Marcinkowska

MDPI • Basel • Beijing • Wuhan • Barcelona • Belgrade

MDPI

Special Issue Editors
Geoffrey Brown
University of Birmingham
UK

Ewa Marcinkowska
University of Wroclaw
Poland

Editorial Office
MDPI AG
St. Alban-Anlage 66
Basel, Switzerland

This edition is a reprint of the Special Issue published online in the open access journal *IJMS* (ISSN 1422-0067) from 2016–2017 (available at: http://www.mdpi.com/journal/ijms/special_issues/ biology_treatment_myeloid_leukaemias).

For citation purposes, cite each article independently as indicated on the article page online and as indicated below:

Lastname, F.M.; Lastname, F.M. Article title. *Journal Name* **Year,** *Article number*, page range.

First Edition 2018

ISBN 978-3-03842-795-7 (Pbk)
ISBN 978-3-03842-796-4 (PDF)

Cover photo courtesy of Geoffrey Brown and Ewa Marcinkowska.

Articles in this volume are Open Access and distributed under the Creative Commons Attribution (CC BY) license, which allows users to download, copy and build upon published articles even for commercial purposes, as long as the author and publisher are properly credited, which ensures maximum dissemination and a wider impact of our publications. The book taken as a whole is © 2018 MDPI, Basel, Switzerland, distributed under the terms and conditions of the Creative Commons license CC BY-NC-ND (http://creativecommons.org/licenses/by-nc-nd/4.0/).

Table of Contents

About the Special Issue Editors

Geoffrey Brown is the Reader in Cellular Immunology at the College of Medical and Dental Sciences at the University of Birmingham, UK. He received a BSc in microbiology from Queen Elizabeth College, University of London and a Ph.D. in tumour biology from University College, London. Postdoctoral research was undertaken at the MRC Immunochemistry Unit, Oxford and the Nuffield Department of Clinical Medicine, Oxford. At Oxford, Geoffrey was also the IBM Fellow, University of Oxford and Research Lecturer of the House, Christ Church College. His early work described human B and T lymphocyte antigens leading to the discovery of the common acute lymphoblastic leukaemia antigen (CD10). This led to designation of the childhood leukaemia as common acute lymphoblastic leukaemia (cALL). His research for many years has concerned the development of blood cells. Geoffrey has proposed the pair-wise model for blood cell development and was the Director of the EU-funded Marie Curie FP7 Initial Training Network and consortium DECIDE (Decision-making within cells and differentiation entity therapies).

Ewa Marcinkowska is a Professor in Cell Biology at the Faculty of Biotechnology at the University of Wrocaw, Poland. She received an MD degree from Wrocaw Medical University in 1991. After obtaining a Ph.D. from the Ludwik Hirszfeld Institute of Immunology and Experimental Therapy in Wrocaw in 1998, she undertook brief postdoctoral training at the Norwegian Radium Hospital in Oslo. In 2003, she was appointed as Assistant Professor at the Faculty of Biotechnology at the University of Wrocaw, and in 2012 she was granted the title of Professor. She supervised four Ph.D. and 12 Master degree students. Her research concentrates on intracellular mechanisms of cell differentiation induced by 1,25-dihydroxyvitamin D. In 1997, Ewa and her colleagues evidenced that 1,25-dihydroxyvitamin D activates MAP kinase pathways in human leukemia cells. Later, she concentrated on vitamin D receptor transcriptional activities which are necessary for cell differentiation and on the crosstalk between vitamin D and retinoic acid receptors. She was Partner and Meeting Coordinator of the EU-funded Marie Curie FP7 Initial Training Network and consortium DECIDE (Decision-making within cells and differentiation entity therapies).

Preface to "The Biology and Treatment of Myeloid Leukaemias"

Acute myeloid leukaemia (AML) is the most prevalent acute haematological malignancy and around 90% of patients aged $\geq$60 years, and more than half aged <60 years die from the disease. There is a pressing need for new, tolerable and effective therapies, particularly for elderly AML patients. Patients co-morbidities preclude aggressive chemotherapy for older people. The survival of patients who are unable to tolerate chemotherapy is dismal, at only 5 to 19 months, and complete remission rates for relapsed patients are <25%. This International Journal of Molecular Sciences (IJMS) Special Issue deals with some of the new potential therapeutic targets and strategies, including vitamin D and all-trans retinoic acid differentiation therapy. This issue also examines the outcomes from the use of current treatments for acute myeloid leukaemia.

Geoffrey Brown, Ewa Marcinkowska
Special Issue Editors

International Journal of
Molecular Sciences

MDPI

Editorial

Acute Myeloid Leukaemia: New Targets and Therapies

Geoffrey Brown [1,2,*] **and Ewa Marcinkowska** [3]

1   Institute of Clinical Science, College of Medical and Dental Sciences, University of Birmingham, Edgbaston, Birmingham B15 2TT, UK
2   Institute of Immunology and Immunotherapy, College of Medical and Dental Sciences, University of Birmingham, Edgbaston, Birmingham B15 2TT, UK
3   Laboratory of Protein Biochemistry, Faculty of Biotechnology, University of Wroclaw, Joliot-Curie 14a, 50-383 Wroclaw, Poland; ema@cs.uni.wroc.pl
*   Correspondence: g.brown@bham.ac.uk; Tel.: +44-0121-414-4082

Received: 30 October 2017; Accepted: 29 November 2017; Published: 30 November 2017

The most common acute hematological malignancy in adults is acute myeloid leukaemia (AML), accounting for more than 80% of cases in patients over 60 years of age [1]. AML is the second most common acute leukaemia in children, accounting for 15–20% of leukaemia cases [2,3]. Morphological classification of AML into eight sub-types (FAB M0–M7) based on the type of cell from which the leukemia developed, and on its degree of maturity, was in use until the end of the last century [4]. A new World Health Organization (WHO) classification introduced four main AML groups, based on cytogenetic abnormalities, and is important to the approach used to treat this disease [5]. In particular, acute promyelocytic leukaemia (APL), a distinct M3 subtype of AML, was once an incurable disease and the use of all-*trans* retinoic acid (ATRA)-based differentiation therapy and anthracycline-based chemotherapy now provides a cure rate above 80% (reviewed in [6]). However, aside from APL, around 90% of older patients (aged $\geq$ 60 years) and more than half of young adult patients (aged < 60 years) die from their disease [7], and AML accounts for the largest number of annual deaths due to leukaemias in the US [2]. The survival of older patients who are unable to tolerate chemotherapy is dismal, at only five to 10 months [8], and complete remission rates for relapsed patients are <25% [9,10].

The Fms-like tyrosine kinase 3 (FLT3), a class III receptor tyrosine kinase, and FLT3 ligand (FL) play an important role in the proliferation, differentiation, and survival of hematopoietic cells [11]. Mutated FLT3 is expressed in a subset of AML patients and confers a poor prognosis. The most common mutation, which occurs in about 25% of AML patients, is an internal tandem duplication (ITD) in the juxtamembrane region of FLT3, which causes ligand-independent dimerization and constitutive receptor activation [12]. Point mutations in the tyrosine kinase domain (TKD), FLT3-TKD, occur in approximately 7% of AMLs [13]. FLT3 tyrosine kinase inhibitors are well tolerated, as a monotherapy and with intensive chemotherapy, and second generation inhibitors have shown significant promise as a treatment for relapsed and refractory AML. The paper by Mooney and colleagues [14] and the review by Tsapogas and colleagues [15] advance the knowledge of FLT3 and FL. Mooney and colleagues have examined the expression of mRNA FLT3 (mRNA and cell surface protein) by hematopoietic stem cells (HSC) and various progenitor cells (HPC). A sub-population of short-term and long-term HSC express FLT3. As expected, expression by HPC was observed for these cells with lymphoid, granulocytic, and myelomomocytic potential. Regarding FLT3 expression by HSC, Tsapogas provides evidence to support an instructive role of FL signalling at early stages of hematopoiesis, in addition to a role in promoting cell survival and proliferation. The precise pattern of expression of FLT3 by HSC/HPC and the roles of FL in normal hematopoiesis are critical to a better understanding of AML subtypes, the process of disease progression, as well as for the development of therapeutic strategies to target FLT3-mutated AML.

Regarding other kinase inhibitors, T315, an inhibitor of integrin-linked kinase, has been shown to suppress the proliferation of breast and stomach cancer cells and chronic lymphocytic leukaemia cells.

Chiu and collegues [16] report that this agent is cytotoxic against the AML cell lines HL-60 and THP-1 and primary leukaemia cells from AML patients. T315 also suppresses the growth of THP-1 xenografts. Chiu and collegues decribe aspects of the mechanism of action of T315, in driving apoptosis and autophagic cell death, and suggest further assessment of the use in treating AML and other leukaemias. However, there is the need to modify T315 to increase efficacy and reduce toxicity. Apoptosis can also be induced in human leukaemia cells by the curcumin analogue EF-24. In their study, Skoupa and colleagues [17] investigate the mechanism by which EF-24 induces cell death in K562 cells. They propose that EF-24 may be suitable as an anticancer agent, specifically in cases of drug resistance.

The oncogenic or chromosomal mutations that are present in a patient's AML cells at diagnosis are important to personalized treatment. The best known is a balanced translocation between chromosomes 15 and 17 [t(15;17)(q22;q21)] resulting in the formation of promyelocytic leukemia (PML) and retinoic acid receptor alpha (RARα) fusion protein [18]. Leukaemic blasts which carry this mutation are susceptible to ATRA-induced cell differentiation. Watts and collegues [19] report a salutary lesson from the treatment outcome of a case of AML. The case is characterised by a new t(4;15)(q31;q22) translocation, resulting in the expression of the TMEM154-RASGRF1 fusion protein. The patient who was treated with ATRA as a part of a clinical study died from rapid disease progression. An increase in the expression of RARγ was observed upon treatment of the patient's cells ex vivo with ATRA, and they proliferated in response to ATRA and a RARγ agonist. The disease progression could be related to an increase in RARγ, which plays a role in hematopoietic stem cell self-renewal and proliferation. Furthermore, there are implications for the use of retinoid-based differentiation therapy for certain cases.

In addition to ATRA, the differentiating agent 1,25-dihydroxyvitamin D (1,25D) has been reported to have beneficial effects in combination therapy for cancer. The paper by Janik and colleagues [20] examines the regulation of the vitamin D receptor (*VDR*) gene by 1,25D and ATRA in blood cells at early stages of their differentiation. ATRA, but not 1,25D, upregulates the expression of *VDR* in human early-stage blood cells. As to early-stage mouse cells, *VDR* is upregulated by 1,25D, but not by ATRA. Hence, *VDR* regulation in humans and mice is different, which is germane to testing combinations of agents for use in differentiation therapy. The findings also bring to attention that the level of expression of VDR protein is low in patients' AML blasts that do not respond to 1,25D. The level can be upregulated by ATRA treatment and this is relevant to the combined therapeutic use of ATRA and 1,25D.

There is an urgent need for new treatments for AML, which will include the need to identify new molecules to target. A number of recurrent mutations in AML involve genes concerned with regulating the epigenetic landscape [21]. Gain or loss of function of the gene encoding the EZH2 methyltransferase occurs in various malignancies. Sbirkov and colleagues [22] describe the use of affinity-purification mass spectrometry to identify new partners of EZH2 and potential new non-histone substrates. Of particular note is that EZHZ has a role in regulating translation, via interacting with RNA binding proteins and methylating eEF1A1. eEF1A1 is a component of protein synthesis, highly expressed in tumours, and therapeutic targeting is a possibility in AML. The review by Gbolahan and colleagues [23] focusses on the benefit of immunotherapeutic interventions to the long-term control of AML, including the use of hypomethylating agents. The results from early phase immunotherapeutic interventions, for example, the use of a monoclonal antibody to CD33 which is highly expressed on AML blasts, are encouraging. The interleukin-3 receptor α chain (CD123), FcγRI (CD64) and C-type lectin-like molecule 1 are also promising targets. The review considers the use of hypomethylating agents to increase the antigenicity of AML cells and their role as immunomodulatory drugs.

The articles by Player et al., Khan et al., and Almeida et al. describe the outcomes from clinical studies. Player and colleagues [24] examine azacitidine for front-line therapy of patients with AML, by reference to international phase 3 trial data and data from the Austrian Azacytidine registry. The authors report clinically-meaningful improvement in overall survival for patients treated with front-line azacytidine versus conventional regimens, and conlude that azacytidine appears efficacious as a front-line treatment for WHO-AML patients. Khan and colleagues [25] examine the clinical outcomes in patients with *RUNX1*-mutated AML. The response of older patients to treatment with hypomethylating agents and survival were found to be independent of *RUNX1* status, leading the authors to conclude that future studies should focus on the prognostic value of *RUNX1* mutations relative to co-occuring mutations. Almeida and colleages [26] examine the outcomes from the various types of treatment for acute erythroleukemia (AEL). AEL typically has a poor prognosis. From a comparison of patients treated with hypomethylating agents, front-line, and intensive chemotherapy, the results provide support to the use of hypomethylating agents to treat AEL. However, high-risk patients treated first-line with hypomethylating agents live just a few months longer (13.3 versus 7.5 for patients treated with intensive chemotherapy).

In summary, there is quite some way to go to extend the success in treating APL to other types of AML. As mentioned above, 90% of older patients die from their disease. In populations of developed countries, cancers are turning into prevalent diseases of the aged. According to epidemiological data, about 80% of cancers are diagnosed in patients older than 55, with the median age at diagnosis above 60 years [27]. This is also the case for AML for which the median age at diagnosis is 67 years [28]. Many older AML patients are not able to receive intensive chemotherapy. Hence, there is the need for new approaches to the treatment of AML, and other malignancies, in elderly patients. New therapeutic targets are important, as is consideration of differentiation therapy, including agents that modify the epigenetic status of cells, combined with gentler chemotherapy. A means of preventing AML relapse and holding in check for life will provide a better quality of life to elderly patients.

Acknowledgments: This project has received funding from the European Union's Seventh Framework Programme for research, technological development and demonstration under grant agreement No. 315902. Geoffrey Brown and Ewa Marcinkowska are Partners within the Marie Curie Initial Training Network DECIDE (Decision-making within cells and differentiation entity therapies).

Author Contributions: Both authors wrote the text, revised the manuscript and gave final approval.

Conflicts of Interest: The authors declare no conflict of interest.

References

1. Pollyea, D.; Kohrt, H.; Medeiros, B. Acute myeloid leukaemia in the elderly: A review. *Br. J. Haematol.* **2011**, *152*, 524–542. [CrossRef] [PubMed]
2. O'Donnell, M.; Tallman, M.; Abboud, C.; Altman, J.; Appelbaum, F.; Arber, D.; Attar, E.; Borate, U.; Coutre, S.; Damon, L.; et al. National Comprehensive Cancer, N., Acute myeloid leukemia, version 2.2013. *J. Natl. Compr. Cancer Netw.* **2013**, *11*, 1047–1055. [CrossRef]
3. Rubnitz, J. How I treat pediatric acute myeloid leukemia. *Blood* **2012**, *119*, 5980–5988. [CrossRef] [PubMed]
4. Bennett, J.; Catovsky, D.; Daniel, M.; Flandrin, G.; Galton, D.; Gralnick, H.; Sultan, C. Proposals for the classification of the acute leukaemias. French-American-British (FAB) co-operative group. *Br. J. Haematol.* **1976**, *33*, 451–458. [CrossRef] [PubMed]
5. Vardiman, J.; Harris, N.; Brunning, R. The World Health Organization (WHO) classification of the myeloid neoplasms. *Blood* **2002**, *100*, 2292–2302. [CrossRef] [PubMed]
6. Lo-Coco, F.; Cicconi, L.; Breccia, M. Current standard treatment of adult acute promyelocytic leukaemia. *Br. J. Haematol.* **2016**, *172*, 841–854. [CrossRef] [PubMed]
7. Ferrara, F.; Schiffer, C. Acute myeloid leukaemia in adults. *Lancet* **2013**, *381*, 484–495. [CrossRef]
8. Dohner, H.; Weisdorf, D.; Bloomfield, C. Acute Myeloid Leukemia. *N. Engl. J. Med.* **2015**, *373*, 1136–1152. [CrossRef] [PubMed]
9. Leopold, L.; Willemze, R. The treatment of acute myeloid leukemia in first relapse: A comprehensive review of the literature. *Leuk. Lymphoma* **2002**, *43*, 1715–1727. [CrossRef] [PubMed]
10. Greenberg, P.; Lee, S.; Advani, R.; Tallman, M.; Sikic, B.; Letendre, L.; Dugan, K.; Lum, B.; Chin, D.; Dewald, G.; et al. Mitoxantrone, etoposide, and cytarabine with or without valspodar in patients with relapsed or refractory acute myeloid leukemia and high-risk myelodysplastic syndrome: A phase III trial (E2995). *J. Clin. Oncol.* **2004**, *22*, 1078–1086. [CrossRef] [PubMed]
11. Rosnet, O.; Schiff, C.; Pébusque, M.; Marchetto, S.; Tonnelle, C.; Toiron, Y.; Birg, F.; Birnbaum, D. Human FLT3/FLK2 gene: CDNA cloning and expression in hematopoietic cells. *Blood* **1993**, *82*, 1110–1119. [PubMed]

12. Yokota, S.; Kiyoi, H.; Nakao, M.; Iwai, T.; Misawa, S.; Okuda, T.; Sonoda, Y.; Abe, T.; Kahsima, K.; Matsuo, Y.; et al. Internal tandem duplication of the FLT3 gene is preferentially seen in acute myeloid leukemia and myelodysplastic syndrome among various hematological malignancies. A study on a large series of patients and cell lines. *Leukemia* **1997**, *11*, 1605–1609. [CrossRef] [PubMed]

13. Callens, C.; Chevret, S.; Cayuela, J.; Cassinat, B.; Raffoux, E.; de Botton, S.; Thomas, X.; Guerc, I.A.; Fegueux, N.; Pigneux, A.; et al. Prognostic implication of FLT3 and Ras gene mutations in patients with acute promyelocytic leukemia (APL): A retrospective study from the European APL Group. *Leukemia* **2005**, *19*, 1153–1160. [CrossRef] [PubMed]

14. Mooney, C.; Cunningham, A.; Tsapogas, P.; Toellner, K.-M.; Brown, G. Selective Expression of Flt3 within the mouse hematopoietic stem cell compartment. *Int. J. Mol. Sci.* **2017**, *18*, 1037. [CrossRef] [PubMed]

15. Tsapogas, P.; Mooney, C.; Brown, G.; Rolink, A. The cytokine Flt3-ligand in normal and malignant hematopoiesis. *Int. J. Mol. Sci.* **2017**, *18*, 1115. [CrossRef] [PubMed]

16. Chiu, C.-F.; Weng, J.-R.; Jadhav, A.; Wu, C.-Y.; Sargeant, A.; Bai, L.-Y. T315 decreases acute myeloid leukemia cell viability through a combination of apoptosis induction and autophagic cell death. *Int. J. Mol. Sci.* **2016**, *17*, 1337. [CrossRef] [PubMed]

17. Skoupa, N.; Dolezel, P.; Ruzickova, E.; Mlejnek, P. Apoptosis induced by curcumin analogue EF-24 is neither mediated by oxidative stress related mechanism nor affected by expression of main drug transporters ABCB1 and ABCG2 in human leukemia cells. *Int. J. Mol. Sci.* **2017**, *18*, 2289. [CrossRef] [PubMed]

18. Kakizuka, A.; Miller, W.J.; Umesono, K.; Warrell, R.J.; Frankel, S.; Murty, V.; Dmitrovsky, E.; Evans, R. Chromosomal translocation t(15;17) in human acute promyelocytic leukemia fuses RAR alpha with a novel putative transcription factor, PML. *Cell* **1991**, *66*, 663–674. [CrossRef]

19. Watts, J.; Perez, A.; Pereira, L.; Fan, Y.-S.; Brown, G.; Vega, F.; Petrie, K.; Swords, R.; Zelent, A. A case of AML characterized by a novel t(4;15)(q31;q22) translocation that confers a growth-stimulatory response to retinoid-based therapy. *Int. J. Mol. Sci.* **2017**, *18*, 1492. [CrossRef] [PubMed]

20. Janik, S.; Nowak, U.; Łaszkiewicz, A.; Satyr, A.; Majkowski, M.; Marchwicka, A.; Śnieżewski, Ł.; Berkowska, K.; Gabryś, M.; Cebrat, M.; et al. Diverse REgulation of vitamin d receptor gene expression by 1,25-dihydroxyvitamin D and atra in murine and human blood cells at early stages of their differentiation. *Int. J. Mol. Sci.* **2017**, *18*, 1323. [CrossRef] [PubMed]

21. Conway O'Brien, E.; Prideaux, S.; Chevassut, T. The epigenetic landscape of acute myeloid leukemia. *Adv. Hematol.* **2014**, *2014*, 103175. [CrossRef] [PubMed]

22. Sbirkov, Y.; Kwok, C.; Bhamra, A.; Thompson, A.; Gil, V.; Zelent, A.; Petrie, K. Semi-quantitative mass spectrometry in AML cells identifies new non-genomic targets of the EZH2 methyltransferase. *Int. J. Mol. Sci.* **2017**, *18*, 1440. [CrossRef] [PubMed]

23. Gbolahan, O.; Zeidan, A.; Stahl, M.; Abu Zaid, M.; Farag, S.; Paczesny, S.; Konig, H. Immunotherapeutic concepts to target acute myeloid leukemia: Focusin on the role of monoclonal antibodies, hypomethylating agents and the leukemic microenvironment. *Int. J. Mol. Sci.* **2017**, *18*, 1660. [CrossRef] [PubMed]

24. Pleyer, L.; Döhner, H.; Dombret, H.; Seymour, J.; Schuh, A.; Beach, C.; Swern, A.; Burgstaller, S.; Stauder, R.; Girschikofsky, M.; et al. Azacitidine for front-line therapy of patients with AML: Reproducible efficacy established by direct comparison of international phase 3 trial data with registry data from the Austrian Azacitidine Registry of the AGMT study group. *Int. J. Mol. Sci.* **2017**, *18*, 415. [CrossRef] [PubMed]

25. Khan, M.; Cortes, J.; Kadia, T.; Naqvi, K.; Brandt, M.; Pierce, S.; Patel, K.; Borthakur, G.; Ravandi, F.; Konopleva, M.; et al. Clinical outcomes and co-occurring mutations in patients with RUNX1-mutated acute myeloid leukemia. *Int. J. Mol. Sci.* **2017**, *18*, 1618. [CrossRef] [PubMed]

26. Almeida, A.; Prebet, T.; Itzykson, R.; Ramos, F.; Al-Ali, H.; Shammo, J.; Pinto, R.; Maurillo, L.; Wetzel, J.; Musto, P.; et al. Clinical outcomes of 217 patients with acute erythroleukemia according to treatment type and line: A retrospective multinational study. *Int. J. Mol. Sci.* **2017**, *18*, 837. [CrossRef] [PubMed]

27. Marosi, C.; Köller, M. Challenge of cancer in the elderly. *ESMO Open* **2016**, *1*. [CrossRef] [PubMed]

28. Almeida, A.; Ramos, F. Acute myeloid leukemia in the older adults. *Leuk. Res. Rep.* **2016**, *6*, 1–7. [CrossRef] [PubMed]

© 2017 by the authors. Licensee MDPI, Basel, Switzerland. This article is an open access article distributed under the terms and conditions of the Creative Commons Attribution (CC BY) license (http://creativecommons.org/licenses/by/4.0/).

International Journal of
Molecular Sciences

MDPI

Article

Semi-Quantitative Mass Spectrometry in AML Cells Identifies New Non-Genomic Targets of the EZH2 Methyltransferase

Yordan Sbirkov [1], Colin Kwok [2], Amandeep Bhamra [3], Andrew J. Thompson [4],
Veronica Gil [2], Arthur Zelent [5] and Kevin Petrie [6,*]

[1] Theodor Boveri Institute and Comprehensive Cancer Center Mainfranken, Biocenter,
    University of Würzburg, Würzburg 97074, Germany; yordan.sbirkov@uni-wuerzburg.de
[2] Division of Clinical Studies, Institute of Cancer Research, London SW7 3RP, UK; colin.kwok@icr.ac.uk (C.K.);
    veronica.gil@icr.ac.uk (V.G.)
[3] UCL Cancer Institute, Paul O'Gormon Building, London WC1E 6DD, UK; a.bhamra@ucl.ac.uk
[4] Proteomics and Metabolomics Core Facility, Institute of Cancer Research, London SW7 3RP, UK;
    a.thompson@invivaconsulting.com
[5] Miller School of Medicine, University of Miami, Miami, FL 33136, USA; a.zelent@med.miami.edu
[6] Faculty of Natural Sciences, University of Stirling, Stirling FK9 4LA, UK
*   Correspondence: kevin.petrie@stir.ac.uk; Tel.: +44-7809-593990

Received: 10 June 2017; Accepted: 29 June 2017; Published: 5 July 2017

Abstract: Alterations to the gene encoding the EZH2 (KMT6A) methyltransferase, including both gain-of-function and loss-of-function, have been linked to a variety of haematological malignancies and solid tumours, suggesting a complex, context-dependent role of this methyltransferase. The successful implementation of molecularly targeted therapies against EZH2 requires a greater understanding of the potential mechanisms by which EZH2 contributes to cancer. One aspect of this effort is the mapping of EZH2 partner proteins and cellular targets. To this end we performed affinity-purification mass spectrometry in the FAB-M2 HL-60 acute myeloid leukaemia (AML) cell line before and after all-*trans* retinoic acid-induced differentiation. These studies identified new EZH2 interaction partners and potential non-histone substrates for EZH2-mediated methylation. Our results suggest that EZH2 is involved in the regulation of translation through interactions with a number of RNA binding proteins and by methylating key components of protein synthesis such as eEF1A1. Given that deregulated mRNA translation is a frequent feature of cancer and that eEF1A1 is highly expressed in many human tumours, these findings present new possibilities for the therapeutic targeting of EZH2 in AML.

Keywords: acute myeloid leukaemia; EZH2; mass spectrometry; methylation; eEF1A1

1. Introduction

Acute myeloid leukaemia (AML) is the most commonly occurring acute haematological malignancy in adults, representing more than 80% of all leukaemias in patients over 60 years of age [1]. AML also accounts for 15–20% of childhood leukaemia cases, making it the second most common acute leukaemia in children [2–4]. Despite advances in diagnosis, stratification and treatment, the disease remains largely incurable (in 60–65% of patients <60 years and 85–95% of patients >60 years [5]), and overall 5-year survival rates remain poor at only 25% [4,6]. Furthermore, treatment outcomes for relapsed patients are low, with complete remission rates under 25% [7,8]. New treatment options are therefore urgently required.

Epigenetic events play a central role in normal development and differentiation, and it is unsurprising that mutation and/or deregulation of DNA and histone modifiers is a frequent event in cancer [9]. Epigenetic enzymes that have been implicated in the promotion of haematological malignancies include *MLL*, translocations of which occur in approximately 80% of infant leukaemias and 5–10% of adult AML [10], resulting in gain-of-function mutations of the encoded H3K4 methyltransferase. Other recently described examples include mutation of DNMT3A (which occurs in approximately 20% of AML patients) [11], aberrant expression of the LSD1 (KDM1A) demethylase (which is strongly implicated in AML) [12–14], and overexpression of histone deacetylase 9 (HDAC9), which is associated with leukaemia and lymphoma [15,16]. Perhaps the most widely-studied epigenetic modifiers, however, are the Polycomb group of proteins (PcG), which form two distinct multiprotein repressive complexes, PRC1 and PRC2. PRC1 is required for the ubiquitination of histone H2A lysine 119, which proceeds via Ring1a or Ring1b E3 ligases. The core PRC1 subunit BMI1 (also known as PCGF4 or RNF51) is required for cancer stem cell maintenance [17] and its overexpression has been implicated in leukaemia and lymphoma. The canonical function of PRC2 is to catalyse the methylation of histone H3 lysine 27 (H3K27) and contains either the EZH1 or EZH2 methyltransferases. EZH2 has been the subject of intense research in recent years due to its role in a wide range of cancers and to its potential as a therapeutic target [18].

In addition to being overexpressed in a number of solid tumours, EZH2 is also frequently mutated in haematological malignances. In contrast to the selection for activating point mutations (Y641 and A677) in the SET (Suppressor of variegation, Enhancer of Zeste, Trithorax) domain of EZH2 described in B-cell malignancies [19–21], a range of loss-of-function aberrations (including in other PRC2 members) have been found in 25% of T-cell acute lymphoblastic leukaemia (ALL) cases, 3% of primary AML, 29% of secondary AML, and 15% of myeloproliferative disorders [22–24]. Importantly, all these perturbations have been shown to lead to poor prognosis and diminished overall survival [23,25,26]. In addition to mutation of EZH2 in AML, evidence strongly suggests its aberrant expression can promote self-renewal of leukaemic stem cells and block differentiation [27–29]. Deregulation of EZH2/PRC2 function can therefore occur in a number of ways, pointing to a complex role that may be dependent on the cell type and on the stage of hematopoietic development at which mutations or deregulation of expression occur.

In order to improve both our understanding of the dynamic functions of EZH2 as well as its potential as a therapeutic target, it is important to identify proteins that interact with the core PRC2 complex, as well as non-histone enzymatic substrates. Here, it is important to note that none of the four core PRC2 components (EZH2, SUZ12, EED, and AEBP1/2) has a DNA binding domain and that the PRC2 complex depends on factors such as JARID2, YY1, or AEBP2 [30], and/or non-coding RNAs such as *XIST* and *HOTAIR* to be recruited to specific loci [31]. Non-canonical EZH2-containing PRC2 activities identified thus far have included scaffold functions and lysine modifications of non-histone proteins, including EZH2 bridging of β-catenin and TCF7 (formerly named Tcf1, T cell factor 1) [32], interaction with RelA and RelB [33], methylation of androgen receptor [34], as well as interactions with STAT3 [35,36], GATA4 [37], RORα [38], and Talin1 [39]. Therefore, using the well-characterised HL-60 cell line as a model system [40], we performed mass spectrometry (MS) as an unbiased approach to quantitatively investigate potential modulators (recruiters or co-repressors) and enzymatic targets of EZH2 in the context of AML. We additionally investigated whether these interactions were modulated in response to all-*trans* retinoic acid (AtRA)-induced myeloid differentiation.

2. Results

We performed five separate MS experiments that comprised a total of five IgG controls and seven EZH2 immunoprecipitations (IP), and one IP with an antibody raised against pan-methyl lysine. Good peptide coverage of EZH2 was achieved in the MS runs (going up to 50% coverage based on peptide assignment of at least 95% peptide threshold using the Peptide Prophet algorithm [41]) and all core components of the PRC2 complex were found to be significantly enriched (highlighted in blue) (Figure 1).

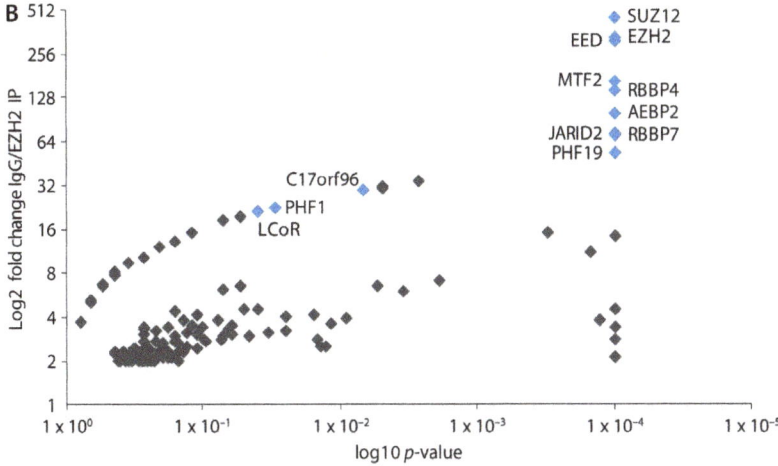

A

EZH2_HUMAN (100%), 85,364.4 Da
Histone-lysine N-methyltransferase EZH2
27 exclusive unique peptides, 37 exclusive unique spectra, 49 total spectra, 373/746 amino acids (50% coverage)

B

Figure 1. Affinity-purification mass spectrometry coverage of EZH2 and enrichment of PRC2 member proteins. (**A**) Sequence coverage of EZH2 (detected peptides are highlighted in yellow and amino acids with post-translational modifications are shown in green) from a representative mass spectrometry run (peptide threshold 95%); (**B**) Scatter plot of hits from all five mass spectrometry experiments passing a 2-fold change cut-off showing enrichment of core PRC2 components (shown in blue).

2.1. The EZH2 Interactome in HL-60 Cells

The first part of this study focussed on building the first EZH2 interactome in AML cells and on quantifying potential changes upon short (overnight) stimulation with AtRA. For this analysis, we considered all seven EZH2 IP and compared the protein abundance to the one found in all five IgG IP (a stringent 2-fold change cut-off was applied in this case) identifying 181 proteins interacting with EZH2. After further filtering (CRAPome [42] and SAINTexpress [43]) and comparison to recently published PcG complexome data [44] (Figure S1), we identified 143 proteins that co-IP with EZH2 in AML cells at high confidence (>0.8 SAINTexpress score). In order to highlight potential molecular complexes among the EZH2 interactors, clustering analysis was then performed with the Cluster maker [45] (MCL cluster), MCODE [46], and ClusterONE [47] plugins in Cytoscape [48], showing one main cluster of proteins and two smaller ones (Figures 2 and S2).

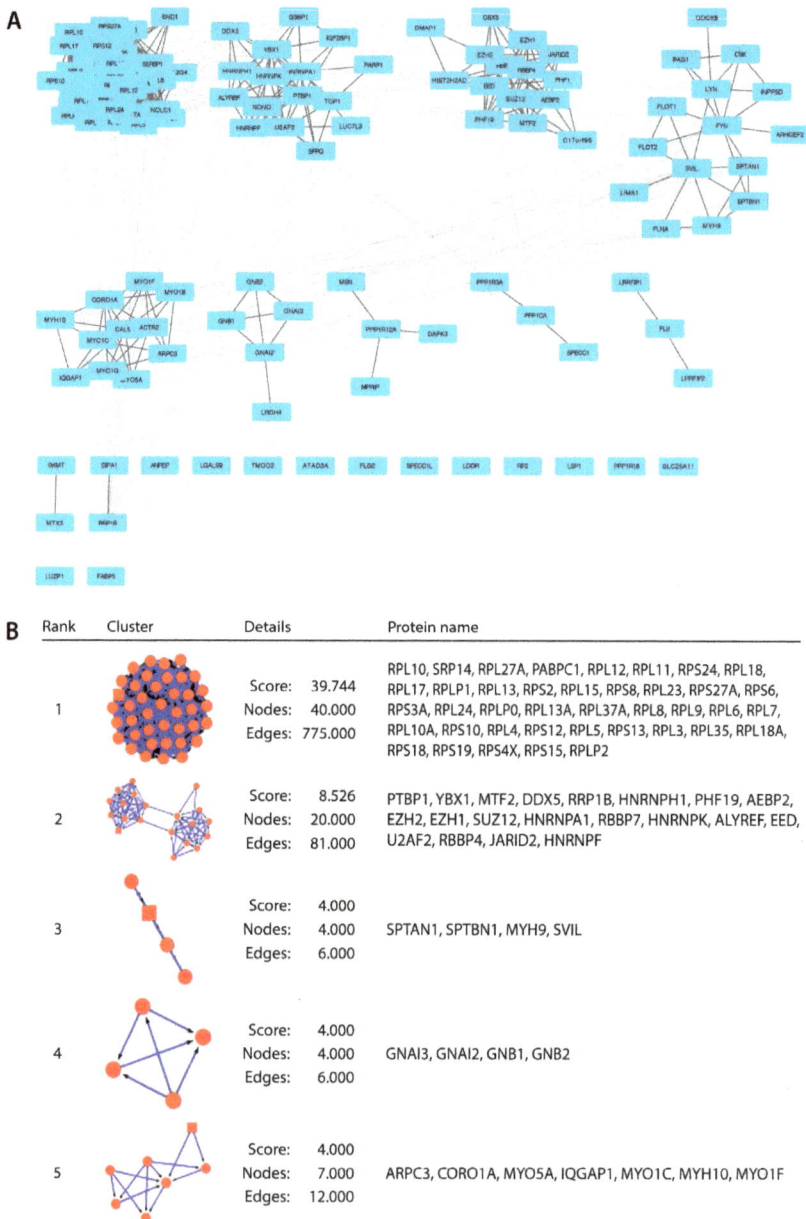

Figure 2. The EZH2 interactome in HL-60 acute myeloid leukaemia (AML) cells. (**A**) Network analysis of EZH2 interactome (2-fold change over IgG and 0.8 SAINTexpress score) showing related groups of proteins as clustered by Cytoscape (Makarov Clustering Algorithm in clusterMaker). Groups of PRC2 interacting proteins (third node from the left), as well as for RNA binding proteins involved in translation and splicing (first and second nodes on the top) can be distinguished; (**B**) MCODE clusters with protein symbols in each node confirming strong enrichment for RNA binding proteins (cluster 1) and known PRC2 associated members (cluster 2).

Further gene ontology (GO) analysis of hits using BinGO [49], David [50], and STRING [51] revealed a strong enrichment for proteins involved in RNA processing (splicing, mRNA metabolic processes) and translation (ribosomal proteins) located in the biggest cluster and one of the two smaller ones (Figure 3A). These results are in agreement with, and expand upon the most comprehensive proteomics dataset on PcG proteins published to date [44]. Importantly, this analysis also identified a major cluster of interacting proteins that are involved in gene expression and chromatin modifications consisting primarily of PRC2 members (the second small cluster from Figure 2 and Figure S2B,C), as well as several ribosomal proteins. Among those, there were several EZH2-recruiting proteins that have not been previously demonstrated to interact with EZH2 in the context of AML, such as PHF1, PHF19, LCOR, and EPOP (also known as C17orf96). Of note, several other proteins implicated in transcriptional control were also identified including THOC4/ALYREF, DDX5, DMAP1, HNRNPK, NONO, SFPQ, FLII, PARP1, PABP1, U2AFA, PTBP1, and YBX1 (Table 1).

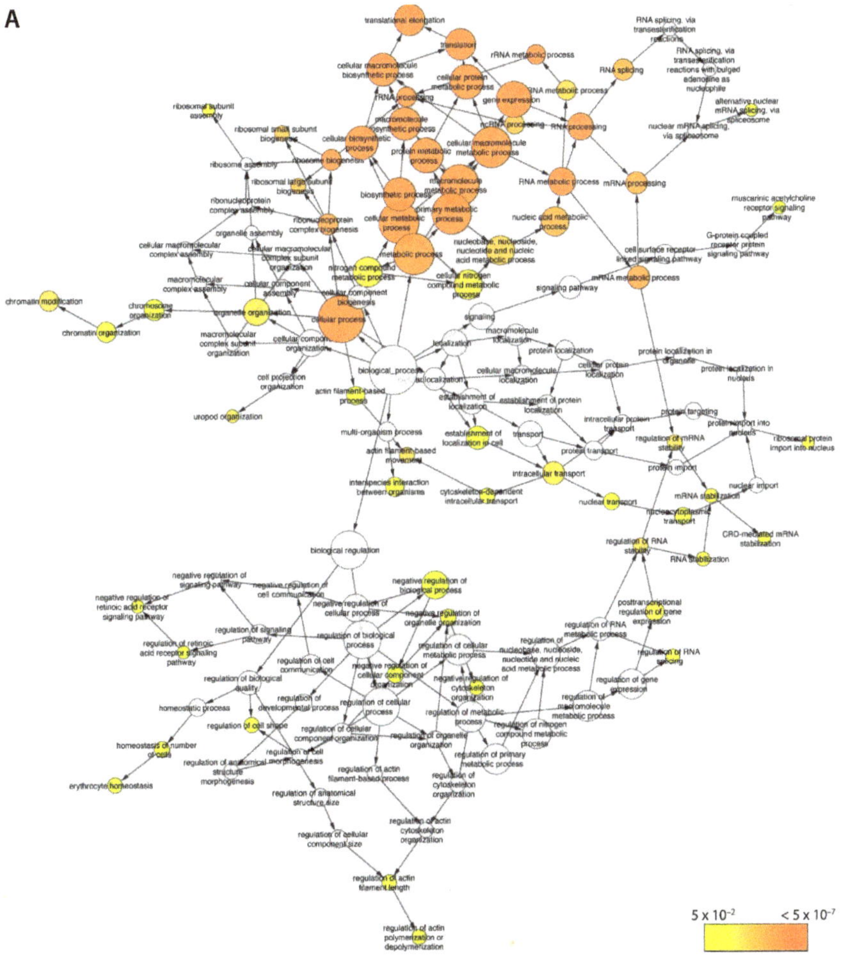

Figure 3. *Cont.*

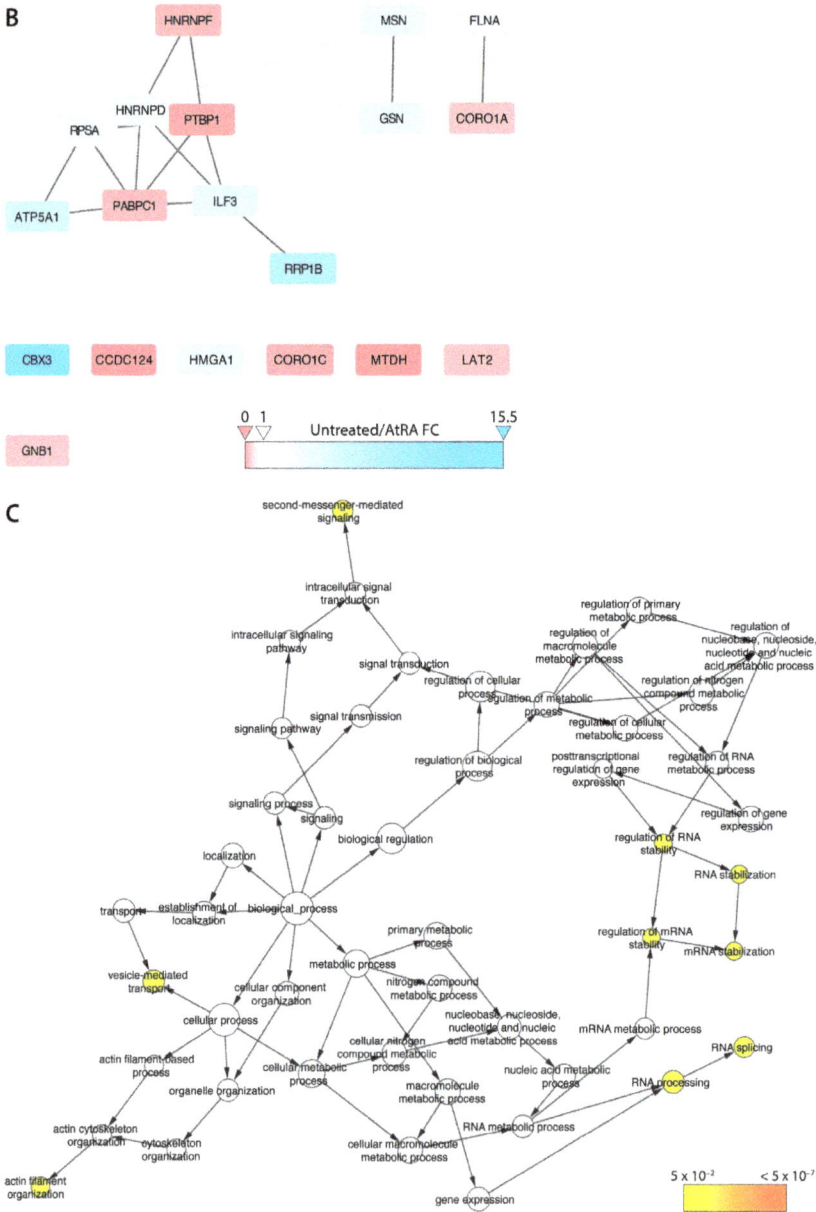

Figure 3. Gene ontology analysis of EZH2 interactome. (**A**) Gene ontology (Cytoscape–BinGO) showing significant enrichment for proteins involved in translation, RNA splicing, and gene expression. Node colour corresponds to *p*-value (see scale). Gene ontology analysis of EZH2 interactome (cont.); (**B**) Quantitative network analysis of EZH2-interacting proteins changing their relative frequency of interaction upon differentiation. Blue-coloured proteins represent ones found to co-IP with EZH2 at higher frequency in untreated HL-60 cells, whereas pink-coloured "preys" were found enriched in cells treated with AtRA; (**C**) Gene ontology (GO) analysis of proteins that change their interaction frequency with EZH2 upon AtRA-driven myeloid differentiation suggesting involvement in RNA metabolism, splicing, and control of gene expression.

Table 1. STRING gene ontology.

GO Category	Description	FDR	Protein Name
GO.0006415	Translational termination	1.27×10^{-43}	RPL10, RPL10A, RPL11, RPL12, RPL13, RPL13A, RPL15, RPL17, RPL18, RPL18A, RPL23, RPL24, RPL27A, RPL3, RPL35, RPL37A, RPL4, RPL5, RPL6, RPL7, RPL8, RPL9, RPLP0, RPLP1, RPLP2, RPS10, RPS12, RPS13, RPS15, RPS19, RPS2, RPS24, RPS27A, RPS3A, RPS4X, RPS6, RPS8
GO.0006414	Translational elongation	4.54×10^{-42}	RPL10, RPL10A, RPL11, RPL12, RPL13, RPL13A, RPL15, RPL17, RPL18, RPL18A, RPL23, RPL24, RPL27A, RPL3, RPL35, RPL37A, RPL4, RPL5, RPL6, RPL7, RPL8, RPL9, RPLP0, RPLP1, RPLP2, RPS10, RPS12, RPS13, RPS15, RPS19, RPS2, RPS24, RPS27A, RPS3A, RPS4X, RPS6, RPS8
GO.0006413	Translational initiation	7.98×10^{-41}	PABPC1, RPL10, RPL10A, RPL11, RPL12, RPL13, RPL13A, RPL15, RPL17, RPL18, RPL18A, RPL23, RPL24, RPL27A, RPL3, RPL35, RPL37A, RPL4, RPL5, RPL6, RPL7, RPL8, RPL9, RPLP0, RPLP1, RPLP2, RPS10, RPS12, RPS13, RPS15, RPS19, RPS2, RPS24, RPS27A, RPS3A, RPS4X, RPS6, RPS8
GO.0010467	Gene expression	2.09×10^{-13}	AEBP2, AICDA, ALYREF, ANPEP, CBX3, DAPK3, DDX5, DMAP1, EED, EZH1, EZH2, FLII, FLNA, HNRNPF, HNRNPH1, HNRNPK, IGF2BP1, JARID2, LCOR, LRRFIP1, MTF2, NOLC1, NONO, PA2G4, PABPC1, PHF1, PHF19, RBBP4, RBBP7, RPL10, RPL12, RPL13, RPL17, RPL18, RPL18A, RPL24, RPL27A, RPL3, RPL35, RPL4, RPL5, RPL6, RPL7, RPL9, RPLP0, RPLP1, RPLP2, RPS10, RPS12, RPS13, RPS19, RPS2, RPS24, RPS27A, RPS3A, RPS6, RPS8, RRP1B, SFPQ, SND1, SRP14, SUZ12, U2AF2, YBX1
GO.0045892	Negative regulation of transcription	3.03×10^{-2}	AEBP2, CBX3, DDX5, DMAP1, FLNA, HNRNPK, LCOR, LRRFIP1, MTF2, NONO, PA2G4, PARP1, RBBP7, RPS27A, SFPQ, SUZ12, YBX1

GO, gene ontology; FDR, false discovery rate.

2.2. Changes in the EZH2 Interactome in Response to AtRA

We next examined alterations to EZH2 interactions in response to myeloid differentiation. Of the seven EZH2 IP mentioned above, four were of untreated HL-60 cells and three of HL-60 cells stimulated with 0.1 μM AtRA. All runs were again merged and relative quantitation was performed based on exclusive spectrum count. Initially, 82 proteins with a greater than 2-fold difference between the conditions were identified, which were found to be at least 1.5 times more abundant in all 7 EZH2 IP versus all five IgG IP. Nevertheless, after filtering of the hits, 19 proteins were identified as selectively interacting with EZH2 upon AtRA-induced differentiation (Table 2 and Figure 3B). GO and network analyses (Figure 3C) showed strong enrichment for RNA binding proteins (13 out of 19, FDR = 4.86×10^{-8}), with a number of these hits being involved in mRNA splicing (HNRNPD, HNRNPF, PABPC1, PTBP1, CCDC124) and transcription regulation (CBX3, HMGA, ILF3, MTDH). Of note, none of the genes encoding these proteins were found to display changes in gene expression in accordance with increased or decreased bait-prey interactions following 72 h of treatment with AtRA (unpublished data). Assuming consistent protein stability, this suggests that the changes in the EZH2 interactome identified here are unlikely to result from altered gene expression levels of any of the preys identified.

Table 2. STRING gene ontology for AtRA-regulated EZH2 interacting proteins.

GO Category	Description	FDR	Protein Name
GO.0003723	RNA binding	4.86×10^{-8}	ATP5A1, CCDC124, CORO1A, FLNA, HNRNPD, HNRNPF, ILF3, MSN, MTDH, PABPC1, PTBP1, RPSA, RRP1B

GO, gene ontology; FDR, false discovery rate.

2.3. Identification of Enzymatic Targets of EZH2

Finally, we sought to identify potential non-histone targets of EZH2 in AML. Four of the five MS runs (six EZH2 IP and four IgG control IP) were re-run in order to detect methylated peptides. Relatively few proteins modified by mono-, di-, or tri-methylation were detected (32 in total, 24 with modified Lysine residues, eight with Arginine methylation). After filtering for proteins identified in the EZH2 interactome (Figure 2), potential hits for lysine methylation identified with high confidence (99.9% proteins threshold and 80–95% peptide threshold) included EZH2 itself (K735^{me1}), SUZ12 (K4^{me1}), CBX3 (K142^{me1}), Histone H4 (K20^{me2}), eEF1A1 (K55^{me2}), and ADT2 (also known as SLC25A5) (K51^{me3}). (Tables 3 and 4, Figures 4 and S3).

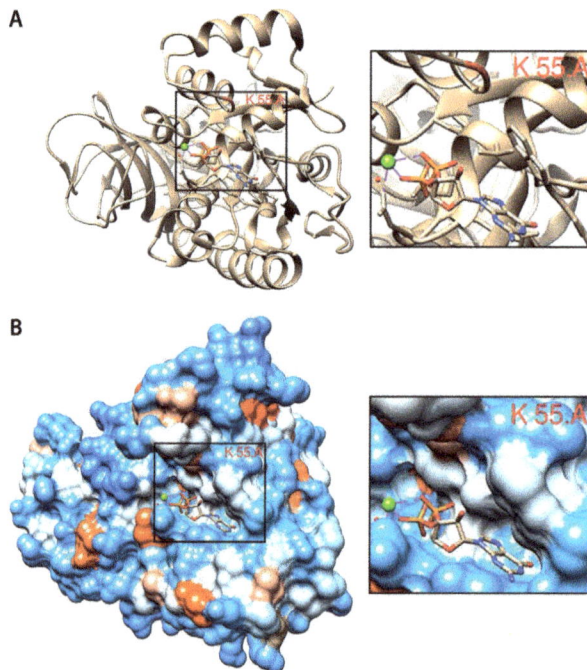

Figure 4. Structure of eEF1A1 and putative EZH2 methylation target site. (**A**) Ribbon structure view of mammalian eEF1A monomer showing close proximity of Lys 55 (highlighted in red) to the GTP/GDP (Guanosine-5′-tri/diphosphate) binding pocket (with a GDP molecule bound in the pocket); (**B**) Hydrophobicity surface view (hydrophobic, red; hydrophilic, blue; neutral, white) of eEF1A1 showing that K55 is in a hydrophilic (blue colour) region close to GTP/GDP binding pocket that is potentially accessible for post-translational modifications and methylation by EZH2.

Table 3. EZH2 interactome lysine methylation hits.

	Protein Name	Peptide Sequence	Peptide Start Index	Peptide Stop Index	Variable Modifications Identified by Spectrum	Methylated Lys −AtRA	Methylated Lys +AtRA
Lys monomethylation	CBX3	WKDSDEADLVLAK	142	154	K2: Methyl	12.5% (1/8)	ND (0/0)
	EZH2	YSQADALKYVGIER	728	741	K8: Methyl	1.5% (2/131)	1.9% (2/106)
	Histone H1.2	KASGPPVSELITK	34	46	K1: Methyl	2.1% (2/96)	5% (1/20)
	Histone H3.1	KSAPATGGVKPHR	28	41	K10: Methyl	0% (0/18)	16.7% (1/6)
	Histone H3.1	EIAQDFKTDLR	74	84	K7: Methyl	5.6% (1/18)	0% (0/6)
	MT1X	MDPNCSCSPVGSCAC-AGSCKCKECKCTSCK	1	30	K22: Methyl	100% (1/1)	ND (0/0)
	RL36L	KQSGYGGQTKPIFR	44	57	K10: Methyl	33.3% (11/33)	44.4% (8/18)
	SUZ12	APQKHGGGGGGSGPSAGS-GGGGHGGSAAVAAATASGGK	2	40	K4: Methyl	1.3% (2/154)	0% (0/138)
Lys dimethylation	eEF1A1	GSFKYAWVLDK	52	62	K4: Dimethyl	11.9% (7/59)	5.9% (1/17)
	eEF1A1	MDSTEPPYSQKR	155	166	K11: Dimethyl	0% (0/59)	5.9% (1/17)
	H3.1	KSAPATGGVKKPHR	28	41	K1: Dimethyl	5.6% (1/18)	50% (3/6)
	Histone H4	KVLRDNIQGITKPAIR	21	36	K1: Dimethyl	0% (0/86)	4.3% (1/23)
	MYO1D	SKDTCIVISGESGAGKTEASK	93	113	K2: Dimethyl	ND (0/0)	100% (3/3)
	RBP56	GPMTGSSGGIDRGGFK	196	210	K15: Dimethyl	36.4% (8/22)	0% (0/6)
	TR150	DSRPSQAAGDNQGDEAKEQ-TFSGGTSQDTK	186	215	K17: Dimethyl	21.9% (7/32)	ND (0/0)
Lys trimethylation	ADT2	QYKGIIDCVVR	50	60	K3: Trimethyl	19.1% (4/21)	7.7% (1/13)
	HNRPQ	GGNVGGKR	558	565	K7: Trimethyl	25% (1/4)	16.7% (1/6)
	MT1X	MDPNCSCSPVGSCACAGSC-KCKECKCTSCK	1	30	K20: Trimethyl, K25: Trimethyl, K30: Trimethyl	100% (1/1)	ND (0/0)
	ALYREF	ADKMDMSLDDIIK	2	14	K3: Trimethyl	0% (0/27)	5.2% (1/19)

K, methylated lysine; ND, not detected.

In order to validate candidate EZH2 targets, another pull-down was performed with a pan-methylated lysine antibody in parallel with the EZH2 IP (data not shown). Although relatively few proteins (25 proteins that had at least 1.5 times more total peptides than in the IgG control) were found to co-IP with this antibody (at high confidence thresholds), this experiment confirmed that eEF1A1 and ADT2 are indeed methylated, suggesting that these two proteins may be direct EZH2 methylation targets (i.e., they co-IP with EZH2 in at least two experiments, they are immunoprecipitated with the pan-methyl lysine antibody, and methylation sites were found by mass spectrometry analysis). Of note, there were six other potential hits from the anti-pan-methylated antibody pull-down that were also found to co-IP with EZH2 (among the final 145 hits)—three other ribosomal proteins (RPL24, RPL35, RPS27A), as well as SPTB2, LUC7L3, and SFPQ (a transcriptional co-repressor). Here, however, mass spectrometry failed to identify specific methylation sites.

Table 4. Summary of potential non-histone EZH2 enzymatic targets.

Protein Name	Gene Symbol	Methylation Site	Protein Function
ADT2 (ADP/ATP Translocase 2)	*SLC25A5*	K52^{me3}	ADP/ATP mitochondrial translocase
CBX3 (Chromobox 3)	*CBX3*	K142^{me1}	Heterochromatin binding
eEF1A1 (Elongation factor 1-α1)	*EEF1A1*	K55^{me2}, K165^{me2}	Regulation of elongation
EZH2 (Enhancer of zeste homology 2)	*EZH2*	K735^{me1}	Protein lysine methyltransferase
SUZ12 (Polycomb Repressive Complex 2 Subunit)	*SUZ12*	K4^{me1}	Regulation of H3K27 methylation and gene expression
ALYREF (Aly/REF Export Factor)	*ALYREF*	K4^{me3}	Chaperone of basic-region leucine zipper (bZIP) proteins

3. Discussion

In this study, we sought to enhance our understanding of the complex biological roles of the PRC2 complex in AML by analysing the EZH2 interactome (Figure 5). We found that previously established interactions with PRC2-recruiting proteins such as PHF1, PHF19, LCOR, JARID2, and EPOP are conserved in HL-60 cells and identified several proteins involved in transcriptional regulation including ALYREF, SFPQ, FLII, PARP1, and YBX1. We also found that EZH2 interacts with a number of RNA binding and processing proteins, including several that are implicated in stem cell maintenance such as YBX1 and DDX family proteins [52], suggesting a role for EZH2 in translational control. With regard to non-histone enzymatic substrates for EZH2, the identification of specifically methylated lysine residues and co-IP of the respective proteins with anti-EZH2 and anti-pan methyl lysine antibodies revealed several potential new targets. Importantly, many of the methylated lysine residues described in this study have been identified elsewhere [53,54], but without pointing to EZH2 as a candidate methyltransferase.

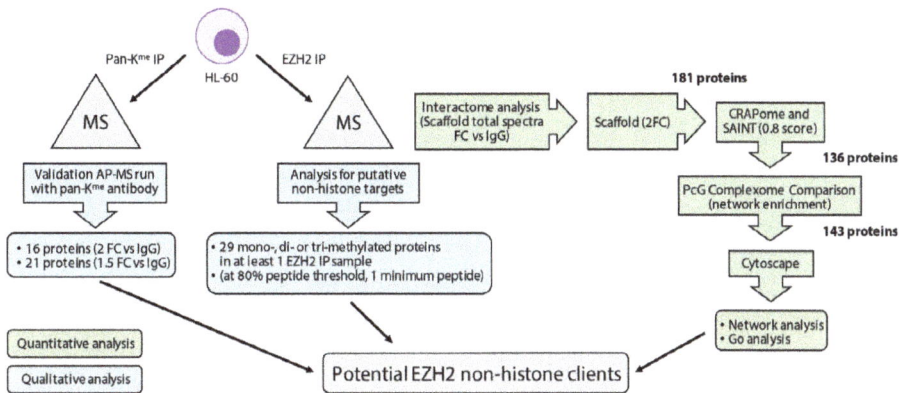

Figure 5. Summary of mass spectrometry (MS) strategy to identify potential non-histone clients in HL-60 cells.

Although the exact biological role of methylation of ribosomal proteins remains poorly characterised, this post-translational modification is likely significant given that it is conserved in all three animal kingdoms [55]. It is well-established that both the 60S and the 40S subunits contain methylated residues and that a number of ribosomal proteins such as RPS2, 3, 9, 10, 12, 14, 25, and 27 as well as RPL23, 29, and 40 can also be methylated including on lysine residues [55]. Such modifications have been suggested to affect protein-protein interactions and ribosome assembly, RNA binding or translation accuracy [55]. In this context, the identification the eEF1A1 translation elongation factor as a substrate for EZH2 warrants further investigation. eEF1A1 lysine methylation has been described for several different residues, including K55 observed here [53,54]. Studies of EF-Tu (elongation factor thermo unstable), an *Escherichia coli* orthologue of eEF1A1, suggest that this conserved residue (K56 in *E. coli*) due to its location in the GTPase switch-1 region may enhance translational accuracy through attenuating GTP hydrolysis [55]. Given that PRC2 complexes have well-established functions in transcriptional regulation [56], our finding that EZH2 may be involved in translational control therefore expands the roles that PRC2 complexes play in the control of gene expression.

To date, therapeutic strategies targeting EZH2 have focused almost exclusively on inhibiting cofactor binding by the enzyme [18]. Even though effective when activating point mutations in the SET domain are driving lymphomas [57,58] or in the background of tumours harbouring other epigenetic perturbations [59], targeting the catalytic activity of EZH2 has thus far failed to elicit the expected response in a number of other malignancies [60,61]. Given that aberrant eEF1A1 activities have been implicated in a number of cancers [62–64] and the translation machinery represents an important area of oncology research [65], our findings suggest new possibilities for combinatorial therapeutic approaches.

4. Methods

4.1. Cell Culture and AtRA Treatment

HL-60 AML cells were cultured in RPMI-1640 (Gibco, Waltham, MA, USA) supplemented with penicillin (50 units/mL)/streptomycin (50 µg/mL) (Gibco, Waltham, MA, USA) and 10% fetal calf serum (Sigma-Aldrich, St. Louis, MO, USA). Cells were maintained at 37 °C and 5% CO_2 in humidified atmosphere. AtRA was purchased from Sigma and was diluted in 1:1 ethanol and dimethyl sulfoxide (DMSO).

4.2. Affinity-Purification and Sample Preparation for Mass Spectrometry

HL-60 cells were centrifuged and washed with phosphate-buffered saline (PBS). Cell lysates prepared using High-salt Lysis Buffer (20 mM HEPES (4-(2-hydroxyethyl)piperazin-1-ylethanesulfonic acid) pH 7.9; 1.5 mM $MgCl_2$; 0.3 mM NaCl; 0.2 mM EDTA (ethylenediaminetetraacetic acid); 0.1% Triton X100; 0.1 mM DTT (dithiothreitol); 25% Glycerol). Cell lysates were pre-cleared with protein A/G magnetic beads (Pierce) and incubated overnight at 4 °C with: rabbit polyclonal anti-EZH2 antibody (Abcam, ab186006, Cambridge, MA, USA); mouse monoclonal anti-EZH2 antibody (Cell Signaling Technology, clone AC22, Danvers, MA, USA); rabbit polyclonal anti-methyl lysine antibody (Abcam, ab7315); mouse IgG (Abcam, ab37355, Cambridge, MA, USA) or rabbit IgG (Abcam, ab27478, Cambridge, MA, USA) isotype controls. Samples were then incubated for 2 h with protein A/G magnetic beads. Bound antibody-protein complexes were washed three times with Low-salt Lysis Buffer (20 mM HEPES pH 7.9; 1.5 mM $MgCl_2$; 0.1 mM NaCl; 0.2 mM EDTA; 0.1% Triton X100; 0.1 mM DTT, three times with 50 mM TEAB (triethylammonium bicarbonate), and eluted with 5% formic acid. Any residual formic acid was neutralised with 1 M TEAB and samples were dried *in vacuo*. Samples were re-dissolved in 5% acetonitrile/50 mM TEAB and then reduced with TCEP (tris(2-carboxyethyl)phosphine, 5 mM final concentration). Free cysteines were alkylated with 2-choloroacetamide (10 mM final concentration). Proteins were digested with trypsin (Promega, Madison, WI, USA) and quenched with neat formic acid after 4 h. An aliquot of these solutions was taken for direct analysis by liquid chromatography tandem-mass spectrometry (LC-MS/MS).

4.3. LC-MS/MS

Reversed phase chromatography was performed using an HP1200 platform (Agilent, Santa Clara, CA, USA). Peptides were resolved on a 75 μm I.D. 15 cm C18 packed emitter column (3 μm particle size; Nikkyo Technos, Tokyo, Japan) over 30 min or 60 min using a linear gradient of 96:4 to 50:50 Buffer A:Buffer B [Buffer A (1% acetonitrile/3% dimethyl sulfoxide/0.1% formic acid), Buffer B (80% acetonitrile/3% dimethyl sulfoxide/0.1% formic acid)] at 250 nL/min. Peptides were ionised by electrospray ionisation using 1.8 kV applied immediately pre-column via a microtee built into the nanospray source. Samples were infused into an LTQ Velos Orbitrap mass spectrometer (Thermo Fisher Scientific, Waltham, MA, USA) directly from the end of the tapered tip silica column (6–8 μm exit bore). MS/MS data were acquired using data dependent acquisition based on a full Fourier Transform mass spectrometry (FT-MS) scan (30,000 resolution, inject time set to 500 milliseconds and Automatic gain control (AGC) set to 1,000,000 with preview mode disabled) and internal lock mass calibration against the ion 401.922718 m/z. The top 20 most intense precursor ions were fragmented by collision-induced dissociation and analysed using normal ion trap scans (AGC set to 30,000, normalised collision energy was set to 35% with an activation time of 10 milliseconds). Precursor ions with unknown or single charge states were excluded from selection. Peptides were measured in the orbitrap at 30,000 resolution (automatic gain control—AGC of 1,000,000). Peptides were then fragmented in the ion trap where they were measured at low resolution (AGC—30,000). Full FT-MS maximum inject time was 500 milliseconds and normalised collision energy was set to 35% with an activation time of 10 milliseconds. Wideband activation was used to co-fragment precursor ions undergoing neutral loss of up to −20 m/z from the parent ion, including loss of water/ammonia. MS/MS was acquired for selected precursor ions with a single repeat count acquired after 5-s delay followed by dynamic exclusion with a 10 ppm mass window for 10 s based on a maximal exclusion list of 500 entries.

4.4. Database Searching

Raw MS/MS data were submitted for database searching using Proteome Discoverer v1.4 and Mascot V2.3. The following Mascot search parameters were used: SwissProt Database,

SwissProt_040511a (526,969 Sequences); taxonomy filter, Homo sapiens (20,305 sequences); enzyme specificity, trypsin (KR) 2 missed cleavages; mass tolerance; precursor 5 ppm, fragment 0.60 Da; variable modifications, acetyl (protein N-term); carbamidomethyl (C); oxidation of methionine pyro-Glu (peptide N-term Q); phosphorylation (STY); methylation (KR); dimethylation (KR); Trimethylation (K). MS/MS-based peptide and protein identifications were grouped and validated using Scaffold v4 (Proteome Software Inc. Portland, OR, USA). Protein identifications were automatically accepted if they contained at least two unique peptides assigned with at least 95% confidence by Peptide Prophet [66].

4.5. Data Analysis

All MS runs were merged in Scaffold 4.0 and assigned to two categories—IgG and EZH2 IP (or three categories for the initial analysis: IgG, EZH2 IP, and EZH2 IP in AtRA-treated cells). Fold change difference compared to IgG control (all IgG samples versus all EZH2 IP samples) were calculated using total spectra as a relative quantitation method (minimum value was set to 0.1 in case spectra were absent e.g., in IgG controls) and then exclusive spectrum count to eliminate potential spectra belonging to more than one protein (Table S1) [67]. Fisher's exact test (with Benjamini-Hoechberg correction) was run in order to highlight potential hits. For the +/−AtRA analysis, a less stringent filter of 1.5-fold change over IgG was set and then the ratio of all four Untreated samples versus all three AtRA samples (two untreated versus two AtRA were matched biological samples) was compared. Proteins altering their relative abundance by a 2-fold change factor (either up or down) were also compared in the two matched +/−AtRA runs (in order to avoid false positive hits that pass the 2-fold change cut off, but have different direction in the two matched runs) and were finally checked to see if they were present in more than one sample. Hits from the global EZH2 interactome (all seven EZH2 IP samples versus all IgG controls) were further processed using the online analysis platform CRAPome [42]. The integrated Significance Analysis of INTeractome (SAINTexpress [43]) module was implemented to remove common contaminants (IgG selected as a criterion since the IP were of native non-overexpressed and non-tagged EZH2 protein) and to provide a relatively stringent confidence filter (score ≥ 0.8) for the probability of the bait-prey interactions. The resulting list of hits (136 proteins) was then enriched using a published PcG complexome MS data set [44] and the resulting data were processed, visualised, and analysed using Cytoscape (Cluster maker, MCODE, ClusterONE, and BinGO) as described previously [68]. USCF Chimera viewer [69] was used to visualise, process, and annotate the crystal structure of EZH2 (pdb: 5HYN) and eEF1A (pdb:4C0S).

5. Conclusions

In this study, we have generated the first EZH2 interactome in AML using an unbiased mass spectrometry approach. We identified EZH2 interactions with several ribosomal proteins, some of which are subject to change upon AtRA-induced myeloid differentiation. Our results strongly suggest that EZH2 is responsible for eEF1A1 K55 di-methylation, indicating a regulatory role in translation/mRNA processing that may present therapeutic opportunities.

Supplementary Materials: Supplementary materials can be found at www.mdpi.com/1422-0067/18/7/1440/s1.

Acknowledgments: Yordan Sbirkov was supported by a Bloodwise (formerly Leukaemia and Lymphoma Research UK) Gordon Piller PhD Studentship (10053). Colin Kwok and Kevin Petrie were supported by Bloodwise Specialist Programme Grants (09001 and 11046). We thank Rowan Watt for a critical reading of the manuscript.

Author Contributions: Yordan Sbirkov, Arthur Zelent, and Kevin Petrie conceived and designed the study. Yordan Sbirkov and Colin Kwok performed experiments. Amandeep Bhamra and Andrew Thompson performed LC-MS/MS and carried out data processing. Colin Kwok, Amandeep Bhamra, Andrew Thompson, and Arthur Zelent analysed data and edited the manuscript. Yordan Sbirkov, Veronica Gil, and Kevin Petrie analysed data and wrote the manuscript.

Conflicts of Interest: The authors declare no conflict of interest.

References

1. Pollyea, D.A.; Kohrt, H.E.; Medeiros, B.C. Acute myeloid leukaemia in the elderly: A review. *Br. J. Haematol.* **2011**, *152*, 524–542. [CrossRef] [PubMed]
2. O'Donnell, M.R.; Tallman, M.S.; Abboud, C.N.; Altman, J.K.; Appelbaum, F.R.; Arber, D.A.; Attar, E.; Borate, U.; Coutre, S.E.; Damon, L.E.; et al. Acute myeloid leukemia, version 2.2013. *J. Natl. Compr. Cancer Netw.* **2013**, *11*, 1047–1055. [CrossRef]
3. Rubnitz, J.E. How I treat pediatric acute myeloid leukemia. *Blood* **2012**, *119*, 5980–5988. [CrossRef] [PubMed]
4. Conway O'Brien, E.; Prideaux, S.; Chevassut, T. The epigenetic landscape of acute myeloid leukemia. *Adv. Hematol.* **2014**, *2014*, 103175. [CrossRef] [PubMed]
5. Dohner, H.; Weisdorf, D.J.; Bloomfield, C.D. Acute Myeloid Leukemia. *N. Engl. J. Med.* **2015**, *373*, 1136–1152. [CrossRef] [PubMed]
6. Ferrara, F.; Schiffer, C.A. Acute myeloid leukaemia in adults. *Lancet* **2013**, *381*, 484–495. [CrossRef]
7. Leopold, L.H.; Willemze, R. The treatment of acute myeloid leukemia in first relapse: A comprehensive review of the literature. *Leuk. Lymphoma* **2002**, *43*, 1715–1727. [CrossRef] [PubMed]
8. Greenberg, P.L.; Lee, S.J.; Advani, R.; Tallman, M.S.; Sikic, B.I.; Letendre, L.; Dugan, K.; Lum, B.; Chin, D.L.; Dewald, G.; et al. Mitoxantrone, etoposide, and cytarabine with or without valspodar in patients with relapsed or refractory acute myeloid leukemia and high-risk myelodysplastic syndrome: A phase III trial (E2995). *J. Clin. Oncol.* **2004**, *22*, 1078–1086. [CrossRef] [PubMed]
9. Vogelstein, B.; Papadopoulos, N.; Velculescu, V.E.; Zhou, S.; Diaz, L.A., Jr.; Kinzler, K.W. Cancer genome landscapes. *Science* **2013**, *339*, 1546–1558. [CrossRef] [PubMed]
10. Yang, W.; Ernst, P. SET/MLL family proteins in hematopoiesis and leukemia. *Int. J. Hematol.* **2017**, *105*, 7–16. [CrossRef] [PubMed]
11. Ley, T.J.; Ding, L.; Walter, M.J.; McLellan, M.D.; Lamprecht, T.; Larson, D.E.; Kandoth, C.; Payton, J.E.; Baty, J.; Welch, J.; et al. DNMT3A mutations in acute myeloid leukemia. *N. Engl. J. Med.* **2010**, *363*, 2424–2433. [CrossRef] [PubMed]
12. Harris, W.J.; Huang, X.; Lynch, J.T.; Spencer, G.J.; Hitchin, J.R.; Li, Y.; Ciceri, F.; Blaser, J.G.; Greystoke, B.F.; Jordan, A.M.; et al. The histone demethylase KDM1A sustains the oncogenic potential of MLL-AF9 leukemia stem cells. *Cancer Cell* **2012**, *21*, 473–487. [CrossRef] [PubMed]
13. Lynch, J.T.; Harris, W.J.; Somervaille, T.C. LSD1 inhibition: A therapeutic strategy in cancer? *Expert Opin. Ther. Targets* **2012**, *16*, 1239–1249. [CrossRef] [PubMed]
14. Schenk, T.; Chen, W.C.; Gollner, S.; Howell, L.; Jin, L.; Hebestreit, K.; Klein, H.U.; Popescu, A.C.; Burnett, A.; Mills, K.; et al. Inhibition of the LSD1 (KDM1A) demethylase reactivates the all-trans-retinoic acid differentiation pathway in acute myeloid leukemia. *Nat. Med.* **2012**, *18*, 605–611. [CrossRef] [PubMed]
15. Gil, V.S.; Bhagat, G.; Howell, L.; Zhang, J.; Kim, C.H.; Stengel, S.; Vega, F.; Zelent, A.; Petrie, K. Deregulated expression of HDAC9 in B cells promotes development of lymphoproliferative disease and lymphoma in mice. *Dis. Models Mech.* **2016**, *9*, 1483–1495. [CrossRef] [PubMed]
16. Suzuki, K.; Okuno, Y.; Kawashima, N.; Muramatsu, H.; Okuno, T.; Wang, X.; Kataoka, S.; Sekiya, Y.; Hamada, M.; Murakami, N.; et al. MEF2D-BCL9 Fusion Gene Is Associated With High-Risk Acute B-Cell Precursor Lymphoblastic Leukemia in Adolescents. *J. Clin. Oncol.* **2016**, *34*, 3451–3459. [CrossRef] [PubMed]
17. Sahasrabuddhe, A.A. BMI1: A Biomarker of Hematologic Malignancies. *Biomark. Cancer* **2016**, *8*, 65–75. [CrossRef] [PubMed]
18. Kim, K.H.; Roberts, C.W. Targeting EZH2 in cancer. *Nat. Med.* **2016**, *22*, 128–134. [CrossRef] [PubMed]
19. Morin, R.D.; Johnson, N.A.; Severson, T.M.; Mungall, A.J.; An, J.; Goya, R.; Paul, J.E.; Boyle, M.; Woolcock, B.W.; Kuchenbauer, F.; et al. Somatic mutations altering EZH2 (Tyr641) in follicular and diffuse large B-cell lymphomas of germinal-center origin. *Nat. Genet.* **2010**, *42*, 181–185. [CrossRef] [PubMed]
20. McCabe, M.T.; Graves, A.P.; Ganji, G.; Diaz, E.; Halsey, W.S.; Jiang, Y.; Smitheman, K.N.; Ott, H.M.; Pappalardi, M.B.; Allen, K.E.; et al. Mutation of A677 in histone methyltransferase EZH2 in human B-cell lymphoma promotes hypertrimethylation of histone H3 on lysine 27 (H3K27). *Proc. Natl. Acad. Sci. USA* **2012**, *109*, 2989–2994. [CrossRef] [PubMed]
21. Bodor, C.; O'Riain, C.; Wrench, D.; Matthews, J.; Iyengar, S.; Tayyib, H.; Calaminici, M.; Clear, A.; Iqbal, S.; Quentmeier, H.; et al. EZH2 Y641 mutations in follicular lymphoma. *Leukemia* **2011**, *25*, 726–729. [CrossRef] [PubMed]

22. Nikoloski, G.; Langemeijer, S.M.; Kuiper, R.P.; Knops, R.; Massop, M.; Tonnissen, E.R.; van der Heijden, A.; Scheele, T.N.; Vandenberghe, P.; de Witte, T.; et al. Somatic mutations of the histone methyltransferase gene EZH2 in myelodysplastic syndromes. *Nat. Genet.* **2010**, *42*, 665–667. [CrossRef] [PubMed]

23. Ernst, T.; Chase, A.J.; Score, J.; Hidalgo-Curtis, C.E.; Bryant, C.; Jones, A.V.; Waghorn, K.; Zoi, K.; Ross, F.M.; Reiter, A.; et al. Inactivating mutations of the histone methyltransferase gene EZH2 in myeloid disorders. *Nat. Genet.* **2010**, *42*, 722–726. [CrossRef] [PubMed]

24. Makishima, H.; Jankowska, A.M.; Tiu, R.V.; Szpurka, H.; Sugimoto, Y.; Hu, Z.; Saunthararajah, Y.; Guinta, K.; Keddache, M.A.; Putnam, P.; et al. Novel homo- and hemizygous mutations in EZH2 in myeloid malignancies. *Leukemia* **2010**, *24*, 1799–1804. [CrossRef] [PubMed]

25. Ernst, P.; Fisher, J.K.; Avery, W.; Wade, S.; Foy, D.; Korsmeyer, S.J. Definitive hematopoiesis requires the mixed-lineage leukemia gene. *Dev. Cell* **2004**, *6*, 437–443. [CrossRef]

26. Grimwade, D.; Hills, R.K.; Moorman, A.V.; Walker, H.; Chatters, S.; Goldstone, A.H.; Wheatley, K.; Harrison, C.J.; Burnett, A.K.; National Cancer Research Institute Adult Leukaemia Working Group. Refinement of cytogenetic classification in acute myeloid leukemia: Determination of prognostic significance of rare recurring chromosomal abnormalities among 5876 younger adult patients treated in the United Kingdom Medical Research Council trials. *Blood* **2010**, *116*, 354–365. [PubMed]

27. Thiel, A.T.; Feng, Z.; Pant, D.K.; Chodosh, L.A.; Hua, X. The trithorax protein partner menin acts in tandem with EZH2 to suppress C/EBPα and differentiation in MLL-AF9 leukemia. *Haematologica* **2013**, *98*, 918–927. [CrossRef] [PubMed]

28. Tanaka, S.; Miyagi, S.; Sashida, G.; Chiba, T.; Yuan, J.; Mochizuki-Kashio, M.; Suzuki, Y.; Sugano, S.; Nakaseko, C.; Yokote, K.; et al. EZH2 augments leukemogenicity by reinforcing differentiation blockage in acute myeloid leukemia. *Blood* **2012**, *120*, 1107–1117. [CrossRef] [PubMed]

29. Shi, J.; Wang, E.; Zuber, J.; Rappaport, A.; Taylor, M.; Johns, C.; Lowe, S.W.; Vakoc, C.R. The Polycomb complex PRC2 supports aberrant self-renewal in a mouse model of MLL-AF9;Nras^{G12D} acute myeloid leukemia. *Oncogene* **2013**, *32*, 930–938. [CrossRef] [PubMed]

30. Mills, A.A. Throwing the cancer switch: Reciprocal roles of polycomb and trithorax proteins. *Nat. Rev. Cancer* **2010**, *10*, 669–682. [CrossRef] [PubMed]

31. Gupta, R.A.; Shah, N.; Wang, K.C.; Kim, J.; Horlings, H.M.; Wong, D.J.; Tsai, M.C.; Hung, T.; Argani, P.; Rinn, J.L.; et al. Long non-coding RNA HOTAIR reprograms chromatin state to promote cancer metastasis. *Nature* **2010**, *464*, 1071–1076. [CrossRef] [PubMed]

32. Shi, B.; Liang, J.; Yang, X.; Wang, Y.; Zhao, Y.; Wu, H.; Sun, L.; Zhang, Y.; Chen, Y.; Li, R.; et al. Integration of estrogen and Wnt signaling circuits by the polycomb group protein EZH2 in breast cancer cells. *Mol. Cell. Biol.* **2007**, *27*, 5105–5119. [CrossRef] [PubMed]

33. Lee, S.T.; Li, Z.; Wu, Z.; Aau, M.; Guan, P.; Karuturi, R.K.; Liou, Y.C.; Yu, Q. Context-specific regulation of NF-κB target gene expression by EZH2 in breast cancers. *Mol. Cell* **2011**, *43*, 798–810. [CrossRef] [PubMed]

34. Xu, K.; Wu, Z.J.; Groner, A.C.; He, H.H.; Cai, C.; Lis, R.T.; Wu, X.; Stack, E.C.; Loda, M.; Liu, T.; et al. EZH2 oncogenic activity in castration-resistant prostate cancer cells is Polycomb-independent. *Science* **2012**, *338*, 1465–1469. [CrossRef] [PubMed]

35. Kim, E.; Kim, M.; Woo, D.H.; Shin, Y.; Shin, J.; Chang, N.; Oh, Y.T.; Kim, H.; Rheey, J.; Nakano, I.; et al. Phosphorylation of EZH2 activates STAT3 signaling via STAT3 methylation and promotes tumorigenicity of glioblastoma stem-like cells. *Cancer Cell* **2013**, *23*, 839–852. [CrossRef] [PubMed]

36. Dasgupta, M.; Dermawan, J.K.; Willard, B.; Stark, G.R. STAT3-driven transcription depends upon the dimethylation of K49 by EZH2. *Proc. Natl. Acad. Sci. USA* **2015**, *112*, 3985–3990. [CrossRef] [PubMed]

37. He, A.; Shen, X.; Ma, Q.; Cao, J.; von Gise, A.; Zhou, P.; Wang, G.; Marquez, V.E.; Orkin, S.H.; Pu, W.T. PRC2 directly methylates GATA4 and represses its transcriptional activity. *Genes Dev.* **2012**, *26*, 37–42. [CrossRef] [PubMed]

38. Lee, J.M.; Lee, J.S.; Kim, H.; Kim, K.; Park, H.; Kim, J.Y.; Lee, S.H.; Kim, I.S.; Kim, J.; Lee, M.; et al. EZH2 generates a methyl degron that is recognized by the DCAF1/DDB1/CUL4 E3 ubiquitin ligase complex. *Mol. Cell* **2012**, *48*, 572–586. [CrossRef] [PubMed]

39. Gunawan, M.; Venkatesan, N.; Loh, J.T.; Wong, J.F.; Berger, H.; Neo, W.H.; Li, L.Y.; La Win, M.K.; Yau, Y.H.; Guo, T.; et al. The methyltransferase EZH2 controls cell adhesion and migration through direct methylation of the extranuclear regulatory protein talin. *Nat. Immunol.* **2015**, *16*, 505–516. [CrossRef] [PubMed]

40. Breitman, T.R.; Selonick, S.E.; Collins, S.J. Induction of differentiation of the human promyelocytic leukemia cell line (HL-60) by retinoic acid. *Proc. Natl. Acad. Sci. USA* **1980**, *77*, 2936–2940. [CrossRef] [PubMed]

41. Ma, K.; Vitek, O.; Nesvizhskii, A.I. A statistical model-building perspective to identification of MS/MS spectra with PeptideProphet. *BMC Bioinform.* **2012**, *13* (Suppl. S16). [CrossRef] [PubMed]

42. Mellacheruvu, D.; Wright, Z.; Couzens, A.L.; Lambert, J.P.; St-Denis, N.A.; Li, T.; Miteva, Y.V.; Hauri, S.; Sardiu, M.E.; Low, T.Y.; et al. The CRAPome: A contaminant repository for affinity purification-mass spectrometry data. *Nat. Methods* **2013**, *10*, 730–736. [CrossRef] [PubMed]

43. Teo, G.; Liu, G.; Zhang, J.; Nesvizhskii, A.I.; Gingras, A.C.; Choi, H. SAINTexpress: Improvements and additional features in Significance Analysis of INTeractome software. *J. Proteom.* **2014**, *100*, 37–43. [CrossRef] [PubMed]

44. Hauri, S.; Comoglio, F.; Seimiya, M.; Gerstung, M.; Glatter, T.; Hansen, K.; Aebersold, R.; Paro, R.; Gstaiger, M.; Beisel, C. A High-Density Map for Navigating the Human Polycomb Complexome. *Cell Rep.* **2016**, *17*, 583–595. [CrossRef] [PubMed]

45. Morris, J.H.; Apeltsin, L.; Newman, A.M.; Baumbach, J.; Wittkop, T.; Su, G.; Bader, G.D.; Ferrin, T.E. clusterMaker: A multi-algorithm clustering plugin for Cytoscape. *BMC Bioinform.* **2011**, *12*, 436. [CrossRef] [PubMed]

46. Bader, G.D.; Hogue, C.W. An automated method for finding molecular complexes in large protein interaction networks. *BMC Bioinform.* **2003**, *4*, 2. [CrossRef]

47. Nepusz, T.; Yu, H.; Paccanaro, A. Detecting overlapping protein complexes in protein-protein interaction networks. *Nat. Methods* **2012**, *9*, 471–472. [CrossRef] [PubMed]

48. Shannon, P.; Markiel, A.; Ozier, O.; Baliga, N.S.; Wang, J.T.; Ramage, D.; Amin, N.; Schwikowski, B.; Ideker, T. Cytoscape: A software environment for integrated models of biomolecular interaction networks. *Genome Res.* **2003**, *13*, 2498–2504. [CrossRef] [PubMed]

49. Maere, S.; Heymans, K.; Kuiper, M. BiNGO: A Cytoscape plugin to assess overrepresentation of gene ontology categories in biological networks. *Bioinformatics* **2005**, *21*, 3448–3449. [CrossRef] [PubMed]

50. Huang da, W.; Sherman, B.T.; Lempicki, R.A. Systematic and integrative analysis of large gene lists using DAVID bioinformatics resources. *Nat. Protoc.* **2009**, *4*, 44–57. [CrossRef] [PubMed]

51. Szklarczyk, D.; Franceschini, A.; Wyder, S.; Forslund, K.; Heller, D.; Huerta-Cepas, J.; Simonovic, M.; Roth, A.; Santos, A.; Tsafou, K.P.; et al. STRING v10: Protein-protein interaction networks, integrated over the tree of life. *Nucleic Acids Res.* **2015**, *43*, D447–D452. [CrossRef] [PubMed]

52. Wang, Y.; Arribas-Layton, M.; Chen, Y.; Lykke-Andersen, J.; Sen, G.L. DDX6 Orchestrates Mammalian Progenitor Function through the mRNA Degradation and Translation Pathways. *Mol. Cell* **2015**, *60*, 118–130. [CrossRef] [PubMed]

53. Guo, A.; Gu, H.; Zhou, J.; Mulhern, D.; Wang, Y.; Lee, K.A.; Yang, V.; Aguiar, M.; Kornhauser, J.; Jia, X.; et al. Immunoaffinity enrichment and mass spectrometry analysis of protein methylation. *Mol. Cell. Proteom.* **2014**, *13*, 372–387. [CrossRef] [PubMed]

54. Cao, X.J.; Arnaudo, A.M.; Garcia, B.A. Large-scale global identification of protein lysine methylation in vivo. *Epigenetics* **2013**, *8*, 477–485. [CrossRef] [PubMed]

55. Polevoda, B.; Sherman, F. Methylation of proteins involved in translation. *Mol. Microbiol.* **2007**, *65*, 590–606. [CrossRef] [PubMed]

56. Laugesen, A.; Hojfeldt, J.W.; Helin, K. Role of the Polycomb Repressive Complex 2 (PRC2) in Transcriptional Regulation and Cancer. *Cold Spring Harb. Perspect. Med.* **2016**, *6*. [CrossRef] [PubMed]

57. McCabe, M.T.; Ott, H.M.; Ganji, G.; Korenchuk, S.; Thompson, C.; Van Aller, G.S.; Liu, Y.; Graves, A.P.; Della Pietra, A.; Diaz, E.; et al. EZH2 inhibition as a therapeutic strategy for lymphoma with EZH2-activating mutations. *Nature* **2012**, *492*, 108–112. [CrossRef] [PubMed]

58. Knutson, S.K.; Wigle, T.J.; Warholic, N.M.; Sneeringer, C.J.; Allain, C.J.; Klaus, C.R.; Sacks, J.D.; Raimondi, A.; Majer, C.R.; Song, J.; et al. A selective inhibitor of EZH2 blocks H3K27 methylation and kills mutant lymphoma cells. *Nat. Chem. Biol.* **2012**, *8*, 890–896. [CrossRef] [PubMed]

59. Knutson, S.K.; Warholic, N.M.; Wigle, T.J.; Klaus, C.R.; Allain, C.J.; Raimondi, A.; Porter Scott, M.; Chesworth, R.; Moyer, M.P.; Copeland, R.A.; et al. Durable tumor regression in genetically altered malignant rhabdoid tumors by inhibition of methyltransferase EZH2. *Proc. Natl. Acad. Sci. USA* **2013**, *110*, 7922–7927. [CrossRef] [PubMed]

60. Wee, Z.N.; Li, Z.; Lee, P.L.; Lee, S.T.; Lim, Y.P.; Yu, Q. EZH2-mediated inactivation of IFN-γ-JAK-STAT1 signaling is an effective therapeutic target in MYC-driven prostate cancer. *Cell Rep.* **2014**, *8*, 204–216. [CrossRef] [PubMed]

61. Kim, W.; Bird, G.H.; Neff, T.; Guo, G.; Kerenyi, M.A.; Walensky, L.D.; Orkin, S.H. Targeted disruption of the EZH2-EED complex inhibits EZH2-dependent cancer. *Nat. Chem. Biol.* **2013**, *9*, 643–650. [CrossRef] [PubMed]

62. Hamrita, B.; Nasr, H.B.; Hammann, P.; Kuhn, L.; Guillier, C.L.; Chaieb, A.; Khairi, H.; Chahed, K. An Elongation factor-like protein (EF-Tu) elicits a humoral response in infiltrating ductal breast carcinomas: An immunoproteomics investigation. *Clin. Biochem.* **2011**, *44*, 1097–1104. [CrossRef] [PubMed]

63. Rehman, I.; Evans, C.A.; Glen, A.; Cross, S.S.; Eaton, C.L.; Down, J.; Pesce, G.; Phillips, J.T.; Yen, O.S.; Thalmann, G.N.; et al. iTRAQ identification of candidate serum biomarkers associated with metastatic progression of human prostate cancer. *PLoS ONE* **2012**, *7*, e30885. [CrossRef]

64. Huang, Y.; Hu, J.D.; Qi, Y.L.; Wu, Y.A.; Zheng, J.; Chen, Y.Y.; Huang, X.L. Effect of knocking down eEF1A1 gene on proliferation and apoptosis in Jurkat cells and its mechanisms. *Zhongguo Shi Yan Xue Ye Xue Za Zhi* **2012**, *20*, 835–841. [PubMed]

65. Bhat, M.; Robichaud, N.; Hulea, L.; Sonenberg, N.; Pelletier, J.; Topisirovic, I. Targeting the translation machinery in cancer. *Nat. Rev. Drug Discov.* **2015**, *14*, 261–278. [CrossRef] [PubMed]

66. Keller, A.; Nesvizhskii, A.I.; Kolker, E.; Aebersold, R. Empirical statistical model to estimate the accuracy of peptide identifications made by MS/MS and database search. *Anal. Chem.* **2002**, *74*, 5383–5392. [CrossRef] [PubMed]

67. Liu, H.; Sadygov, R.G.; Yates, J.R., III. A Model for Random Sampling and Estimation of Relative Protein Abundance in Shotgun Proteomics. *Anal. Chem.* **2004**, *76*, 4193–4201. [CrossRef] [PubMed]

68. Morris, J.H.; Knudsen, G.M.; Verschueren, E.; Johnson, J.R.; Cimermancic, P.; Greninger, A.L.; Pico, A.R. Affinity purification-mass spectrometry and network analysis to understand protein-protein interactions. *Nat. Protoc.* **2014**, *9*, 2539–2554. [CrossRef] [PubMed]

69. Pettersen, E.F.; Goddard, T.D.; Huang, C.C.; Couch, G.S.; Greenblatt, D.M.; Meng, E.C.; Ferrin, T.E. UCSF Chimera—A visualization system for exploratory research and analysis. *J. Comput. Chem.* **2004**, *25*, 1605–1612. [CrossRef] [PubMed]

© 2017 by the authors. Licensee MDPI, Basel, Switzerland. This article is an open access article distributed under the terms and conditions of the Creative Commons Attribution (CC BY) license (http://creativecommons.org/licenses/by/4.0/).

International Journal of
Molecular Sciences

MDPI

Article

Diverse Regulation of Vitamin D Receptor Gene Expression by 1,25-Dihydroxyvitamin D and ATRA in Murine and Human Blood Cells at Early Stages of Their Differentiation

Sylwia Janik [1,†], Urszula Nowak [2,†], Agnieszka Łaszkiewicz [1], Anastasiia Satyr [2], Michał Majkowski [1], Aleksandra Marchwicka [2], Łukasz Śnieżewski [1], Klaudia Berkowska [2], Marian Gabryś [3], Małgorzata Cebrat [1] and Ewa Marcinkowska [2,*]

[1] Laboratory of Molecular and Cellular Immunology, Department of Tumor Immunology, Institute of Immunology and Experimental Therapy, Polish Academy of Science, Weigla 12, 53-114 Wrocław, Poland; sw90@interia.pl (S.J.); bijbo@interia.pl (A.Ł.); michal.majkowski@iitd.pan.wroc.pl (M.M.); lukasz.sniezewski@iitd.pan.wroc.pl (Ł.Ś.); cebrat@iitd.pan.wroc.pl (M.C.)

[2] Laboratory of Protein Biochemistry, Faculty of Biotechnology, University of Wrocław, Joliot-Curie 14a, 50-383 Wrocław, Poland; urszula.nowak.bio@gmail.com (U.N.); anastasiia.satyr@gmail.com (A.S.); alexandramarchwicka@interia.pl (A.M.); airealphiel@wp.pl (K.B.)

[3] First Department of Obstetrics and Gynecology, Wrocław Medical University, Chałubińskiego 3, 50-368 Wrocław, Poland; mst_gabrys@post.pop.pl

* Correspondence: ema@cs.uni.wroc.pl; Tel.: +48-71-375-2929

† These authors contributed equally to this work.

Academic Editor: Anthony Lemarié

Received: 2 May 2017; Accepted: 14 June 2017; Published: 21 June 2017

Abstract: Vitamin D receptor (VDR) is present in multiple blood cells, and the hormonal form of vitamin D, 1,25-dihydroxyvitamin D (1,25D) is essential for the proper functioning of the immune system. The role of retinoic acid receptor α (RARα) in hematopoiesis is very important, as the fusion of RARα gene with PML gene initiates acute promyelocytic leukemia where differentiation of the myeloid lineage is blocked, followed by an uncontrolled proliferation of leukemic blasts. RARα takes part in regulation of *VDR* transcription, and unliganded RARα acts as a transcriptional repressor to *VDR* gene in acute myeloid leukemia (AML) cells. This is why we decided to examine the effects of the combination of 1,25D and all-*trans*-retinoic acid (ATRA) on *VDR* gene expression in normal human and murine blood cells at various steps of their development. We tested the expression of *VDR* and regulation of this gene in response to 1,25D or ATRA, as well as transcriptional activities of nuclear receptors VDR and RARs in human and murine blood cells. We discovered that regulation of *VDR* expression in humans is different from in mice. In human blood cells at early stages of their differentiation ATRA, but not 1,25D, upregulates the expression of *VDR*. In contrast, in murine blood cells 1,25D, but not ATRA, upregulates the expression of *VDR*. VDR and RAR receptors are present and transcriptionally active in blood cells of both species, especially at early steps of blood development.

Keywords: blood cells; vitamin D receptor; retinoic acid receptors; expression; *CYP24A1*; *CYP26A1*; differentiation; hematopoietic stem cells

---

1. Introduction

Retinoic acid (RA), an active metabolite of vitamin A, and the hormonal form of vitamin D, 1,25-dihydroxyvitamin D (1,25D), are very active compounds, which regulate many important cellular processes, such as differentiation and proliferation [1,2]. RA and 1,25D are the ligands for nuclear receptors, which act as transcription factors after binding the ligand. A dominating RA metabolite in humans is all-*trans*-RA (ATRA), which binds with high affinity to all retinoic acid receptors (RARα, β and γ). A less abundant metabolite, which is nevertheless present in almost all tissues, is 13-*cis*-RA, which most probably serves as a depot for isomerisation to ATRA or to 9-*cis*-RA. 9-*cis*-RA is hard to detect in human tissues, and it binds predominantly to retinoid X receptors (RXRα, β and γ) [3,4]. 1,25D is a ligand to only one vitamin D receptor (VDR), which in its active state forms heterodimers with RXRs [5]. After ligation, VDR and RARs undergo conformational changes that induce binding to specific sequences in the promoter regions of target genes. These sequences are called vitamin D response elements (VDRE) and retinoic acid response elements (RARE) [6]. The role of RARα in blood development is very important, as the fusion of the RARα gene with the PML gene, caused by a translocation t(15;17), initiates acute promyelocytic leukemia where differentiation of the myeloid lineage is blocked at a promyelocyte stage and followed by an uncontrolled proliferation of leukemic blasts [7]. It is also well known and widely accepted that VDR is present in multiple blood cells, and that the correct levels of 1,25D are essential for proper functioning of the immune system [8]. Both compounds, ATRA and 1,25D, can be used in therapy to induce differentiation of acute myeloid leukemia (AML) blasts. ATRA induces differentiation of these blasts to granulocyte-like cells, and 1,25D to monocyte-like cells [9,10]. Past research indicated a synergistic cancer differentiation effect by using a combination of 1,25D and ATRA [11]. However, our recent experiments have revealed that in AML cell lines the effect of combination treatment varies, due to either down- or upregulation of *VDR* expression in response to ATRA, depending on the AML cell line examined [12,13].

Since beneficial effects of 1,25D and ATRA combination treatment in anticancer therapy have been reported and their wider use postulated [14], the effects of such combination towards normal cells should be addressed. Hematopoiesis seems to be the most relevant process which might be influenced by ATRA and 1,25D. The roles of vitamin A and its most active metabolites during hematopoiesis have been extensively studied and are well appreciated [15]. The actions of RA are multiple, and they start as early as in embryonal yolk sac and aorta-gonad-mesonephros, where RA causes the appearance of hematopoietic progenitors from the hemogenic endothelium [16]. In adult hematopoiesis, RA is important for granulopoiesis, and it controls differentiation of B and T lymphocytes [15]. However, it should be remembered that, due to difficulties in the use of human models of hematopoiesis, mice models have often been used in the experiments [15]. The role of 1,25D in hematopoiesis is less well documented than that of ATRA; moreover, some of the data come from zebrafish models. It should be remembered that, in contrast to humans and mice, there are two forms of VDR in zebrafish [17]. However, the available data show that the correct levels of 1,25D are necessary to maintain hematopoietic stem and progenitor cells (HSPCs) [18]. It was also shown that in human hematopoietic stem cells (HSCs) exposed to physiological concentrations of 1,25D, markers of monocytic differentiation are induced [19].

The gene encoding human VDR is located on chromosome 12. This gene is composed of 14 exons, and translation of VDR protein starts from the exon 2. Region 5′ of human *VDR* gene is very complex, and is composed of the seven exons 1a–g. These exons, together with corresponding promoter regions, are alternatively used for *VDR* transcription in different tissues. Transcripts starting from exon 1a and from exon 1d are regulated by the common promoter upstream to exon 1a, and the exons 1f and 1c have separate upstream promoters [20]. Our recent experiments have revealed a new exon, 1g, regulated from the promoter of exon 1a. Exon 1g is used in *VDR* transcripts present in AML cells [13]. Multiple publications confirm that *VDR* expression in humans is regulated in response to ATRA [12,21–23], while there are conflicting reports concerning regulation of human *VDR* by 1,25D [13,24–26].

The murine *VDR* gene is located on chromosome 15, and its composition is less complex than in humans. In the 5′ UTR region of *VDR* gene, exons 1 and 2 were identified, which show strong homology to human 1a and 1c, respectively [27]. Although exon 1d is well conserved (1d-like), transcripts containing this exon have not been reported in mice. The sequence similarity of the exons 1f and 1b is low between man and mice. Translation of mouse VDR protein starts from exon 3 [28]. It has been shown that transcription of *VDR* is upregulated in response to 1,25D in murine osteoblasts [29,30].

This is why we decided to examine the effects of the 1,25D and ATRA combination on *VDR* gene expression in blood cells at various steps of their development. We were interested in discovering if, in normal human blood cells, transcriptional variants of the *VDR* gene are as multiple as in AML cells, and if they are regulated in response to ATRA and 1,25D. Since the availability of human hematopoietic cells for experiments is very limited, we decided to examine whether human cells could be replaced in this type of studies with murine blood model.

An important question in studies concerning nuclear receptors is whether or not they are transcriptionally active in the cells. It is therefore important to be able to study expression of the genes that are specific targets of regulation by either VDR or RARs. In the case of VDR, expression of the gene which encodes 24-hydroxylase of 1,25D (*CYP24A1*) is the best measure of VDR's activity. 24-Hydroxylase of 1,25D is the central enzyme in the catabolism of 1,25D to calcitroic acid. *CYP24A1* is the most strongly regulated out of all 1,25D-target genes [31]. This is why 1,25D-dependent upregulation of *CYP24A1* confirms that VDR protein is expressed and active in cells. *CYP24A1* is upregulated in response to 1,25D, but not in response to ATRA [13], and its expression can also be detected in human HSCs [18]. We decided to search for a similar sensor of RARs transcriptional activity. The effective concentration of the most active metabolite of vitamin A, ATRA is strictly controlled in the human body. In order to maintain safe concentrations of ATRA, its catabolism is also strictly regulated [32]. The major catabolizing enzyme is retinoic acid 4-hydroxylase (CYP26A1), whose transcription is upregulated in response to ATRA [33] by activating RARE, which is present in the *CYP26A1* promoter region [34]. Thus, our first aim was to verify that *CYP26A1* is regulated by ATRA in RAR-positive cells, and that it is not regulated by 1,25D in VDR-positive cells.

2. Results

2.1. Activation Of Retinoic Acid 4-Hydroxylase (CYP26A1) Gene Expression in Acute Myeloid Leukemia (AML) Cells

In a search for the sensor of transcriptional activity of RARs, we decided to concentrate on isoforms of CYP26. According to the available literature, CYP26A1 is the most ubiquitously expressed isoform [32], present in HSCs [35], and exhibits the greatest increase upon stimulation with ATRA [36]. Therefore, we wanted to verify that the gene encoding CYP26A1 is upregulated in AML cells in response to ATRA-treatment, and what the kinetics of its activation are. Our initial experiments conducted on HL60 cells revealed that the kinetics of *CYP26A1* upregulation are: a slow, significant increase can be seen after 72 h of treatment, and after 96 h, expression is about 200–250 times higher than in vehicle-treated cells. Then we treated HL60 cells with ATRA at concentrations ranging from 10 nM to 1 µM for 96 h. It appeared that at 100 nM we could detect noticeable upregulation of *CYP26A1* expression, but in order to obtain significant upregulation, cells had to be exposed to 1 µM ATRA. In our next experiments, HL60 cells were exposed to either 1 µM ATRA or to 10 nM 1,25D. From our past experiments, we know that 10 nM 1,25D significantly upregulates the gene encoding CYP24A1, and that ATRA does not regulate this gene in many different AML cell lines [13]. Now we wanted to determine if 1,25D has any influence on *CYP26A1* expression. HL60 cells are very sensitive to 1,25D [12], and are moderately sensitive to ATRA. We also tested how *CYP26A1* is regulated in KG1 cells. These cells have higher expression levels of *RARA* and *RARB* genes, and a higher level of RARα protein than HL60 [13,37], but they have low expression of *VDR* [13]. Our experiments showed that the expression of *CYP26A1* could be upregulated in HL60 cells by ATRA by up to 250 times, compared to the control cells, but is not regulated by 1,25D. In KG1 cells, ATRA induced upregulation of *CYP26A1* as well, and again this gene was not regulated by 1,25D (Figure 1). Therefore we decided to measure levels of *CYP26A1* mRNA in order to detect transcriptional responses to ATRA in our next experiments performed using blood cells.

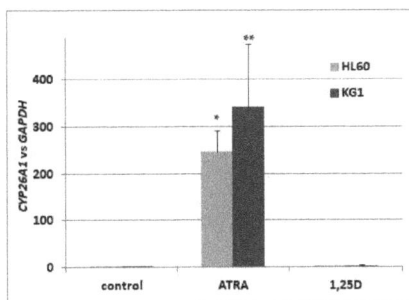

Figure 1. Expression of retinoic acid 4-hydroxylase gene (*CYP26A1*) in acute myeloid leukemia (AML) cell lines exposed to all-*trans*-retinoic acid (ATRA) or to 1,25-dihydroxyvitamin D (1,25D). HL60 and KG1 cells were exposed to 1 μM ATRA or to 10 nM 1,25D and after 96 h the expression of *CYP26A1* mRNA was measured by Real-time polymerase chain reaction (PCR). The bars represent mean values (±standard error of the mean (SEM)) of the fold changes in mRNA levels relative to glyceraldehyde 3-phosphate dehydrogenase (*GAPDH*) mRNA levels. Values significantly different from these obtained for respective control cells are marked with asterisks (* $p < 0.01$, ** $p < 0.05$).

2.2. Detection of Transcriptional Variants of Vitamin D Receptor Gene (VDR) and Their Regulation in Response to All-trans-Retinoic Acid (ATRA) in Normal Human Blood Cells

In our previous experiments, we found that in HL60 cells *VDR* transcripts originated from exon 1a, which was spliced to exon 2. KG1 cells express a more varied set of *VDR* transcripts than HL60 cells. In unstimulated KG1 cells, transcript variants originating from exon 1a, 1d and 1g were detected. In these cells, transcripts containing exon 1b present in transcripts originating from exon 1a or 1g were also found. Transcripts originating from exons 1a and 1g appeared to be regulated by ATRA [13]. Now we wanted to verify whether, in normal human blood cells, such a large variety of *VDR* transcripts could also be detected. For this reason, we used variant-specific primers in order to detect the above transcripts and to check if they are regulated in response to ATRA. Primer sequences for general *VDR* transcripts and for variant-specific transcripts are given in the Materials and Methods section, and graphical representation of polymerase chain reaction (PCR) products is shown in Figure A1.

For our experiments, we used three different sets of blood cells. The first set consisted of the mononuclear cells from the peripheral blood of healthy adults (PBM), isolated using Histopaque-based centrifugation. The second (UCB) consisted of mononuclear cells from human umbilical cord blood, which is strongly enriched in HSCs and progenitor cells in comparison to peripheral blood [38]. The third set (HSCs) was up to 95% enriched in CD34+ cells isolated from UCB (efficacy of CD34+ cells isolation is presented in Figure A2). All three transcriptional variants were present in human blood cells, however only in UCB and HSC cells were *VDR* and its transcriptional variants regulated by ATRA (Figure 2a–c). Our results suggest that there also exist other transcriptional variants of *VDR* in addition to the tested ones. This is because, the regulation of total *VDR* transcripts in response to ATRA is stronger than the regulation of specific transcripts identified in our earlier studies.

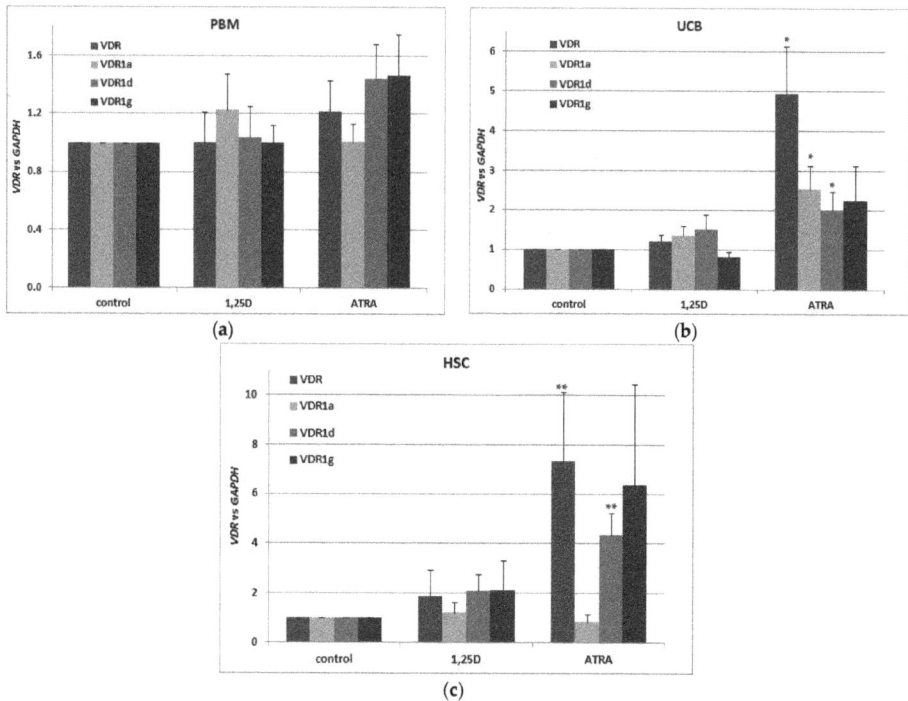

Figure 2. Regulation of *VDR* by 1,25D or ATRA in human blood cells. Human blood cells were isolated as described in Materials and Methods. The cells were exposed ex vivo to 10 nM 1,25D or to 1 μM ATRA for 96 h. Then mRNA was isolated and expression of *VDR* transcriptional variants was measured by Real-time PCR for peripheral blood of healthy adults (PBM) (**a**), mononuclear cells from human umbilical cord blood (UCB) (**b**) and hematopoietic stem cells (HSC) (**c**). The bars represent mean values (±SEM) of the fold changes in mRNA levels relative to *GAPDH* mRNA levels. Values that differ significantly from these obtained for respective control are marked with asterisks (* $p < 0.01$, ** $p < 0.05$).

2.3. Detection Of Transcriptional Activities of VDR and Retinoic Acid Receptors (RARs) in Normal Human Blood Cells

Next, we wanted to verify whether *VDR* transcripts in these cells are translated into a transcriptionally active VDR protein. This is why we exposed the blood cells to 10 nM 1,25D and/or 1 μM ATRA and tested expression of *CYP24A1* in these cells. As presented in Figure 3a, 1,25D caused an upregulation of *CYP24A1* mRNA levels, but co-treatment with 1 μM ATRA did not cause further upregulation. This shows that ATRA-induced upregulation of *VDR* transcription was not followed by an increase in transcriptionally active VDR protein. Then, we wanted to verify that ATRA did, indeed, activate RARs in these cells. This is why we checked the expression of *CYP26A1* in blood cells exposed to 1 μM ATRA for 96 h. As presented in Figure 3b, transcriptional activity in response to ATRA was very high in HSCs, and moderate in UCB cells. In PBM cells from healthy adults, ATRA did not cause upregulation of *CYP26A1*, which might indicate that RAR receptors are not present in these cells. This might also explain the lack of regulation of *VDR* in response to ATRA in PBM cells presented in Figure 2a. *CYP26A1* in normal blood cells was not upregulated in response to 1,25D (Figure 3b).

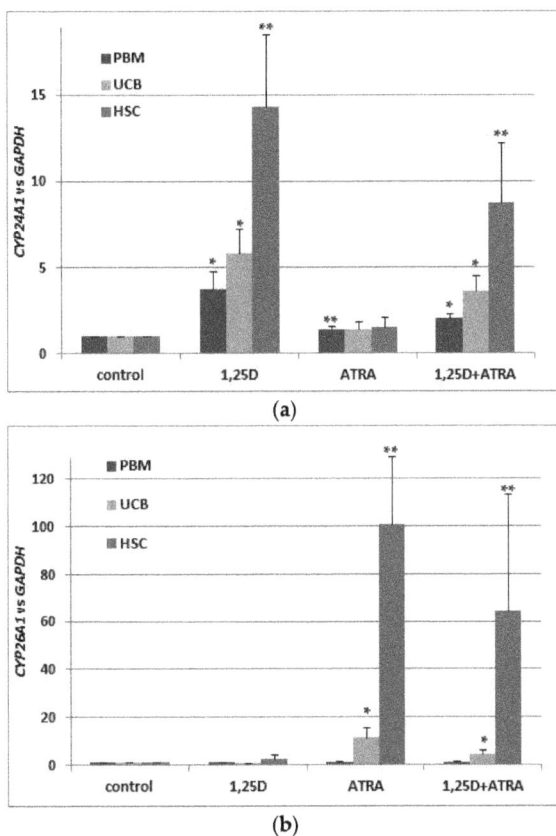

Figure 3. Transcriptional activity of VDR (a) and RARs (b) in human blood cells. Human blood cells were isolated as described in Materials and Methods. The cells were exposed ex vivo to 10 nM 1,25D $\pm$ 1 µM ATRA for 96 h. Then mRNA was isolated and *CYP24A1* (a) or *CYP26A1* (b) expression was measured by Real-time PCR. The bars represent mean values ($\pm$SEM) of the fold changes in mRNA levels relative to *GAPDH* mRNA levels. Values that are significantly higher than from these obtained for respective control are marked with asterisks (* $p < 0.01$, ** $p < 0.05$).

2.4. Detection of Transcriptional Variants of VDR in Murine Blood Cells Using 5′-Rapid Amplification of cDNA Ends (RACE) Assay

The murine *VDR* transcript consists of 10 exons, of which exons 3–10 contain the open reading frame. The organization of the murine *VDR* locus is similar to human, however less non-coding 5′ exons has been identified in the murine locus (Figure 4 top). We used the 5′-RACE method to identify the 5′-ends and transcriptional start sites of *VDR* transcripts in the selected murine tissues: intestine, kidney and bone marrow. We have found that the transcription start sites of murine *VDR* are localized in a single region spanning 45 nucleotides, and identified only one splice variant of the *VDR* transcript (Figure 4 bottom). We did not find any strong tissue-specific preference in the transcription start sites used. This observation strongly suggests that, in contrast to the complex regulation of *VDR* transcription in human cells controlled by several promoters, *VDR* transcription in mice is controlled by a single promoter region.

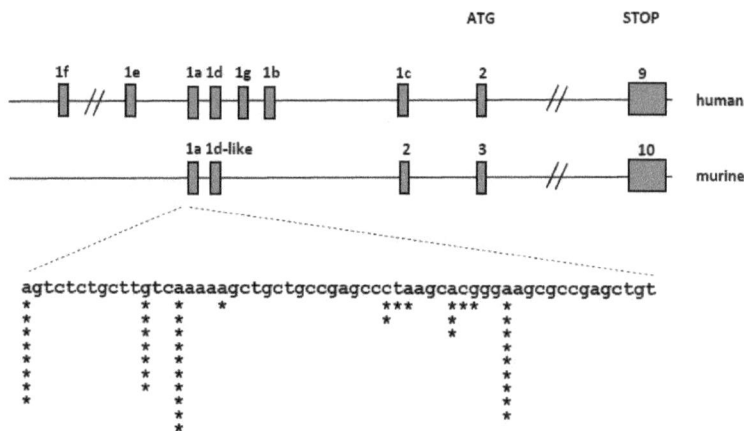

Figure 4. Organization of the human (**upper panel**) and murine (**middle panel**) *VDR* locus. Gray boxes represent exons. The expanded view (**bottom panel**) represents the genomic sequence surrounding the transcription start sites of the murine transcripts—asterisks represent the localization of the start sites identified during 5′-RACE, the number of asterisks—the number of transcripts starting at a given site.

2.5. Basal Expression of VDR in Murine Kidneys and in Blood Cells

Knowing that there is only one transcriptional variant of *VDR* gene present in C57BL/6 mice, we decided to test its basic expression in different tissues. We used kidneys as a reference 1,25D-responsive organ, therefore we normalized expression of *VDR* in remaining tissues to the expression measured for kidneys. We used pooled bone marrow cells (BM), bone marrow granulocytes isolated using CD45/SSC-based sorting (BM granulocytes), hematopoietic stem and progenitor cells isolated from bone marrow (BM HSPC), pooled thymocytes isolated from thymus, pooled spleen cells, spleen T cells isolated using anti-CD3 and spleen B cells isolated using anti-CD19 antibody. Indeed, the expression of *VDR* appeared to be the highest in kidneys, and in all other tissues tested it was significantly lower except of BM HSPC. The results are presented in Figure 5.

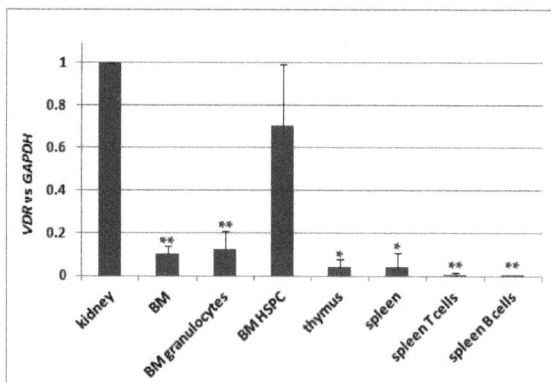

Figure 5. Expression of *VDR* in C57BL/6 mice. Tissues from 2 to 9 mice were isolated as described in Materials and Methods. mRNA was isolated and *VDR* expression was measured by Real-time PCR. The bars represent mean values (±SEM) of the fold changes in mRNA levels relative to *GAPDH* mRNA levels. Values that are significantly different from these obtained for kidney are marked with asterisks (* $p < 0.01$, ** $p < 0.05$).

2.6. Regulation of VDR Expression by 1,25D and ATRA in Murine Kidneys and Blood Cells

Our next goal was to verify whether, in murine blood cells, *VDR* expression is regulated by ATRA in a similar manner to human blood cells. Since previous reports indicated a 1,25D-induced regulation of *VDR* in murine osteoblasts [29,30], we decided to also test this compound. Therefore we took different fractions of cells from bone marrow, and kidney cells as a control of 1,25D-responsive tissue, and exposed them ex vivo to either 10 nM 1,25D or to 1 µM ATRA for 96 h, and then we measured the levels of *VDR* mRNA relative to glyceraldehyde 3-phosphate dehydrogenase (*GAPDH*) expression. In kidneys, expression of *VDR* was not regulated to a significant degree by either of the compounds used. In pooled BM cells, there was significant upregulation of *VDR* expression in response to 1,25D, which was not present in BM granulocytes. The strongest and the most significant upregulation of *VDR* expression in response to 1,25D was detected in BM HSPC. ATRA did not regulate *VDR* expression in all of the cells tested (Figure 6).

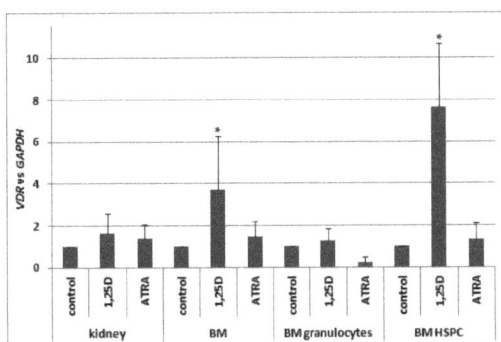

Figure 6. Regulation of *VDR* by 1,25D or ATRA in murine cells. Tissues from 3 to 10 mice were isolated as described in Materials and Methods. The cells were exposed ex vivo to 10 nM 1,25D or to 1 µM ATRA for 96 h. Then mRNA was isolated and *VDR* expression was measured by Real-time PCR. The bars represent mean values (±SEM) of the fold changes in mRNA levels relative to *GAPDH* mRNA levels. Values that differ significantly ($p < 0.01$) from those obtained for respective controls are marked with asterisks.

2.7. Detection of Transcriptional Activities of RARs and VDR in Murine Blood Cells

Since in murine blood cells an expression of *VDR* was not regulated by ATRA as in human blood cells, we hypothesized that RARs might not be present in these cells. Since RARs are ligand-activated transcription factors, and ATRA is a universal ligand for all RAR isoforms, we decided to test expression of *CYP26A1* as a RAR-target gene. Again, we took different populations of cells from bone marrow and kidney cells, and exposed them ex vivo to either 10 nM 1,25D or/and to 1 µM ATRA for 96 h. After ex vivo stimulation, mRNA was isolated from the cells, the expression of *CYP26A1* was tested, and *GAPDH* expression was used as a reference. Relative expression of *CYP26A1* in control cells (exposed to vehicle) was calculated as 1. As presented in Figure 7a, *CYP26A1* was upregulated in all samples exposed to ATRA; the highest and most significant upregulation was observed in pooled BM cells and in BM HSPC. These results indicate that RARs are expressed in BM and specifically in HSPC, and are transcriptionally active.

Our next goal was to verify whether the *VDR* mRNA detected in murine blood cells was effectively translated into a transcriptionally active protein. Similarly as with human cells, we decided to test this activity by measuring the expression of its most strongly regulated target gene in the cells exposed to 10 nM 1,25D. As we used the cDNA obtained for the experiment described above, we tested expression of *CYP24A1* in the cells exposed ex vivo to either 10 nM 1,25D or/and to 1 µM ATRA for 96 h. Again *GAPDH* was used as a reference gene, and relative expression of *CYP24A1* in control cells (exposed to vehicle) was calculated as 1. As presented in Figure 7b, 1,25D-induced upregulation of *CYP24A1* is very high (more than 10^3 times) and significant in kidney cells. However, 1,25D-induced expression of *CYP24A1* in BM HSPC is even higher, reaching a level of almost 10^6 times that of vehicle-treated cells. This shows that in HSPCs exposed to 1,25D, *VDR* is present and is transcriptionally active. An unexpected result was obtained in kidney cells and in BM HSPCs exposed to ATRA. In these cells, we observed significant increase of *CYP24A1* expression. The molecular mechanism of this effect will be studied in our future experiments.

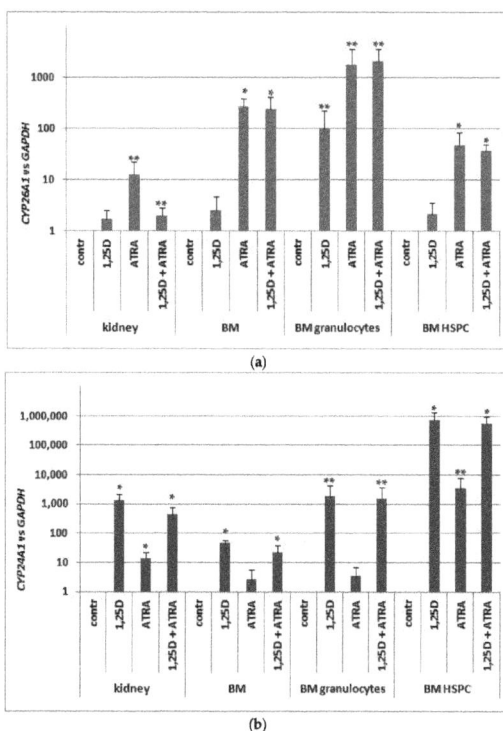

(a)

(b)

Figure 7. Transcriptional activity of RARs (**a**) and VDR (**b**) in murine cells. Tissues from 3 to 8 mice were isolated as described in Materials and Methods. The cells were exposed ex vivo to 10 nM 1,25D ± 1 μM ATRA for 96 h. Then mRNA was isolated and *CYP26A1* (**a**) or *CYP24A1* (**b**) expression was measured by Real-time PCR. The bars represent mean values (±SEM) of the fold changes in mRNA levels relative to *GAPDH* mRNA levels. Values that are significantly higher than these obtained for respective control are marked with asterisks (* $p < 0.01$, ** $p < 0.05$).

3. Discussion

The findings from studies of leukemia cell lines support the use of 1,25D as an anticancer agent, since 1,25D causes the growth arrest and differentiation of a wide variety of AML cell lines [39]. Our results revealed that there are patients whose AML blasts respond to 1,25D analogs with differentiation, while the blasts of others are resistant [39,40]. The possible reason of these differences may lie in the expression level of the *VDR* gene and the VDR protein level in AML cells. We have recently found that low *VDR* expression levels may be upregulated using ATRA and that unliganded RARα acts as transcriptional repressor to *VDR* [13]. The possibility of using RA analogs to induce differentiation of blasts was investigated for many years and in the case of ATRA it was successfully introduced into clinics to treat one subtype of AML [41]. Unfortunately, in other subtypes of leukemia, ATRA is not effective.

This is why the combined use of 1,25D and ATRA, or a combination of their more active analogs, was postulated [14]. In AML cells, the *VDR* gene is transcribed in multiple variants, and some of them are transcriptionally regulated by ATRA. We thus wanted to ascertain whether similar transcriptional variants of VDR can be found in normal human blood cells. Our experiments revealed that transcripts *VDR1a*, *VDR1d* and *VDR1g* can also be detected in normal blood cells, however they are upregulated in response to ATRA, in a manner similar to KG1 cell line, only in blood cells found in UCB. *VDR* protein in UCB cells is transcriptionally active after exposure to 1,25D, and in PBM cells this transcriptional activity is much lower. Moreover, in contrast to AML cell lines, upregulation of *VDR* expression is not followed by an increased transcriptional activity of VDR protein. This might suggest that possible side effects of combination treatment using 1,25D and ATRA in normal human blood cells do not synergize.

It is widely accepted that the eventual cell fate during hematopoiesis is governed by spatiotemporal fluctuations in transcription factor concentrations, which either cooperate or compete in driving target gene expression [42]. Nuclear receptors for vitamin A and vitamin D do not belong to the set of the most important hematopoietic regulators, but their roles in blood cell differentiation are becoming apparent and appreciated. The availability of human HSC cells is limited, and their differentiation can't be studied in vivo. This is why the majority of the available data about blood cell formation comes from murine models, however one should be aware that human hematopoiesis does not reflect murine hematopoiesis in all aspects [43]. This might also concern the roles of VDR and RARs in blood cell formation. As mentioned in the Introduction, the organization and regulation of the *VDR* locus in humans and mice are different, and the 5' UTR region composition is less complex in mice. Despite the high resemblance of the symptoms of *VDR* knock-out in mice and inactivating mutations of *VDR* in humans, not all findings concerning vitamin D endocrine systems in mice are present in humans [44]. This is why we addressed regulation of *VDR* transcription in murine blood cells in parallel to human cells.

Our results indicate that there are differences in the regulation of *VDR* transcription between mice and man. In contrast to human cells, in murine cells ATRA does not influence *VDR* expression, even though RARs are present and transcriptionally active. On the other hand, the *VDR* gene in murine blood is positively auto-regulated by the VDR ligand, 1,25D. This positive feedback causes transcriptional activity of VDR to be particularly high in murine HSPC exposed to 1,25D. This might indicate that VDR is important for murine blood cells at early stages of their commitment. Such auto-regulation of VDR does not occur in human blood cells. Since there is no good alternative for testing the toxicity of potential drugs other than rodents, we suggest that observations concerning toxic effects of 1,25D in mice should be translated to humans with caution.

Another issue to consider is the fact that there are differences in the vitamin D system between mice and man. 1,25D is a steroid hormone, which is entirely produced by organisms from 7-dehydrocholesterol, and biologically activated by subsequent hydroxylations at carbons C25 and C1 [45]. This requires exposure to the UVB light, and vitamin D must only be delivered with food for people who live in regions deficient in sunlight. Mice are much less likely to produce vitamin D in high amounts from exposure to sunlight due to their fur, as well as their nocturnal and underground activities. Therefore, the role and regulation of *VDR* is likely to be different in these two distinct species, and should be taken into consideration when vitamin D compounds are being tested in mice. Very high positive auto-regulation of *VDR* expression in murine blood cells at their early steps of development, which does not occur in humans, might cause unwanted side-effects of 1,25D, or of its highly active analogues, to be more pronounced in mice than in humans.

4. Materials and Methods

4.1. Chemicals

1,25D was purchased from Cayman Europe (Tallinn, Estonia) and ATRA was from Sigma-Aldrich (St. Louis, MO, USA). The compounds were dissolved in an ethanol to reach $1000\times$ final concentrations, and subsequently diluted in the culture medium to the concentration required for experiments.

4.2. Cell Lines and Normal Cells

HL60 cells were acquired from the cell bank at the Institute of Immunology and Experimental Therapy in Wrocław, Poland and KG1 cells were purchased from the German Resource Center for Biological Material (DSMZ GmbH, Braunschweig, Germany). The cells were cultured in RPMI-1640 medium (Biowest, Nuaillé, France) with 10% fetal bovine serum (FBS), 2 mM L-glutamine, 100 units/mL penicillin and 100 µg/mL streptomycin (all from Sigma-Aldrich) and maintained at standard cell culture conditions.

Human UCB was obtained post-delivery at the First Department of Obstetrics and Gynecology, Wrocław Medical University (Wrocław, Poland) from mothers who gave informed consent for this study. The study was accepted by the local Ethical Committee. Two to eight mL of cord blood were diluted with PBS in 1:1 ratio. Diluted blood was carefully layered onto the equal volume of Histopaque 1077 (Sigma-Aldrich), and centrifuged at $400 \times g$ for 30 min. The opaque interface containing mononuclear cells was moved to fresh sterile tube, and washed three times with PBS. The cells were transferred to Biotarget-1 (Biological Industries, Kibbutz Beit-Haemek, Israel) medium containing 4 mM L-glutamine, 100 units/mL penicillin and 100 µg/mL streptomycin and maintained at standard cell culture conditions.

The experiments using animals were performed according to the procedures approved by the First Local Ethical Commission for Animal Experimentation in Wrocław at the Institute of Immunology and Experimental Therapy (permit numbers 21/2016/W, 21/2016/U, 20/2016/U issued on 5 January 2016). Cell suspensions from 8 week old C57BL/6 mice were prepared as follows: bone marrow cells were isolated by washing the femur and tibia with ice-cold PBS stream. Spleen and thymus were washed with ice-cold PBS and strained through 30-µm mesh. Kidneys were washed to remove blood, cut into small $1–2$ mm^2 pieces and incubated in 0.5 mL with colagenase type II (1 mg/mL) and DNAse I (10 units/mL) at 37 °C for 40 min. Cells isolated from bone marrow, spleen and thymus were treated with red blood cell lysis buffer (155 mM NH_4Cl, 10 mM $KHCO_3$, 0.1 mM ethylenediaminetetraacetic acid) to remove erythrocytes. All tissues and cells were mechanically dissociated by the syringe trituration, washed twice with PBS by centrifugation (400 rcf, 5 min, 4 °C) and resuspended in PBS supplemented with 5% FBS. Single cell suspension was filtered through 70-µm mesh.

4.3. Sorting of Blood Cells and Flow Cytometry

Human HSCs were sorted from cord blood mononuclear cells using the Miltenyi MACS CD34 Isolation Kit (Miltenyi Biotec, Bergisch Gladbach, Germany) in accordance with the manufacturer's instructions. Briefly, cord blood mononuclears were resuspended in Separation Buffer (PBS with 10% bovine serum albumin (BSA)) and incubated with FcR Blocking Reagent and magnetic microbeads conjugated to monoclonal mouse anti-human CD34 antibody for 30 min at 4 °C. Labeled cell suspension was sorted using magnetic separator. After three washes with Rinsing Solution (PBS supplemented with 2 mM EDTA and 0.5% BSA), the column was removed from the separator and labeled CD34+ cells were eluted with 1 mL of Rinsing Solution. To determine the purity of CD34+ cell fraction, 1×10^5 cells were stained with phycoerythrin (PE)-conjugated anti-CD34 (Becton Dickinson, San Jose, CA, USA) monoclonal antibody for 60 min on ice. Isotype-identical monoclonal antibodies served as controls. Next, the stained cells were analyzed using flow cytometry (BD Accuri™ C6, Becton Dickinson, San Jose, CA, USA). The purity ranged from 92% to 95%, and sample staining is presented in Figure A1. CD34+ cells were grown in Stemline Hematopoietic Stem Cell Expansion Medium with 4 mM L-glutamine, 100 units/mL penicillin and 100 µg/mL streptomycin, recombinant human cytokines (all from ImmunoTools, Friesoythe, Germany): stem cell factor (100 ng/mL), thrombopoietin (100 ng/mL) and granulocyte colony-stimulating factor (100 ng/mL).

Hematopoietic stem and progenitor cells were isolated from murine bone marrow using Mouse Hematopoeitic Progenitor Cell Isolation Kit (Stemcell, Cologne, Germany) according to the manufacturer's recommendations. Briefly, the cells were resuspended in PBS (with 2% FBS and 1 mM EDTA, rat serum 50 µL/mL at the density of 1×10^8 cells/mL) and incubated with EasySep Mouse

Hematopoietic Progenitor Cell Isolation Cocktail (50 µL/mL) for 15 min at 4 °C. Next, EasySep™ Streptavidin RapidSpheres (75 µL/mL) were added and after 10 min of incubation the cell suspension was sorted with magnets. The purity of the obtained population was monitored by flow cytometry (FACS-Calibur, Becton Dickinson, San Jose, CA, USA) using anti-c-kit-APC (eBioscience, Vienna, Austria) and anti-Sca-1-FITC (eBioscience) staining. According to the manufacturer, the lineage antigen-negative cell content of the isolated fraction typically ranges from 60% to 84%. In our experiments c-kit+ cells constituted 65% of sorted population. Stemline Hematopoietic Stem Cell Expansion Medium (Sigma-Aldrich) and recombinant murine cytokines (all from ImmunoTools): stem cell factor (50 ng/mL), Flt3-ligand (50 ng/mL), thrombopoietin (50 ng/mL) and interleukin-6 (10 ng/mL) were used for further ex vivo culture of the isolated cells.

Murine spleen cells were stained with anti-CD3-APC and anti-CD19-PE antibodies (Becton Dickinson) to isolate mature T- and B-cells, respectively (Figure A3a). Bone marrow cells were stained with anti-CD45-FITC antibody (Becton Dickinson) to isolate granulocytes, using CD45/SSC-based sorting criteria (Figure A3b). Cells were stained in 0.5 mL PBS supplemented with 2% FBS using 1 µg of each antibody for 30 min on ice. Cells were sorted using FACS-Aria (Becton Dickinson).

4.4. 5'-RACE Assay

In order to identify the transcriptional start sites for murine *VDR* transcript(s), 5'-RACE was used [46]. Ten micrograms of total RNA were isolated from intestine, kidney or bone marrow of C57BL/6 mice and then processed as before: digested with calf alkaline phosphatase (CIP, New England Biolabs, Ipswich, MA, USA) in the presence of RiboLock RNAse inhibitor for 1 h at 37 °C and purified by extraction with TRI Reagent (Sigma-Aldrich). Half of the CIP-digested RNA was treated with tobacco acid pyrophosphatase (TAP, Epicentre, Madison, WI, USA) for 1 h at 37 °C in 10 µL reaction mixture containing TAP buffer, 0.5 units of TAP and 20 units of RiboLock RNAse inhibitor. 2 µL of TAP-treated RNA was ligated with a RNA oligonucleotide (5'-GCUGAUGGCGAUGAAUGAACACUGCGUUUGCUGGCUUUGAUGAAA-3') for 1 h at 37 °C in a 10 µL reaction mixture containing 0.3 µg of the oligonucleotide, 5 U of RNA ligase (New England Biolabs), RNA ligase buffer and 20 U of RiboLock RNase inhibitor. 2 µL of RNA was then reverse transcribed using SuperScript III reverse transcriptase (Invitrogen, Carlsbad, CA, USA) and random hexamers. The cDNA was amplified in nested PCR reactions (2×20 cycles, annealing temp. 52 °C, in the presence of 1.2 M betaine) using primers complementary to 5'-adapter (5'-GCTGATGGCGATGAAT GAACACTG-3', 5'-CGCGGATCCGAACACTGCGTTTGCTGGCTTTGATG-3') and exon 5 and 4 of *VDR* gene (5'-TCTGTGAGGATGAACTCCTTCATC-3', 5'-TCCTTGGTGATGCGGCAATCTC-3'). The amplification products were directly cloned into pGEMT-easy vector. The individual plasmid clones were sequenced using SP6 primer (5'-ATTTAGGTGACACTATAG-3') and BigDye 3.1 Terminator Cycle Sequencing Kit (Life Technologies, Carlsbad, CA, USA). The sequencing reaction was analyzed using ABI Prism 310 Genetic Analyzer (Applied Biosystems, Foster City, CA, USA). The obtained sequences of *VDR* transcripts were aligned with the genomic sequence of *VDR* gene using Spidey software (https://www.ncbi.nlm.nih.gov/spidey/ National Center for Biotechnology Information, Bethesda, MD, USA) to identify exons and transcriptional start sites.

4.5. cDNA Synthesis and Real-Time PCR

For PCR analyses, the cells were stimulated with 10 nM 1,25D and/or 1 µM ATRA for 96 h. RNA from unstimulated and stimulated cells was isolated using either TRI Reagent (Sigma-Aldrich), Extractme Total RNA Kit (DNA-Gdańsk, Gdańsk, Poland) (for $>10^6$ cells) or PicoPure RNA Isolation Kit (ThermoFisher Scientific, Waltham, MA, USA) (for $<10^6$ cells) according to manufacturer's recommendations. Reverse transcription of 100 ng of total RNA (34 ng in case of human CD34+ cells) was done with High-Capacity cDNA Reverse Transcription Kit (ThermoFisher Scientific) using random hexamers. The Real-time PCR analysis was performed using Real-time PCR–PowerUp™

SYBR Green Master Mix (Applied Biosystems) or SensiFAST SYBR® No-ROX Kit (Bioline, London, UK). For murine samples the reaction consisted of 40 cycles (95 °C for 15 s and 60 °C for 60 s), preceded by uracil-DNA glycosylase and AmpliTaq DNA polymerase activation at 50 °C for 120 s and 95 °C for 120 s, respectively, and was performed on BioRad CFX Connect apparatus (Bio-Rad Laboratories Inc., Hercules, CA, USA). The thermal profile for human samples consisted of 45 cycles (95 °C for 5 s, 54/58 °C for 10 s, 72 °C for 5 s) followed by one step at 95 °C for 2 min, the reaction was performed using CFX Real-time PCR System (Bio-Rad).

The following primer pairs were used:

hGAPDH: forward 5'-CATGAGAAGTATGACAACAGCCT-3', reverse 5'-AGTCCTTCCACGATA CCAAAGT-3';

hVDR: forward 5'-CCTTCACCATGGACGACATG-3', reverse 5'-CGGCTTTGGTCACGTCACT-3';

hVDR1a: forward 5'-GCGGAACAGCTTGTCCACCC-3', reverse 5'-GAAGTGCTGGCCGC CATTG-3';

hVDR1d: forward 5'-GCTCAGAACTGCTGGAGTGG-3', reverse 5'-GAAGTGCTGGCCGCC ATTG-3';

hVDR1g: forward 5'-TTGCTCATCCAGCTTCCCAGAC-3', reverse 5'-GAAGTGCTGGCCG CCATTG-3';

hCYP24A1: forward 5'-CTCATGCTAAATACCCAGGTG-3', reverse 5'-TCGCTGGCAAAACGCG ATGGG-3';

hCYP26A1: forward 5'-CGCATCGAGCAGAACATTCG-3', reverse 5'-GCTTTAGTGCCTGC ATGT-3';

mGAPDH: forward 5'-AACTTTGGCATTGTGGAAGG-3', reverse 5'-ACACATTGGGGGTAGG AACA-3';

mVDR: forward 5'-CACCTGGCTGATCTTGTCAGT-3', reverse 5'-CTGGTCATCAGAGGTGA GGTC-3';

mCYP24A1: forward 5'-CACGGTAGGCTGCTGAGATT-3', reverse 5'-CCAGTCTTCGCAGTT GTCC-3';

mCYP26A1: forward 5'-GCAGGCACTAAAACAATCGTC-3', reverse 5'-GCTGTTCCAAAGTT TCCATGTC-3'. Relative quantification of gene expression was analyzed with the $\Delta\Delta C_t$ method using *GAPDH* as the endogenous control.

4.6. Statistical Analysis

The sample distribution was assessed using the Shapiro-Wilk test. For samples with normal distribution, *t*-test was used to assess significance of differences. For the remaining samples, a non-parametric one-way ANOVA test followed by a Mann-Whitney *U* test was used for assessing the significance of the differences in gene expression levels.

5. Conlusions

The main findings of this paper indicate that there are differences in the regulation of *VDR* transcription between mice and man. Our study has revealed that in murine blood stem and progenitor cells expression of *VDR* is auto-regulated by ligand-activated VDR. Such auto-regulation does not occur in human blood stem and progenitor cells. On the contrary, in human cells expression of *VDR* is regulated by ligand-activated RARs, and this other kind of regulation does not occur in murine cells.

Acknowledgments: The research was supported by National Science Centre, Poland (grant No. 2015/17/B/ NZ4/02632).

Author Contributions: Ewa Marcinkowska conceived the general idea of the study, obtained financing, wrote most of the paper, and designed the experiments on human cells; Urszula Nowak and Aleksandra Marchwicka designed, performed and analyzed the experiments on human blood cells; Anastasiia Satyr, Aleksandra Marchwicka and Klaudia Berkowska designed, performed and analyzed the experiments on AML cell lines; Sylwia Janik, Małgorzata Cebrat, Michał Majkowski, Agnieszka Łaszkiewicz conceived and designed the experiments on mice; Sylwia Janik,

Michał Majkowski, Małgorzata Cebrat, Łukasz Śnieżewski performed the experiments on mice; Sylwia Janik, Małgorzata Cebrat, Agnieszka Łaszkiewicz analyzed the data obtained from mice experiments; Małgorzata Cebrat took part in writing the paper, Marian Gabryś selected mothers from whom cord blood was obtained, informed them about the aim of the study and obtained an informed consent to participate in the study. All authors accepted the text of the paper.

Conflicts of Interest: The authors declare no conflicts of interest.

Abbreviations

1,25D	1,25-Dihydroxyvitamin D
AML	Acute myeloid leukemia
ATRA	All-*trans*-retinoic acid
BM	Bone marrow
BSA	Bovine serum albumin
CYP24A1	24-Hydroxylase of 1,25D
CYP26A1	Retinoic acid 4-hydroxylase
FBS	Fetal bovine serum
HSC	Hematopoietic stem cells
HSPC	Hematopoeitic stem and progenitor cells
GAPDH	Glyceraldehyde 3-phosphate dehydrogenase
PCR	Polymerase chain reaction
PBM	Peripheral blood mononuclears
RA	Retinoic acid
RAR	Retinoic acid receptors
RARE	Retinoic acid response element
RXR	Retinoid X receptors
UCB	Umbilical cord blood
UTR	Untranslated region
VDR	Vitamin D receptor
VDRE	Vitamin D response element

Appendix A

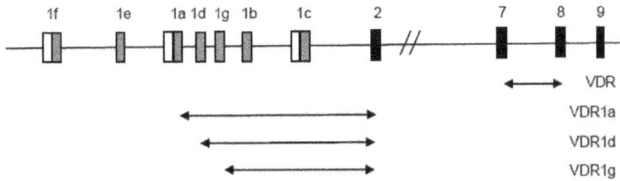

Figure A1. Graphical representation of transcriptional variants of human *VDR* detected in Real-time PCR. Primer sequences are given in Materials and Methods. All *VDR* transcripts were detected using primers which amplify product on the border of exons 7 and 8 (VDR). Transcripts that start at exon 1a were detected using primers which amplify product on the border of exons 1a and 2 (*VDR1a*). Transcripts which start at exon 1d were detected using primers which amplify product on the border of exons 1d and 2 (*VDR1d*) and transcripts which start at exon 1g were detected using primers which amplify product on the border of exons 1g and 2 (*VDR1g*). Black boxes represent coding exons, gray boxes represent noncoding exons and white boxes represent defined promoter regions.

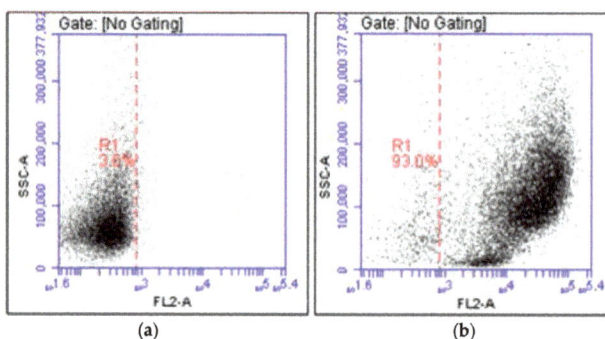

(a)   (b)

Figure A2. Human CD34+ cells were isolated from UCB as described in Materials and Methods. The cells before sorting, and the cells after magnetic sorting were stained using CD34-PE antibody. Cell fluorescence was detected in flow cytometry using Accuri flow cytometer (Becton Dickinson). Red line separates CD34-negative cells (left side of red line) from CD34-positive cells (right side of red line). Sample staining is presented for unsorted (**a**) and sorted (**b**) cells.

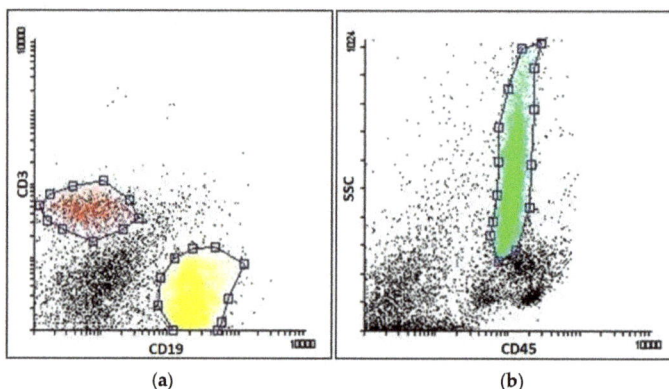

(a)   (b)

Figure A3. Murine BM leukocytes, T cells and B cells were sorted as described in Materials and Methods. (**a**) Murine spleen cells were stained with anti-CD3-APC and anti-CD19-PE antibodies to isolate mature T (red gate) and B (yellow gate) cells, respectively; (**b**) Bone marrow cells were stained with anti-CD45-FITC antibody to isolate granulocytes, using CD45/SSC-based sorting criteria (green gate). Cells were sorted using FACS-Aria (Becton Dickinson).

References

1. Rhinn, M.; Dollé, P. Retinoic acid signalling during development. *Development* **2012**, *139*, 843–858. [CrossRef] [PubMed]
2. Carlberg, C. The physiology of vitamin D-far more than calcium and bone. *Front. Physiol.* **2014**, *5*, 335. [CrossRef] [PubMed]
3. Chambon, P. A decade of molecular biology of retinoic acid receptors. *FASEB J.* **1996**, *10*, 940–954. [PubMed]
4. Kane, M. Analysis, occurrence, and function of 9-*cis*-retinoic acid. *Biochim. Biophys. Acta* **2012**, *1821*, 10–20. [CrossRef] [PubMed]
5. Baker, A.; McDonnell, D.; Hughes, M.; Crisp, T.; Mangelsdorf, D.; Haussler, M.; Pike, J.; Shine, J.; O'Malley, B. Cloning and expression of full-length cDNA encoding human vitamin D receptor. *Proc. Natl. Acad. Sci. USA* **1988**, *85*, 3294–3298. [CrossRef] [PubMed]
6. Aranda, A.; Pascual, A. Nuclear hormone receptors and gene expression. *Physiol. Rev.* **2001**, *81*, 1269–1304. [PubMed]

7. Grignani, G.; Ferrucci, P.; Testa, U.; Talamo, G.; Fagioli, M.; Alcalay, M.; Mencarelli, A.; Grignani, F.; Peschle, C.; Nicoletti, I.; et al. The acute promyelocytic leukemia-specific PML-RARα fusion protein inhibits differentiation and promotes survival of myeloid precursor cells. *Cell* **1993**, *74*, 423–431. [CrossRef]

8. Van Etten, E.; Mathieu, C. Immunoregulation by 1,25-dihydroxyvitamin D₃: Basic concepts. *J. Steroid Biochem. Mol. Biol.* **2005**, *97*, 93–101. [CrossRef] [PubMed]

9. Brackman, D.; Lund-Johansen, F.; Aarskog, D. Expression of cell surface antigens during the differentiation of HL-60 cells induced by 1,25-dihydroxyvitamin D₃, retinoic acid and DMSO. *Leuk. Res.* **1995**, *19*, 57–64. [CrossRef]

10. Tenen, D. Disruption of differentiation in human cancer: AML shows the way. *Nat. Rev. Cancer* **2003**, *3*, 89–101. [CrossRef] [PubMed]

11. Ma, Y.; Trump, D.; Johnson, C. Vitamin D and acute myeloid leukemia. *J. Cancer* **2010**, *3*, 101–107. [CrossRef]

12. Gocek, E.; Marchwicka, A.; Baurska, H.; Chrobak, A.; Marcinkowska, E. Opposite regulation of vitamin D receptor by ATRA in AML cells susceptible and resistant to vitamin D-induced differentiation. *J. Steroid Biochem. Mol. Biol.* **2012**, *132*, 220–226. [CrossRef] [PubMed]

13. Marchwicka, A.; Cebrat, M.; Łaszkiewicz, A.; Śnieżewski, Ł.; Brown, G.; Marcinkowska, E. Regulation of vitamin D receptor expression by retinoic acid receptor α in acute myeloid leukemia cells. *J. Steroid Biochem. Mol. Biol.* **2016**, *159*, 121–130. [CrossRef] [PubMed]

14. Ma, Y.; Trump, D.; Johnson, C. Vitamin D in combination cancer treatment. *J. Cancer* **2010**, *1*, 101–107. [CrossRef] [PubMed]

15. Cañete, A.; Cano, E.; Muñoz-Chápuli, R.; Carmona, R. Role of vitamin A/retinoic acid in regulation of embryonic and adult hematopoiesis. *Nutrients* **2017**, *9*, 159. [CrossRef] [PubMed]

16. Gritz, E.; Hirschi, K. Specification and function of hemogenic endothelium during embryogenesis. *Cell. Mol. Life Sci.* **2016**, *73*, 1547–1567. [CrossRef] [PubMed]

17. Lin, C.; Su, C.; Tseng, D.; Ding, F.; Hwang, P. Action of vitamin D and the receptor, VDRa, in calcium handling in zebrafish (*Danio rerio*). *PLoS ONE* **2012**, *7*, e45650. [CrossRef] [PubMed]

18. Cortes, M.; Chen, M.; Stachura, D.; Liu, S.; Kwan, W.; Wright, F.; Vo, L.; Theodore, L.; Esain, V.; Frost, I.; et al. Developmental vitamin D availability impacts hematopoietic stem cell production. *Cell Rep.* **2016**, *17*, 458–468. [CrossRef] [PubMed]

19. Grande, A.; Montanari, M.; Tagliafico, E.; Manfredini, R.; Marani, T.Z.; Siena, M.; Tenedini, E.; Gallinelli, A.; Ferrari, S. Physiological levels of 1α, 25 dihydroxyvitamin D₃ induce the monocytic commitment of CD34+ hematopoietic progenitors. *J. Leukoc. Biol.* **2002**, *71*, 641–651. [PubMed]

20. Crofts, L.; Hancock, M.; Morrison, N.; Eisman, J. Multiple promoters direct the tissue-specific expression of novel N-terminal variant human vitamin D receptor gene transcripts. *Proc. Natl. Acad. Sci. USA* **1998**, *95*, 10529–10534. [CrossRef] [PubMed]

21. Miyamoto, K.; Kesterson, R.; Yamamoto, H.; Taketani, Y.; Nishiwaki, E.; Tatsumi, S.; Inoue, Y.; Morita, K.; Takeda, E.; Pike, J. Structural organization of the human vitamin D receptor chromosomal gene and its promoter. *Mol. Endocrinol.* **1997**, *11*, 1165–1179. [CrossRef] [PubMed]

22. Byrne, I.; Flanagan, L.; Tenniswood, M.; Welsh, J. Identification of a hormone-responsive promoter immediately upstream of exon 1c in the human vitamin D receptor gene. *Endocrinology* **2000**, *141*, 2829–2836. [CrossRef] [PubMed]

23. Marchwicka, A.; Corcoran, A.; Berkowska, K.; Marcinkowska, E. Restored expression of vitamin D receptor and sensitivity to 1,25-dihydroxyvitamin D₃ in response to disrupted fusion FOP2-FGFR1 gene in acute myeloid leukemia cells. *Cell Biosci.* **2016**, *6*, 7. [CrossRef] [PubMed]

24. Wiese, R.; Uhland-Smith, A.; Ross, T.; Prahl, J.; DeLuca, H. Up-regulation of the vitamin D receptor in response to 1,25-dihydroxyvitamin D₃ results from ligand-induced stabilization. *J. Biol. Chem.* **1992**, *267*, 20082–20086. [PubMed]

25. Gocek, E.; Kielbinski, M.; Wylob, P.; Kutner, A.; Marcinkowska, E. Side-chain modified vitamin D analogs induce rapid accumulation of VDR in the cell nuclei proportionately to their differentiation-inducing potential. *Steroids* **2008**, *73*, 1359–1366. [CrossRef] [PubMed]

26. Pan, P.; Reddy, K.; Lee, S.; Studzinski, G. Differentiation-related regulation of 1,25-dihydroxyvitamin D₃ receptor mRNA in human leukaemia cells HL-60. *Cell Prolif.* **1991**, *24*, 159–170. [CrossRef] [PubMed]

27. Halsall, J.; Osborne, J.; Hutchinson, P.; Pringle, J. In silico analysis of the 5′ region of the vitamin D receptor gene: Functional implications of evolutionary conservation. *J. Steroid Biochem. Mol. Biol.* **2007**, *103*, 352–356. [CrossRef] [PubMed]

28. Gardiner, E.; Esteban, L.; Fong, C.; Allison, S.; Flanagan, J.; Kouzmenko, A.; Eisman, J. Vitamin D receptor b1 and exon 1d: Functional and evolutionary analysis. *J. Steroid Biochem. Mol. Biol.* **2004**, *89–90*, 233–238. [CrossRef] [PubMed]

29. Zella, L.; Kim, S.; Shevde, N.; Pike, J. Enhancers located within two introns of the vitamin D receptor gene mediate transcriptional autoregulation by 1,25-dihydroxyvitamin D_3. *Mol. Endocrinol.* **2006**, *20*, 1231–1247. [CrossRef] [PubMed]

30. Zella, L.; Meyer, M.; Nerenz, R.; Lee, S.; Martowicz, M.; Pike, J. Multifunctional enhancers regulate mouse and human vitamin D receptor gene transcription. *Mol. Endocrinol.* **2010**, *24*, 128–147. [CrossRef] [PubMed]

31. Kahlen, J.; Carlberg, C. Identification of a vitamin D receptor homodimer-type response element in the rat calcitriol 24-hydroxylase gene promoter. *Biochem. Biophys. Res. Commun.* **1994**, *202*, 1366–1372. [CrossRef] [PubMed]

32. Thatcher, J.; Isoherranen, N. The role of CYP26 enzymes in retinoic acid clearance. *Expert Opin. Drug Metab. Toxicol.* **2009**, *5*, 875–886. [CrossRef] [PubMed]

33. Ray, W.; Bain, G.; Yao, M.; Gottlieb, D. CYP26, a novel mammalian cytochrome p450, is induced by retinoic acid and defines a new family. *J. Biol. Chem.* **1997**, *272*, 18702–18708. [CrossRef] [PubMed]

34. Loudig, O.; Maclean, G.; Dore, N.; Luu, L.; Petkovich, M. Transcriptional co-operativity between distant retinoic acid response elements in regulation of CYP26A1 inducibility. *Biochem. J.* **2005**, *392*, 241–248. [CrossRef] [PubMed]

35. Ghiaur, G.; Yegnasubramanian, S.; Perkins, B.; Gucwa, J.; Gerber, J.; Jones, R. Regulation of human hematopoietic stem cell self-renewal by the microenvironment's control of retinoic acid signaling. *Proc. Natl. Acad. Sci. USA* **2013**, *110*, 16121–16126. [CrossRef] [PubMed]

36. Owusu, S.; Ross, A. Retinoid homeostatic gene expression in liver, lung and kidney: Ontogeny and response to vitamin A-retinoic acid (VARA) supplementation from birth to adult age. *PLoS ONE* **2016**, *11*, e0145924. [CrossRef] [PubMed]

37. Brown, G.; Marchwicka, A.; Cunningham, A.; Toellner, K.; Marcinkowska, E. Antagonizing retinoic acid receptors increases myeloid cell production by cultured human hematopoietic stem cells. *Arch. Immunol. Ther. Exp.* **2017**, *65*, 69–81. [CrossRef] [PubMed]

38. Porada, C.; Atala, A.; Almeida-Porada, G. The hematopoietic system in the context of regenerative medicine. *Methods* **2016**, *99*, 44–61. [CrossRef] [PubMed]

39. Gocek, E.; Kielbinski, M.; Baurska, H.; Haus, O.; Kutner, A.; Marcinkowska, E. Different susceptibilities to 1,25-dihydroxyvitamin D_3-induced differentiation of aml cells carrying various mutations. *Leuk. Res.* **2010**, *34*, 649–657. [CrossRef] [PubMed]

40. Baurska, H.; Kiełbiński, M.; Biecek, P.; Haus, O.; Jaźwiec, B.; Kutner, A.; Marcinkowska, E. Monocytic differentiation induced by side-chain modified analogs of vitamin D in ex vivo cells from patients with acute myeloid leukemia. *Leuk. Res.* **2014**, *38*, 638–647. [CrossRef] [PubMed]

41. Ablain, J.; de Thé, H. Retinoic acid signaling in cancer: The parable of acute promyelocytic leukemia. *Int. J. Cancer* **2014**, *135*, 2262–2272. [CrossRef] [PubMed]

42. Göttgens, B. Regulatory network control of blood stem cells. *Blood* **2015**, *125*, 2614–2620. [CrossRef] [PubMed]

43. Doulatov, S.; Notta, F.; Laurenti, E.; Dick, J. Hematopoiesis: A human perspective. *Cell Stem Cell* **2012**, *10*, 120–136. [CrossRef] [PubMed]

44. Bouillon, R.; Bischoff-Ferrari, H.; Willett, W. Vitamin D and health: Perspectives from mice and man. *J. Bone Miner. Res.* **2008**, *23*, 974–979. [CrossRef] [PubMed]

45. Christakos, S.; Dhawan, P.; Verstuyf, A.; Verlinden, L.; Carmeliet, G. Vitamin D: Metabolism, molecular mechanism of action, and pleiotropic effects. *Physiol. Rev.* **2016**, *96*, 365–408. [CrossRef] [PubMed]

46. Laszkiewicz, A.; Cebrat, M.; Miazek, A.; Kisielow, P. Complexity of transcriptional regulation within the RAG locus: Identification of a second NWC promoter region within the RAG2 intron. *Immunogenetics* **2011**, *63*, 183–187. [CrossRef] [PubMed]

© 2017 by the authors. Licensee MDPI, Basel, Switzerland. This article is an open access article distributed under the terms and conditions of the Creative Commons Attribution (CC BY) license (http://creativecommons.org/licenses/by/4.0/).

International Journal of
Molecular Sciences

MDPI

Case Report

A Case of AML Characterized by a Novel t(4;15)(q31;q22) Translocation That Confers a Growth-Stimulatory Response to Retinoid-Based Therapy

Justin M. Watts [1,†], Aymee Perez [1,†] (iD), Lutecia Pereira [1], Yao-Shan Fan [2], Geoffrey Brown [3,4], Francisco Vega [2], Kevin Petrie [5,*] (iD), Ronan T. Swords [1] and Arthur Zelent [1,*]

1   Division of Hematology/Oncology, Department of Medicine, Miller School of Medicine,
    University of Miami, Miami, FL 33136, USA; jxw401@miami.edu (J.M.W.); aperez10@med.miami.edu (A.P.);
    lpereira@med.miami.edu (L.P.); rswords@med.miami.edu (R.T.S.)
2   Department of Pathology, Miller School of Medicine, University of Miami, Miami, FL 33136, USA;
    yfan@med.miami.edu (Y.-S.F.); fvega@med.miami.edu (F.V.)
3   Institute of Immunology and Immunotherapy, University of Birmingham, Birmingham B15 2TT, UK;
    g.brown@bham.ac.uk
4   Institute of Clinical Sciences, University of Birmingham, Birmingham B15 2TT, UK
5   Faculty of Natural Sciences, University of Stirling, Stirling FK9 4LA, UK
*   Correspondence: kevin.petrie@stir.ac.uk (K.P.); a.zelent@med.miami.edu (A.Z.); Tel.: +44-1786-466840 (K.P.);
    +1-305-243-5071 (A.Z.)
†   These authors contributed equally to this study.

Received: 6 June 2017; Accepted: 5 July 2017; Published: 11 July 2017

Abstract: Here we report the case of a 30-year-old woman with relapsed acute myeloid leukemia (AML) who was treated with all-*trans* retinoic acid (ATRA) as part of investigational therapy (NCT02273102). The patient died from rapid disease progression following eight days of continuous treatment with ATRA. Karyotype analysis and RNA-Seq revealed the presence of a novel t(4;15)(q31;q22) reciprocal translocation involving the *TMEM154* and *RASGRF1* genes. Analysis of primary cells from the patient revealed the expression of *TMEM154-RASGRF1* mRNA and the resulting fusion protein, but no expression of the reciprocal *RASGRF1-TMEM154* fusion. Consistent with the response of the patient to ATRA therapy, we observed a rapid proliferation of t(4;15) primary cells following ATRA treatment ex vivo. Preliminary characterization of the retinoid response of t(4;15) AML revealed that in stark contrast to non-t(4;15) AML, these cells proliferate in response to specific agonists of RARα and RARγ. Furthermore, we observed an increase in the levels of nuclear RARγ upon ATRA treatment. In summary, the identification of the novel t(4;15)(q31;q22) reciprocal translocation opens new avenues in the study of retinoid resistance and provides potential for a new biomarker for therapy of AML.

Keywords: acute myeloid leukemia; t(4; 15)(q31; q22); TMEM154-RASGRF1; ATRA; retinoid; NCT02273102

---

1. Introduction

Acute myeloid leukemia (AML) is the most commonly occurring leukemia in adults, accounting for an estimated 33% of all cases and more that 80% of acute cases in 2016 [1]. AML is a disease whose incidence increases with age and—in line with an aging population—this has increased from 3.4 per 100,000 in 2004 to 5.1 per 100,000 in 2013 [2]. Advances in the treatment of AML have dramatically improved treatment outcomes for younger patients, with a relative 5-year survival rate (2005–2011) of approximately 67% (0–14 years) [1]. In the elderly, however (who account for the majority of new cases), the prognosis remains poor. Indeed, AML accounts for an estimated 43% of all leukemia deaths, with an overall average 5-year survival rate of only 27% [2].

The use of combination therapy based on the use all-*trans* retinoic acid (ATRA) and arsenic trioxide (As_2O_3) has revolutionized the clinical outcome of acute promyelocytic leukemia (APL), with complete remission rates of 90% and cure rates of around 80% [3]. However, the success of ATRA-based therapy in AML has not been translated into other non-APL subtypes of AML. Based on in vitro data showing efficacy with cytotoxic chemotherapy agents such as Ara-C and anthracyclines, several clinical studies have investigated the impact of adding ATRA to chemotherapy in patients with non-APL AML [4]. Unfortunately, however, these studies have yielded inconsistent and conflicting results. One possible reason for a lack of response to ATRA treatment in non-APL AML patients is epigenetic repression of the retinoic acid receptor pathway, especially the *RARA* gene [5]. Indeed, combination therapy of ATRA with epi-drugs targeting the histone H3 lysine 4 demethlyase LSD1 (KDM1A) has shown promise in in vivo models of AML [6]. This approach is now being evaluated in a phase I clinical trial where patients are given escalating doses of the LSD1 inhibitor Parnate (tranylcypromine, TCP) with fixed continuous doses of ATRA.

Acute myeloid leukemia is characterized by clonal or oligoclonal hematopoiesis that varies amongst patients. On average, at least four-to-five oncogenic mutations or chromosomal abnormalities are present at diagnosis, along with passenger mutations of unclear significance [7]. Clonal evolution occurs after exposure to chemotherapy, and the secondary mutations that give rise to these clones are more likely to be random, as opposed to recurrent primary abnormalities [8]. Here we report a case of relapsed AML characterized by a previously unidentified chromosomal translocation.

2. Results

2.1. Clinical Presentation

A 30-year-old previously healthy African-American female presented with a white blood cell count (WBC) of 180,000/μL, hemoglobin 7.6 g/dL, platelets 93,000/μL. Physical examination was unremarkable. Bone marrow aspiration and biopsy were consistent with AML with monocytic differentiation. Blasts amounted to 95% of nucleated elements and expressed CD34, CD117, CD13, CD33, CD64 (weak), CD38, HLA-DR, and were negative for terminal deoxynucleotidyl transferase (TdT) and myeloperoxidase (MPO) by flow cytometry. Cytogenetic analysis revealed an abnormal female karyotype: 46,XX,t(9;11)(p22;q23)[13]/46,XX[7]. Additionally, mutational analysis revealed an activating mutation in *NRAS* gene (c.34G > A; p.G12S). Following treatment with standard therapy, the patient achieved complete cytogenetic remission. Post remission therapy followed, with sequential cycles of high-dose cytarabine in the absence of an optimal donor. Following a disease-free interval of 8 months, the patient relapsed. Karyotype at relapse was consistent with clonal evolution: 46,XX,t(4;15)(q31;q22), t(9;11)(p22;q23)[20] (Figure 1A). Mutational analysis confirmed the presence of an *NRAS* mutation as before. The patient failed two lines of salvage chemotherapy. In the setting of chemotherapy refractory disease, the patient was enrolled in the phase I clinical trial (NCT02273102) and given escalating doses of TCP with fixed continuous doses of ATRA. However, the patient died from rapid disease progression after eight days of treatment (WBC at screening was 400/μL and at withdrawal from study WBC was 99,300/μL).

2.2. Molecular and Functional Characterization of the t(4;15) Fusion Gene

To characterize the translocation partners and their breakpoints, we used RNA-Seq and Sanger sequencing, confirming the t(9;11) *MLL(KMT2A)-AF9(MLLT3)* reciprocal translocation present at diagnosis (data not shown). We also defined a novel fusion gene comprising exons 1–6 of the *TMEM154* gene (4q31.3) and exons 15–24 of the *RASGRF1* gene (15q24.2) (Figure 1B). This generates a putative 830 amino acid TMEM154-RASGRF1 fusion protein comprising amino acids 1–178 of TMEM154 and

622–1273 of RASGRF1 (Figure 1C). RT-PCR analysis confirmed the presence of *TMEM154-RASGRF1* mRNA in t(4;15) AML cells but not normal controls (Figure 1D, lanes 1–2 and 5–6). By contrast, controls for *TMEM154* mRNA were present in both t(4;15) AML cells and normal controls (Figure 1D, lanes 4 and 8). Consistent with the RNA-Seq results, expression of mRNA transcripts arising from the reciprocal *RASGRF1-TMEM154* fusion gene were not detected in t(4;15) AML cells (Figure 1D, lane 3).

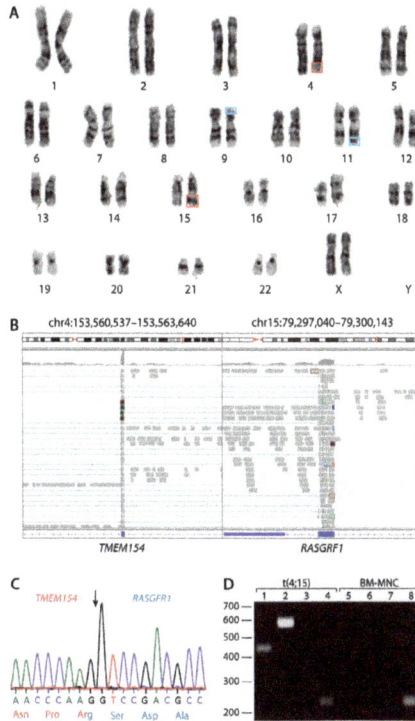

Figure 1. Analysis of the t(4;15)(q31;q22) reciprocal chromosomal translocation. (**A**) Conventional karyotyping showing 46,XX,t(4;15)(q31;q22), outlined in red and t(9;11)(p22;q23), outlined in blue. Chromosome analysis was performed on 20 G-banded metaphase cells from multiple unstimulated cultures. Both translocations were present in all cells examined. (**B**) Breakpoint analysis of the novel fusion transcript produced by the t(4;15) chromosomal translocation. *TMEM154* and *RASGRF1* form a chimeric mRNA transcript with the breakpoint indicated by the black arrow. (**C**) Sanger sequencing chromatogram showing the transcribed sequence surrounding the breakpoint. (**D**) PCR analysis of and *TMEM154-RASGRF1*, *RASGRF1-TMEM154*, and *TMEM154* transcripts. PCR was performed on t(4;15) acute myeloid leukemia (AML) cells (lanes 1–4) and bone marrow mononuclear cells (BM-MNCs) from a healthy donor (lanes 5–8). Bands correspond to a 445 bp product amplified from*TMEM154-RASGRF1* mRNA (lanes 1 and 5), a 589 bp product amplified from *TMEM154-RASGRF1* mRNA (lanes 2 and 6), and a 445 bp product amplified from*TMEM154* mRNA (lanes 4 and 8). No product was amplified using primers corresponding to *RASGRF1-TMEM154* mRNA (lanes 3 and 7).

In agreement with our analysis of mRNA expression, Western blot analysis revealed the expression of endogenous TMEM154 and RASGRF1 proteins, as well as TMEM154-RASGRF1 fusion protein, which could be detected using antibodies directed against both a N-terminal region of TMEM154 as well as a C-terminal region of RASGRF1 (Figure 2A). Interestingly, treatment with ATRA led to an increase in the expression of TMEM154-RASGRF1 fusion protein, but not endogenous TMEM154 or RASGRF1 (Figure 2A). Supporting the clinical history, we observed a rapid proliferation of the patient's blasts following treatment with ATRA ex vivo (Figure 2B). The same phenotype was seen when the cells were exposed to other agonists of the retinoic acid pathway. Primary cells lacking a t(4;15)(q31;q22) rearrangement displayed no growth alteration compared to controls when treated with either ATRA (a pan-RAR isotype agonist) or isotype-selective agonists specific for RARα or RARγ (Figure 2C). To further examine the effects of ATRA on primary t(4;15) AML cells, we performed immunofluorescence analysis of RARα or RARγ. Here, we found that treatment of t(4;15) AML cells with ATRA led to an increase in nuclear RARγ (Figure 2D).

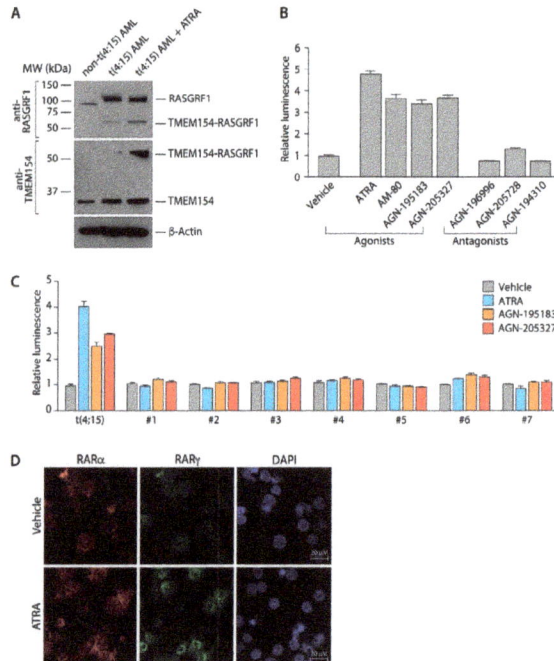

Figure 2. Characterization of t(4;15) AML primary cells. (**A**) Immunoblot analysis of *TEM154-RASGRF1* expression. After 72 h treatment, vehicle control non-t(4;15) AML cells (lane 1), vehicle control t(4;15) AML cells (lane 2) and all-trans retinoic acid (ATRA) t(4;15) AML cells (lane 3) were subjected to immunoblot analysis. Samples were probed with rabbit monoclonal antibody against a C-terminal region of RASGRF1 and mouse monoclonal antibody against an N-terminal region of TMEM154. β-actin was used as a loading control. Molecular weight standards (left) and the identities of bands (right) are indicated. (**B**) Proliferation of t(4;15) AML cells in response to retinoids. Cells were treated with vehicle control, ATRA, RARα agonist (AM-80 or AGN-195183), or RARγ agonist (AGN-205327). Cells were also treated with RARα antagonist (AGN-196996), RARγ antagonist (AGN-205728), or RARα/β/γ antagonist (AGN-194310). All retinoids were used at a concentration of 1 μM. Cell proliferation was determined by CellTiter-Glo luminescent cell viability assay (Promega) after 72 h of treatment. Values shown are normalized to vehicle control. (**C**) Proliferation of t(4;15) and non-t(4;15) AML cells in response to retinoid agonists. Samples from t(4;15) and 7 non-t(4;15) AML patients were incubated with vehicle control, ATRA, RARα agonist (AGN-195183), or RARγ agonist (AGN-205327) at a concentration of 1 μM. Cells were analyzed and normalized as described for (B). (**D**) Treatment of t(4;15) cells with ATRA increases levels of nuclear RARγ. t(4;15) cells were treated with vehicle control or 1 μM ATRA and incubated for 72 h before staining with RARα mouse monoclonal and RARγ rabbit polyclonal antibodies.

3. Discussion

Our data suggest that the t(4;15)(q31;q22) reciprocal translocation confers a growth-stimulatory response to retinoid-based therapy. However, as we did not measure levels of apoptosis at the time the proliferation assays were performed, it is possible that t(4;15) cells also exhibited improved survival ex vivo. Nevertheless, an ex vivo growth-stimulatory response is consistent with the increased WBC in the patient following ATRA treatment. The basis for this response requires further investigation, but could lie—at least in part—in an increase in nuclear RARγ following ATRA treatment. While RARα is required for myeloid differentiation, RARγ has been shown to promote hematopoietic stem and progenitor cell self-renewal and proliferation [9]. Additionally, it is likely that the ATRA-mediated increase in TMEM154-RASGRF1 fusion protein (which is consistent with established ATRA regulation of TMEM154 [10]) cooperates with mutant NRAS G12S (which results in decreased GTPase activity and constitutive RAS signaling). Between 12–19% of AML patients possess gain-of-function mutated *RAS* genes [7], with NRAS being most frequently affected [11]. Furthermore, other components of the RAS pathway signaling cascade are often mutated [12]. RASGRF1 (RAS protein specific nucleotide releasing factor 1) is a guanine nucleotide exchange factor (GEF) similar to the *Saccharomyces cerevisiae CDC25* gene product, and functions to activate RAS by catalyzing the exchange of RAS-bound GDP for GTP [13]. While we have not yet investigated the activity of TMEM154-RASGRF1 towards NRAS, it is noteworthy that the fusion protein (which lacks regulatory domains contained in N-terminal RASGRF1) contains the membrane anchoring domain of TMEM154 and functional domains of RASGRF1. These include the GDP/GTP RAS exchanger and CDC25H motifs of RASGRF1, which have been shown to be sufficient for functional protein [14].

Another aspect of the consequences of the t(4;15) translocation that merits further consideration is the fact that *RASGRF1* is a paternally expressed imprinted gene. Thus, under normal conditions, only the paternal allele of the gene is translated into protein. In t(4;15) AML, both the TMEM154-RASGRF1 fusion protein as well endogenous RASGRF1 are expressed, leading to possible gene dosage effects. Conversely, the lack of expression of the reciprocal *RASGRF1-TMEM154* transcript indicates that the differentially methylated domain (DMD) located 30 kb 5' of the *RASGRF1* transcription initiating site [15] is intact, resulting in imprinted methylation. A functionally important consequence of this could be inappropriate silencing of genes downstream of *TMEM154*. Here, it is noteworthy that the gene immediately downstream of *TMEM154* is *FBXW7*, which encodes an F-box family protein that is a component of the SCF (SKP1-cullin-F-box) ubiquitin protein ligase complex. In the context of AML, FBXW7 has clinically relevant targets including MYC, MCL-1, and Cyclin E [16,17].

4. Materials and Methods

4.1. Patient Samples

Primary AML samples were collected by the Tissue Banking Core Facility (TBCF) at the University of Miami according to an institutional review board approved protocol (20060858). In order to purify the mononuclear fraction, bone marrow aspirates were subjected to density gradient centrifugation using Ficoll Paque Plus (GE Healthcare Life Sciences, Pittsburgh, PA, USA).

4.2. DNA/RNA Isolation

DNA and RNA were isolated from mononuclear cells from patients' bone marrow aspirate. DNA was isolated using QIAamp DNA Mini Kit (Qiagen, Valencia, CA, USA) according to the manufacturer's instructions. RNA was extracted with PureLink RNA kit (Life Technologies, Carlsbad, CA, USA) following the manufacturer's recommendations.

4.3. RNA-Seq for Breakpoint Identification

Preparation of RNA libraries for sequencing on the Illumina HiSeq2500 (Illumina, San Diego, CA, USA) platform was carried out in the John P. Hussman Institute for Human Genomics Center for

Genome Technology at the University of Miami. Briefly, total RNA was quantified and qualified by Agilent Bioanalyzer (Agilent, Santa Clara, CA, USA) to have an RNA integrity score (RIN) of 7. For each of the two samples, 1000 ng of total RNA was used as input for the Illumina TruSeq Stranded Total RNA Library Prep Kit with Ribo-Zero to create ribosomal RNA-depleted sequencing libraries. Each sample was barcoded to allow for multiplexing and was sequencing to ~30 million raw reads in a 2 × 125 paired end sequencing run on an Illumina HiSeq2500. Raw sequence data from the Illumina HiSeq2500 was processed by the on-instrument Real Time Analysis software (v1.8, Illumina, San Diego, CA, USA) to basecall files. These were converted to demultiplexed FASTQ files with the Illumina supplied scripts in the BCL2FASTQ software (v1.8.4, Illumina, San Diego, CA, USA). The quality of the reads was determined with FastQC software for per-base sequence quality, duplication rates, and overrepresented kmers (www.bioinformatics.babraham.ac.uk/projects/fastqc/). Illumina adapters were trimmed from the ends of the reads using Trim Galore! resulting in two trimmed FASTQ files per sample (http://www.bioinformatics.babraham.ac.uk/projects/trim_galore/). Reads were aligned to the human reference genome (hg19) with the STAR aligner (v2.5.0a) [18]. Gene fusions were detected in the aligned BAM files using the Manta software [19].

4.4. Accession Numbers

The Genbank accession number for *TMEM154-RASGRF1* mRNA is MF175878. RNA-Seq FASTQ files were submitted to the sequence read archive (SRA) with the Genbank accession number SRR5681084.

4.5. RT-PCR

Reverse transcription and PCR amplification was performed using One-Step RT-PCR System (Life Technologies) according to manufacturer's instructions. Primers used to amplify *TMEM154-RASGRF1* mRNA were *TMEM154*-Fwd333: 5′-TAGCCAAGGATCTCAGAGTG-3′; and *RASGRF1*-Rev778: 5′-GAATGGCACTGATAGGCTTC-3′ or *RASGRF1*-Rev922: 5′-GTGACGATG TCTTGGTGATG-3′. Primers used to amplify TMEM154 were *TMEM154*-Fwd1762: 5′-TAGCCAAG GATCTCAGAGTG-3′ and *TMEM154*-Rev2272: 5′-TGAGCGCCATTCAGGTTTAG-3′. Primers to amplify *RASGRF1-TMEM154* were RASGRF1-Fwd1762: 5′-CTCATTCAGGTGCCCATGTC-3′ and *TMEM154*-Rev2272: 5′-TGAGCGCCATTCAGGTTTAG-3′. Amplified PCR fragments were subjected to DNA sequence (Eurofins, Louisville, KY, USA) to validate the translocation breakpoint.

4.6. Western Blot

Protein sample concentrations were normalized using DC Protein Assay (Bio-Rad, Hercules, CA, USA). Fifty micrograms of protein were separated by SDS-PAGE using 4–20% Mini-PROTEAN TGX pre-cast gels with Precision Plus ProteinDual Color Standards (Bio-Rad). Primary antibodies used were as follows: rabbit monoclonal antibody against the C-terminal region of RASGRF1 (ab118830, Abcam, Cambridge, UK) and mouse monoclonal antibody against the N-terminal region of TMEM154 (sc-398802, Santa Cruz Biotechnology, Dallas, TX, USA). β-actin was used as a loading control (A5441, Sigma, St. Louis, MO, USA,). Protein bands were detected using HRP-linked Anti-rabbit IgG (#7074, Cell Signaling Technology, Danvers, MA, USA) and HRP-linked Anti-mouse IgG (Cell Signaling Technology, #7076) secondary antibodies.

4.7. Cell Proliferation

Cell proliferation was measured using CellTiter-Glow luminescent cell viability assay (Promega, Madison, WI, USA) following the manufacturer's recommendations. Primary AML mononuclear cells were seeded at a density of 10,000 cells/well in 96-well plate for 24 h, followed by retinoid or vehicle (DMSO) treatment for 72 h. Cell were cultured in RPMI supplemented with 5% Charcoal-dextran fetal bovine serum (FBS, Life Technologies). ATRA was purchased from Sigma. AM-80 was purchased from Tocris (Bio-Techne, Minneapolis, MN, USA). AGN195183, AGN196996, AGN205327, AGN205728, and

AGN194310 were manufactured under contract at the Chinese National Center for Drug Screening, Shanghai Institute of Materia Medica, Shanghai, China.

4.8. Immunofluorescence

AML cells were fixed with 4% paraformaldehyde in 1X phosphate-buffered saline (PBS) for 20 min at room temperature. For immunostaining, cells were permeabilized with 1% Triton X-100 in 1X PBS and blocked with 2.5% bovine serum albumin (BSA) and 1% Triton X-100 in 1X PBS for 1 h shaking at room temperature. Primary antibodies (dilution 1:100) were incubated for 2 h at room temperature in blocking buffer and washed five times with wash buffer (0.1% Triton X-100 in 1X PBS). Primary antibodies used were RARα-9a (mouse monoclonal) and RARγ-453 (rabbit polyclonal) (gift from Cécile Rochette-Egly, L'Institut de Génétique et de Biologie Moléculaire et Cellulaire (IGBMC), France). Secondary antibodies (dilution 1:1000) were added for 1 h at room temperature in blocking buffer and then washed five times. Secondary antibodies used were Alexa Fluor 488-conjugated anti-mouse IgG (H + L), F(ab')2 fragment (Cell Signaling Technology, #4408) and Alexa Fluor 555-conjugated anti-rabbit IgG (H + L), F(ab')2 fragment (Cell Signaling Technology, #4413). Negative controls were performed using primary and secondary antibodies alone. Slides were then mounted with ProLong Gold antifade reagent with DAPI (Molecular Probes, Eugene, OR, USA), following the manufacturer's instructions. Immunofluorescence microscopy was performed using a Leica DFC 310 FX microscope.

5. Conclusions

Here we report a case of relapsed AML characterized by a novel translocation associated with aggressive disease and increased proliferation in the presence of ATRA. We are currently generating a cell line bearing this translocation that will provide a new model system for the study of AML. Although the *TMEM154-RASGRF1* translocation is likely to be a rare occurrence (as has been the case in the past with other translocations), it is predicted that further study of its biology will yield important insights into retinoic acid signaling and resistance to retinoid-based therapy in AML.

Acknowledgments: We thank Omar Lopez, Ana Rodrigues, Veronica Gil, Terrence Bradley, Fernando Vargas, Sandra Algaze and Jennifer Chapman for helpful advice and comments. This study was supported by Grant Number KL2TR000461, Miami Clinical and Translational Science Institute, from the National Center for Advancing Translational Sciences and the National Institute on Minority Health and Health Disparities. Its contents are solely the responsibility of the authors and do not necessarily represent the official views of the NIH.

Author Contributions: Justin M. Watts, Aymee Perez, Ronan T. Swords and Arthur Zelent conceived and designed the study. Justin M. Watts collected the clinical data. Aymee Perez and Lutecia Pereira performed laboratory experiments. Yao-Shan Fan and Francisco Vega performed cytogenetic analysis. Geoffrey Brown provided technical expertise on the manufacture and use of retinoid agonists and antagonists. Justin M. Watts, Aymee Perez and Kevin Petrie analyzed and interpreted the data and wrote the manuscript. Geoffrey Brown, Francisco Vega, Ronan T. Swords and Arthur Zelent analyzed and interpreted the data and edited the manuscript.

Conflicts of Interest: The authors declare no conflict of interest.

References

1. Siegel, R.L.; Miller, K.D.; Jemal, A. Cancer statistics, 2016. *CA Cancer J. Clin.* **2016**, *66*, 7–30. [CrossRef] [PubMed]
2. American Cancer Society. *Cancer Facts & Figures*; American Cancer Society: Atlanta, GA, USA, 2017.
3. Coombs, C.C.; Tavakkoli, M.; Tallman, M.S. Acute promyelocytic leukemia: Where did we start, where are we now, and the future. *Blood Cancer J.* **2015**, *5*, e304. [CrossRef] [PubMed]
4. Forghieri, F.; Bigliardi, S.; Quadrelli, C.; Morselli, M.; Potenza, L.; Paolini, A.; Colaci, E.; Barozzi, P.; Zucchini, P.; Riva, G.; et al. All-trans retinoic acid (ATRA) in non-promyelocytic acute myeloid leukemia (AML): Results of combination of ATRA with low-dose Ara-C in three elderly patients with NPM1-mutated AML unfit for intensive chemotherapy and review of the literature. *Clin. Case Rep.* **2016**, *4*, 1138–1146. [CrossRef] [PubMed]

5. Glasow, A.; Barrett, A.; Petrie, K.; Gupta, R.; Boix-Chornet, M.; Zhou, D.C.; Grimwade, D.; Gallagher, R.; von Lindern, M.; Waxman, S.; et al. DNA methylation-independent loss of *RARA* gene expression in acute myeloid leukemia. *Blood* **2008**, *111*, 2374–2377. [CrossRef] [PubMed]

6. Schenk, T.; Chen, W.C.; Gollner, S.; Howell, L.; Jin, L.; Hebestreit, K.; Klein, H.U.; Popescu, A.C.; Burnett, A.; Mills, K.; et al. Inhibition of the LSD1 (KDM1A) demethylase reactivates the all-trans-retinoic acid differentiation pathway in acute myeloid leukemia. *Nat. Med.* **2012**, *18*, 605–611. [CrossRef] [PubMed]

7. Cancer Genome Atlas Research Network; Ley, T.J.; Miller, C.; Ding, L.; Raphael, B.J.; Mungall, A.J.; Robertson, A.; Hoadley, K.; Triche, T.J., Jr.; Laird, P.W.; et al. Genomic and epigenomic landscapes of adult de novo acute myeloid leukemia. *N. Engl. J. Med.* **2013**, *368*, 2059–2074.

8. Klco, J.M.; Miller, C.A.; Griffith, M.; Petti, A.; Spencer, D.H.; Ketkar-Kulkarni, S.; Wartman, L.D.; Christopher, M.; Lamprecht, T.L.; Helton, N.M.; et al. Association between mutation clearance after induction therapy and outcomes in acute myeloid leukemia. *Jama* **2015**, *314*, 811–822. [CrossRef] [PubMed]

9. Petrie, K.; Zelent, A.; Waxman, S. Differentiation therapy of acute myeloid leukemia: Past, present and future. *Curr. Opin. Hematol.* **2009**, *16*, 84–91. [CrossRef] [PubMed]

10. Simandi, Z.; Balint, B.L.; Poliska, S.; Ruhl, R.; Nagy, L. Activation of retinoic acid receptor signaling coordinates lineage commitment of spontaneously differentiating mouse embryonic stem cells in embryoid bodies. *FEBS Lett.* **2010**, *584*, 3123–3130. [CrossRef] [PubMed]

11. Bacher, U.; Haferlach, T.; Schoch, C.; Kern, W.; Schnittger, S. Implications of NRAS mutations in AML: A study of 2502 patients. *Blood* **2006**, *107*, 3847–3853. [CrossRef] [PubMed]

12. Scholl, C.; Gilliland, D.G.; Frohling, S. Deregulation of signaling pathways in acute myeloid leukemia. *Semin. Oncol.* **2008**, *35*, 336–345. [CrossRef] [PubMed]

13. Fernandez-Medarde, A.; Santos, E. The RasGrf family of mammalian guanine nucleotide exchange factors. *Biochim. Biophys. Acta* **2011**, *1815*, 170–188. [CrossRef] [PubMed]

14. Lenzen, C.; Cool, R.H.; Prinz, H.; Kuhlmann, J.; Wittinghofer, A. Kinetic analysis by fluorescence of the interaction between Ras and the catalytic domain of the guanine nucleotide exchange factor Cdc25Mm. *Biochemistry* **1998**, *37*, 7420–7430. [CrossRef] [PubMed]

15. Shibata, H.; Yoda, Y.; Kato, R.; Ueda, T.; Kamiya, M.; Hiraiwa, N.; Yoshiki, A.; Plass, C.; Pearsall, R.S.; Held, W.A.; et al. A methylation imprint mark in the mouse imprinted gene *Grf1/Cdc25Mm* locus shares a common feature with the *U2afbp-rs* gene: An association with a short tandem repeat and a hypermethylated region. *Genomics* **1998**, *49*, 30–37. [CrossRef] [PubMed]

16. Gores, G.J.; Kaufmann, S.H. Selectively targeting Mcl-1 for the treatment of acute myelogenous leukemia and solid tumors. *Genes Dev.* **2012**, *26*, 305–311. [CrossRef] [PubMed]

17. Takeishi, S.; Nakayama, K.I. Role of Fbxw7 in the maintenance of normal stem cells and cancer-initiating cells. *Br. J. Cancer* **2014**, *111*, 1054–1059. [CrossRef] [PubMed]

18. Dobin, A.; Davis, C.A.; Schlesinger, F.; Drenkow, J.; Zaleski, C.; Jha, S.; Batut, P.; Chaisson, M.; Gingeras, T.R. STAR: Ultrafast universal RNA-seq aligner. *Bioinformatics* **2013**, *29*, 15–21. [CrossRef] [PubMed]

19. Chen, X.; Schulz-Trieglaff, O.; Shaw, R.; Barnes, B.; Schlesinger, F.; Kallberg, M.; Cox, A.J.; Kruglyak, S.; Saunders, C.T. Manta: Rapid detection of structural variants and indels for germline and cancer sequencing applications. *Bioinformatics* **2016**, *32*, 1220–1222. [CrossRef] [PubMed]

© 2017 by the authors. Licensee MDPI, Basel, Switzerland. This article is an open access article distributed under the terms and conditions of the Creative Commons Attribution (CC BY) license (http://creativecommons.org/licenses/by/4.0/).

International Journal of
Molecular Sciences

MDPI

Article

Selective Expression of Flt3 within the Mouse Hematopoietic Stem Cell Compartment

Ciaran James Mooney [1], Alan Cunningham [1], Panagiotis Tsapogas [2], Kai-Michael Toellner [1] and Geoffrey Brown [1,3,*]

[1] Institute of Immunology and Immunotherapy, College of Medical and Dental Sciences, University of Birmingham, Edgbaston, Birmingham B15 2TT, UK; c.mooney@smd15.qmul.ac.uk (C.J.M.); alan.cunningham43@gmail.com (A.C.); k.m.toellner@bham.ac.uk (K.-M.T.)
[2] Developmental and Molecular Immunology, Department of Biomedicine, University of Basel, Basel 4058, Switzerland; panagiotis.tsapogas@unibas.ch
[3] Institute of Clinical Sciences, College of Medical and Dental Sciences, University of Birmingham, Edgbaston, Birmingham B15 2TT, UK
* Correspondence: g.brown@bham.ac.uk; Tel.: +44-0121-414-4082

Academic Editor: Johannes Haybaeck
Received: 20 April 2017; Accepted: 5 May 2017; Published: 12 May 2017

Abstract: The fms-like tyrosine kinase 3 (Flt3) is a cell surface receptor that is expressed by various hematopoietic progenitor cells (HPC) and Flt3-activating mutations are commonly present in acute myeloid and lymphoid leukemias. These findings underscore the importance of Flt3 to steady-state and malignant hematopoiesis. In this study, the expression of Flt3 protein and *Flt3* mRNA by single cells within the hematopoietic stem cell (HSC) and HPC bone marrow compartments of C57/BL6 mice was investigated using flow cytometry and the quantitative reverse transcription polymerase chain reaction. Flt3 was heterogeneously expressed by almost all of the populations studied, including long-term reconstituting HSC and short-term reconstituting HSC. The erythropoietin receptor (EpoR) and macrophage colony-stimulating factor receptor (M-CSFR) were also found to be heterogeneously expressed within the multipotent cell compartments. Co-expression of the mRNAs encoding Flt3 and EpoR rarely occurred within these compartments. Expression of both Flt3 and M-CSFR protein at the surface of single cells was more commonly observed. These results emphasize the heterogeneous nature of HSC and HPC and the new sub-populations identified are important to understanding the origin and heterogeneity of the acute myeloid leukemias.

Keywords: Flt3; growth factor receptors; cytokines; HSC; hematopoiesis; leukemia

1. Introduction

The hematopoietic stem and progenitor cells (HSPC) that give rise to all of the blood and immune cell types are within a small fraction of bone marrow cells that lack lineage markers (Lin⁻) and express Sca-1 and c-kit (LSK). HSPC are heterogeneous and multi-color flow cytometry has greatly aided the sub-fractionation of the LSK compartment, for example, on the basis of CD34 expression [1]. However, how to best phenotypically delineate populations of HSPC with differing biological properties is, as yet, uncertain.

A marker that is used to sub-divide LSK is the cell surface receptor fms-like tyrosine kinase 3 (Flt3/Flk2) which is a class III tyrosine kinase with structural homology to the c-kit and macrophage colony-stimulating factor (M-CSF/CSF1) receptors [2,3]. Flt3 was first identified on murine hematopoietic progenitor cells (HPC) [3] while expression in humans is restricted to CD34⁺ bone marrow cells, which include hematopoietic stem cells (HSC) [4]. Within the LSK compartment, only the Flt3⁻ fraction is capable of long-term myeloid reconstitution which has led to the viewpoint that Flt3 expression is linked to a loss of self-renewal capacity [5–7]. The promoter region of the Flt3 gene (*Flt3*) in Flt3⁻ LSK cells is occupied in a primed state [8] and the use of Flt3-Cre:loxp-eYFP mice has revealed that *Flt3* expression occurs within a phenotypically defined HSC compartment [9]. However, when LSK eYFP⁺ and eYFP⁻ cells from Flt3-Cre: loxp-eYFP mice are transplanted into secondary recipients only the latter provide robust myeloid reconstitution [9]. Boyer and colleagues have confirmed that all hematopoietic cells develop from HSC via a Flt3⁺ progenitor [10]. Together, the above results provide strong evidence to support the viewpoint that Flt3 protein can be first detected at the multipotent progenitor (MPP) stage during murine hematopoiesis. However, Flt3 may be expressed at a low level during earlier developmental stages and it remains unknown whether such expression might mark functionally distinct HSPC.

Dimerization of Flt3 occurs upon binding of its ligand (Flt3L) resulting in auto-phosphorylation of tyrosine residues [11,12], recruitment of the adapter proteins SHC, CBL and GRB [13–15] and signaling via the phosphoinositide 3 kinase (PI3K) and RAS pathways [16,17]. PI3K signaling is important to cell survival and, accordingly, the ligand promotes the survival and growth of hematopoietic progenitors, particularly myeloid and B lymphoid pathway progenitors [18–20]. The use of semi-solid medium assays has revealed that Flt3L influences the formation of granulocyte-macrophage (GM) colonies by human bone marrow CD34⁺ cells [21]. Flt3L also synergizes with other cytokines. The addition of Flt3L to interleukin (IL)-3 or IL-6 doubles the cell number in the colonies derived from mouse Lin⁻ Thy^lo Sca-1⁺ bone marrow cells and FltL combined with IL-3 or granulocyte-macrophage colony-stimulating factor (GM-CSF) enhances the growth of Lin⁻ CD34⁺ CD33⁺ human fetal liver progenitor cells [22]. Flt3L alone has little or no effect on these populations [19,23–26]. Flt3L has also been shown to synergize with the GM-CSF-IL-3 fusion protein Pixy 321 for human HPC [21] and with stem cell factor, GM-CSF, IL-6, IL-7, IL-11 and IL-12 for both murine and human HPC [23–30]. Importantly, Flt3L alone or combined with other appropriate cytokines does not affect the growth of the erythroid (BFU-E and CFU-E) [23,26,28] or megakaryocyte colonies in vitro [25,31,32]. In essence, the range of action of Flt3 is restricted to cells belonging to the lymphoid and GM pathways.

Flt3L⁻/⁻ mice have a reduced bone marrow, spleen and lymph node cellularity, and decreased numbers of dendritic cells (DC), Gr-1⁺ CD11b⁺ myeloid cells and lymphoid cells, including innate lymphoid cells [33,34]. Injection of Flt3L into mice leads to leukocytosis which is mostly due to an elevation in monocytes. The absolute number of LSK in bone marrow, spleen and peripheral blood is increased, lymphocytes are elevated, and there is a significant decrease in the hematocrit value and a 90% reduction in immature TER119⁺ erythroid cells [35]. Ceredig and colleagues injected mice with Flt3L and observed a 50% expansion of Flt3⁺ CD19⁻ B220⁺ CD117^lo cells, termed Early Progenitors with Lymphoid and Myeloid potential, and an increase in the number of DC [36,37]. Similarly, transgenic mice that express supra-physiological levels of human Flt3L (Flt3L-Tg) have increased numbers of Gr-1⁺ CD11b⁺ myeloid cells, NK1.1⁺ cells and DC. Studies of Flt3L-Tg mice have led to the proposition that Flt3L above a certain threshold level instructs myeloid and lymphoid development at the expense of cells developing along the megakaryocytic and erythroid (MegE) pathways, as these mice are anemic, thrombocytopenic and have a 9.7-fold decrease in megakaryocyte-erythrocyte progenitors (MEP) [38].

Blast cells of most cases of acute myeloid leukemia (AML) express Flt3 [39,40] and Flt3L has a strong stimulatory effect on these cells, enhancing colony growth when other cytokines are present at suboptimal levels [41]. Furthermore, around 35% of AML patients harbor a *FLT3* mutation [42,43], which often leads to constitutive activation of Flt3. In frame internal tandem duplications (ITD), in the juxta-membrane part of Flt3, account for 25–35% of the mutations in AML [44] and 5–10% of myelodysplastic syndrome (MDS) cases [45,46]. FLT3-ITD has also been associated with malignant transformation of MDS [45,47] and a poor prognostic outcome in AML [42,44,48–50], with the ratio of mutant to wild-type alleles having an impact [51]. The second most common *FLT3* mutations are missense point mutations in the tyrosine kinase domain which occur in approximately 5–10% of AML,

2–5% of MDS and 1–3% of acute lymphocytic leukemia (ALL) cases [46,51,52]. As to all of the above, selective Flt3 inhibitors are being examined as a means of treating some cases of AML [44,53].

Various populations of murine HSPC can be isolated by the use of cell surface markers, for example, the signaling lymphocytic activation molecule family of markers CD48, CD150 and CD224 [54]. In this study, we have used a combination of the quantitative reverse transcription polymerase chain reaction (qRT-PCR), to examine mRNA within single cells, and flow cytometry, for protein expression, to delineate the extent sub-populations of HSPC express Flt3. The expression of Flt3, or not, has been examined in relation to the expression of the receptors for erythropoietin (Epo) and M-CSF. The use of a triplex qRT-PCR assay and multi-color flow cytometry has revealed substantial heterogeneity of HSPC.

2. Results

2.1. Heterogeneous Expression of Flt3 Transcripts by Single HSPC

Cell surface phenotypes that had been adapted from published profiles were used to identify and isolate the various populations of HSPC from murine bone marrow (Figure 1). The LSK compartment was divided into HSC (LSK CD150$^+$ CD48$^-$), MPP (LSK CD150$^-$ CD48$^-$), HPC-1 (LSK CD150$^-$ CD48$^+$) and HPC-2 (LSK CD150$^+$ CD48$^+$), as published by the Morrison group [54,55]. HSC were divided into LT-HSC (LSK CD150$^+$ CD48$^-$ CD34$^-$) or ST-HSC (LSK CD150$^+$ CD48$^-$ CD34$^+$) based on their expression of CD34 [1]. The gating strategy used to identify HPC-1 is similar to the one used by Adolfsson et al. to identify lymphoid-primed multipotent progenitors (LMPP), as these two cell populations express high levels of Flt3 at their surface [55,56]. Therefore, HPC-1 were divided into Flt3$^{-/lo}$ HPC-1 (LSK CD150$^-$ CD48$^+$ Flt3$^{-/lo}$) and Flt3^{hi} LMPP (LSK CD150$^-$ CD48$^+$ Flt3^{hi}). The common myeloid progenitors (CMP; LS$^-$K IL-7Rα$^-$ CD16/32^{lo} CD34^{hi}), granulocyte-macrophage progenitors (GMP; LS$^-$K IL-7R$^-$ CD16/32^{hi} CD34^{hi}), MEP (LS$^-$K IL-7Rα$^-$ CD16/32^{lo} CD34^{lo}) and common lymphoid progenitors (CLP; LS^{int}K^{int} IL-7R$^+$) were identified according to the strategies published by the Weissman group [57,58].

First, expression of Flt3 mRNA by single cells within each of the above populations was investigated using qRT-PCR (Figure 2). Cells that expressed Flt3 mRNA (Flt3mRNA$^+$) and the endogenous control mRNA, Actb, could be readily identified (Figure 2A,B). In total, 1465 single cells were sorted and Actb mRNA was detected in 1416 samples (96.7%). Samples that did not contain detectable levels of Actb mRNA were removed from analysis.

Within the LT-HSC compartment, 11.6 ± 2.19% of cells (28 of 248 cells) expressed a detectable level of Flt3 mRNA (Figure 2C,D). Flt3 mRNA was expressed by 21 ± 3.4% (56 of 252 cells) of ST-HSC, which was significantly higher than the percentage of Flt3mRNA$^+$ LT-HSC ($p = 0.0476$) (Figure 2C,D). Two hundred ninety-seven MPP were analyzed and 64 ± 2.2% (190 cells) of these cells were Flt3mRNA$^+$ (Figure 2C,D). Analysis of progenitors that are associated with the myeloid pathways revealed a clear decrease in the percentage of Flt3mRNA$^+$ cells as compared to the percentage of Flt3mRNA$^+$ MPP (Figure 2C,D). Eighty-one CMP were analyzed and 21 of these cells expressed Flt3 mRNA (24.9 ± 5.4%). Similarly, 21.4 ± 2.8% (21 of 98 cells) and 9.9 ± 6.0% (8 of 56 cells) of GMP and HPC-2 were Flt3mRNA$^+$, respectively. Only one of the 98 MEP assayed (1 ± 1%) was found to be Flt3mRNA$^+$. On the other hand, the fractions of Flt3mRNA$^+$ cells within the lymphoid pathway-associated progenitor compartments were either comparable or greater than the percentage of Flt3mRNA$^+$ cells within the MPP population (Figure 2C,D). Within the LMPP population, 92.8 ± 2.0 of cells (90 of 97 cells) were Flt3mRNA$^+$, while 57.2 ± 14.1% (57 of 101 cells) and 66.2 ± 3.8% (62 of 94 cells) of Flt3$^{-/lo}$ HPC-1 and CLP were Flt3mRNA$^+$, respectively. Together, these data demonstrate that Flt3 mRNA is expressed by cells within almost all HSPC populations, including HSC.

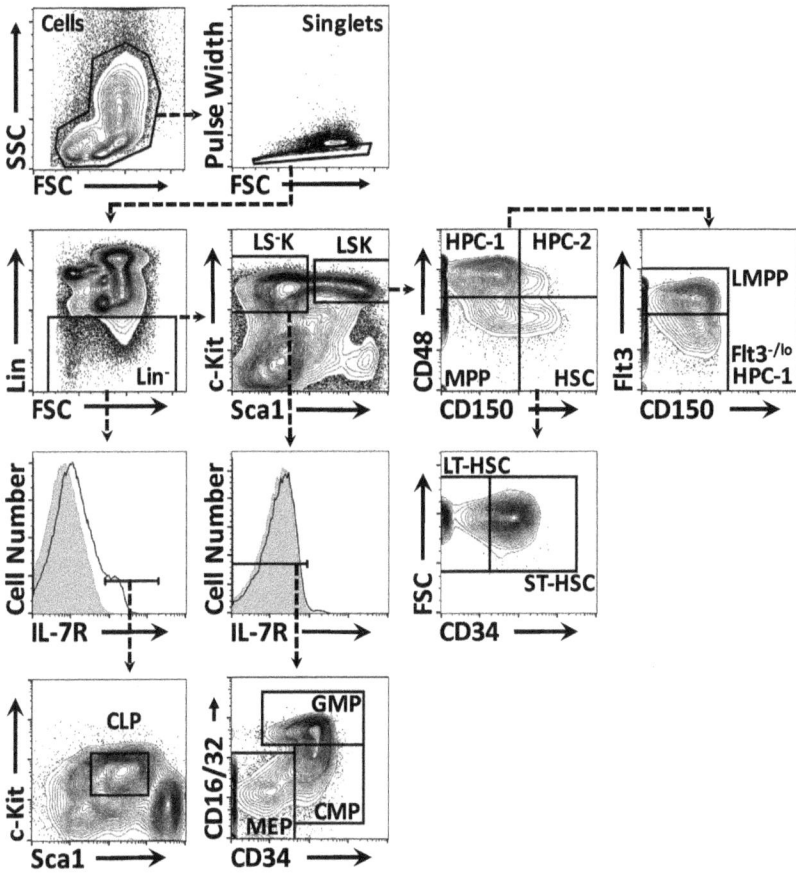

Figure 1. Isolation of hematopoietic stem and progenitor cells (HSPC) from mouse bone marrow. The gating strategies used to identify hematopoietic stem cells (HSC) and various progenitors are shown. The areas delineated by the black boxes in the scatter plots and solid black lines in the histograms indicate the population of cells gated. Dashed arrows point to further gating of these cells. Shaded areas in the histograms depict the isotype control staining. Lineage markers (Lin) included CD3ε, B220, CD11b, Ly-6G and TER-119. HSPC, hematopoietic stem and progenitor cells; LT-HSC, long-term reconstituting hematopoietic stem cell; ST-HSC, short-term reconstituting hematopoietic stem cell; HSC, hematopoietic stem cell; MPP, multipotent progenitor; HPC, hematopoietic progenitor cell; lymphoid-primed multipotent progenitors (LMPP), lymphoid-primed multipotent progenitor; common lymphoid progenitors (CLP), common lymphoid progenitor; common myeloid progenitors (CMP), common myeloid progenitor; GMP, granulocyte-macrophage progenitor; MEP, megakaryocyte-erythrocyte progenitor; LS$^-$K, Lineage marker$^-$ Sca1$^-$ c-Kit$^+$; LSK, Lineage marker$^-$ Sca1$^+$ c-Kit$^+$; Flt3, fms-like tyrosine kinase 3; IL-7R, interleukin-7 receptor. SSC, side light scatter; FSC, forward light scatter.

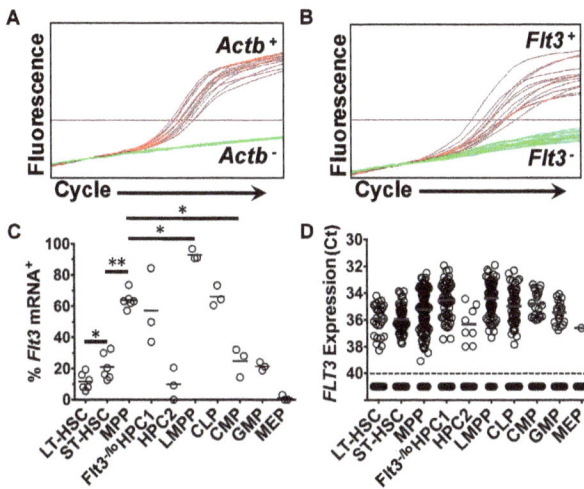

Figure 2. Expression of *Flt3* mRNA transcripts by HSPC. Analysis of *Flt3* mRNA expression by HSPC was carried out using duplex qRT-PCR assays specific for both *Actb* and *Flt3* transcripts. Reactions that did not give rise to a detectable amplification of *Actb* were removed from analysis. Representative amplification of: (**A**) the endogenous control mRNA, *Actb*; and (**B**) *Flt3* mRNA in single HSC are shown, where there is a clear distinction between cells that expressed the target gene (shown in red) and those that did not (shown in green); (**C**) The percentage of cells within each HSPC population that expressed *Flt3* mRNA; and (**D**) the levels of *Flt3* mRNA expressed by single HSPC. *Flt3* mRNA expression was estimated using the cycle at which the PCR signal crossed an arbitrary threshold (Ct) and each individual cell is represented by a single data point. Single cells that gave rise to a detectable amplification of *Actb* mRNA, but not *Flt3* mRNA, are plotted below the dotted lines. Data in (**C**,**D**) are the mean of the values obtained from $n = 3$–6 mice. In total, 248 LT-HSC, 252 ST-HSC, 297 MPP, 101 Flt3$^{-/lo}$ HPC-1, 97 LMPP, 56 HPC-2, 94 CLP, 81 CMP, 98 GMP and 98 MEP were analyzed. p values were obtained by two-tailed non-parametric Student's *t*-test, where * $p < 0.05$; ** $p < 0.005$,. HSC, hematopoietic stem cell; LT-HSC, long-term reconstituting hematopoietic stem cell; ST-HSC, short-term reconstituting hematopoietic stem cell; MPP, multipotent progenitor; LMPP, lymphoid-primed multipotent progenitor; CLP, common lymphoid progenitor; CMP, common myeloid progenitor; GMP, granulocyte-monocyte progenitor; MEP, megakaryocyte-erythrocyte progenitor; Flt3, fms-like tyrosine kinase 3; HSPC, hematopoietic stem and progenitor cell.

2.2. Expression of Flt3 Protein by HSPC Populations

Next, the expression of Flt3 protein on the surface of bone marrow HSPC was investigated by flow cytometry (Figure 3). Flt3 protein was detected on the surface of $4.6 \pm 1\%$ of LT-HSC and $7.7 \pm 1.2\%$ of ST-HSC (Figure 3A,B). Interestingly, the expression of Flt3 protein by HSC correlated with the presence of a low level of CD150 at the cell surface (Figure 3D,E). A low level of expression of CD150 is associated with a reduced potential for self-renewal, and MegE development [59,60]. Within the MPP population, $63.7 \pm 3\%$ of cells had detectable levels of Flt3 at their cell surface (Figure 3A,B). In agreement with the gene expression analysis, the percentages of Flt3$^+$ cells within the myeloid pathway-associated progenitor populations were less as compared to the MPP compartment (Figure 3A). Flt3 was detected on the surface of $36.8 \pm 1.4\%$ of CMP (Figure 3A,B). Within the HPC-2 and GMP populations, $9.3 \pm 1.1\%$ and $15.7 \pm 0.9\%$ of cells were Flt3$^+$, respectively (Figure 3A). Flt3 was virtually absent from the MEP compartment ($0.5 \pm 0.06\%$) (Figure 3A). Lymphoid pathway-associated progenitor compartments contained either comparable or greater percentages of Flt3$^+$ cells as compared to the MPP population. Flt3 was detected on the surface of $80 \pm 2.1\%$ of CLP, and $58.6 \pm 3\%$ of Flt3$^{-/lo}$ HPC-1 (Figure 3A,B).

As to their designated phenotype, 100% of LMPP were Flt3$^+$ (Figure 3A). Similar to the *Flt3* mRNA expression data, these results show that Flt3 protein is expressed by all HSPC populations, apart from MEP, and is most commonly found on the surface of HPC with robust lymphoid potential.

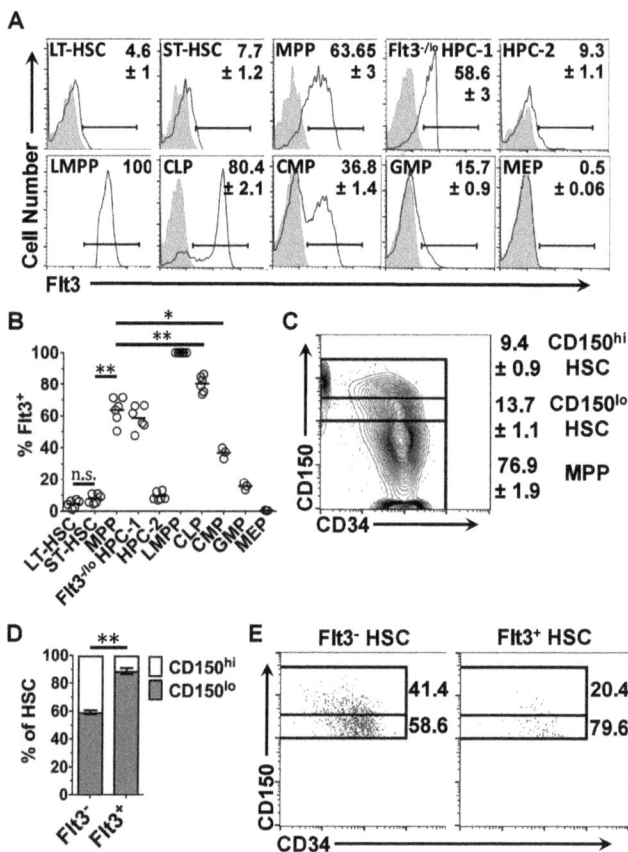

Figure 3. Cell surface expression of Flt3 protein by HSPC: (**A**) the gating strategy used to detect Flt3 on the surface of HSC and HPC populations (solid black line), compared to an isotype control (shaded histogram); (**B**) the percentage of cells within each HSPC compartment that expressed Flt3 protein at their surface; (**C**) the gating strategy used to identify LSK CD48$^-$ cells with high (CD150^{hi}) and low levels of CD150 (CD150^{lo}) expressed at their surface and those lacking CD150 expression (MPP); (**D**) the percentages of Flt3$^+$ and Flt3$^-$ cells within the HSC compartment that express low and high levels of CD150 on their surface; and (**E**) the gating strategy in (**C**) applied to cells within the Flt3$^+$ and Flt3$^-$ HSC populations in a representative sample. Data in (**A**,**B**) and (**C**,**D**) are the mean of the values obtained from $n = 3$–6 and $n = 6$ mice, respectively. Values depicted in (**A**,**C**,**E**) represent the percentage of cells within each gate. Solid arrows in (**A**,**C**,**E**) indicate increasing cell number or the signal intensity in the designated channel. p values obtained by two-tailed non-parametric Student's t-test, where ** $p < 0.005$. HSC, hematopoietic stem cell; LT-HSC, long-term reconstituting hematopoietic stem cell; ST-HSC, short-term reconstituting hematopoietic stem cell; MPP, multipotent progenitor; HPC, hematopoietic progenitor cell; LMPP, lymphoid-primed multipotent progenitor; CLP, common lymphoid progenitor; CMP, common myeloid progenitor; GMP, granulocyte-monocyte progenitor; MEP, megakaryocyte-erythrocyte progenitor; Flt3, fms-like tyrosine kinase 3; LSK, Lineage marker$^-$ Sca1$^+$ c-Kit$^+$.

2.3. Effects of Flt3 Stimulation on the Phosphorylation of Intracellular Ribosomal Protein S6 by HSC and MPP

Murine HSC are thought not to express Flt3 and we identified a small fraction of cells within the LT-HSC and ST-HSC compartments that express both *Flt3* mRNA and Flt3 protein. To determine whether Flt3$^+$ HSC and MPP respond to Flt3L in vitro, we adapted a previously described phospho-flow protocol for detecting phosphorylated S6 (pS6) in HSPC, as both the PI3K and RAS pathways converge to stimulate protein translation *via* phosphorylation of S6 [16,61–63]. Even though CD150 staining of whole bone marrow is compromised following cell fixation and permeabilization with acetone [63], CD150 staining was observed to be maintained within the LSK compartment when compared to untreated controls (Figure 4A,B). As such, we were able to monitor the levels of pS6 in the HSC and MPP populations.

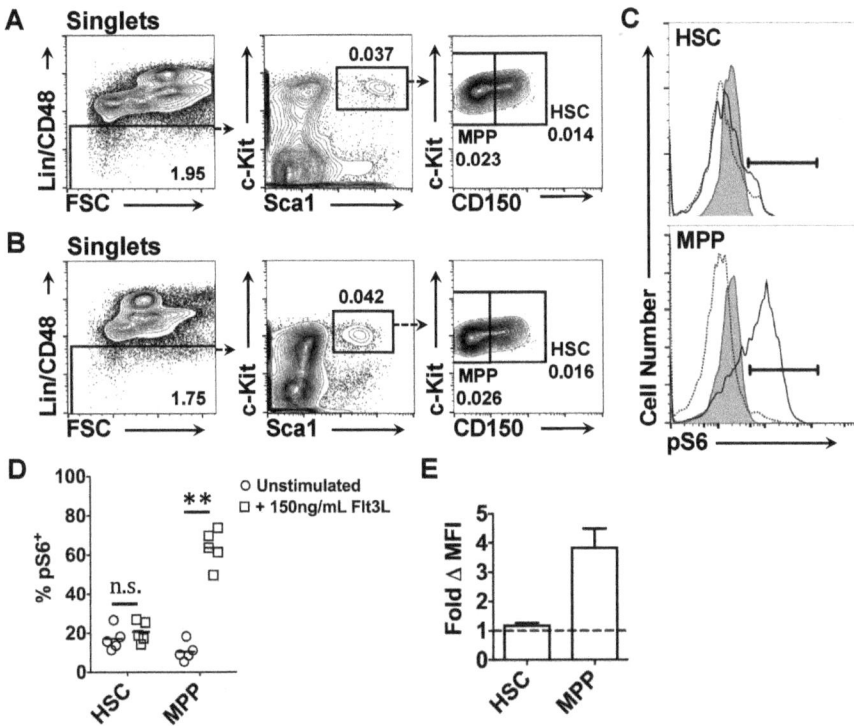

Figure 4. Changes in the phosphorylation of the ribosomal protein S6 in HSC and MPP following treatment with Flt3 ligand. Representative gating strategies used to analyze phosphorylated ribosomal protein S6 (pS6) showing: (**A**) unfixed; and (**B**) fixed and permeabilized samples. Gating of pS6$^+$ HSC and MPP for treated (solid black line) and untreated cells (dotted black line) are shown in (**C**) and values for the percentages of positive cells are given in the text. Shaded histograms depict the isotype control; (**D**) The percentage of pS6$^+$ cells in untreated and treated HSC and MPP; (**E**) The fold change in the MFI of pS6 staining in treated HSC and MPP compared to untreated cells. The areas delineated by solid black boxes in the scatter plots and solid black lines in the histograms indicate the population of cells gated. Dashed arrows point to further gating of these cells. Solid arrows in (**A,B,C**) indicate increasing cell number or signal intensity in the designated channel. Values depicted in (**A,B**) represent the percentage of total bone marrows cells within each gate. Data in (**D,E**) are the mean of the values obtained from *n* = 5 mice. *p* values obtained by two-tailed non-parametric Student's *t*-test, where ** *p* < 0.005; n.s., non-significant. HSC, hematopoietic stem cell; MPP, multipotent progenitor; Flt3, fms-like tyrosine kinase 3; Flt3L, flt3 ligand; MFI, mean fluorescence intensity.

After starvation in serum-free medium, HSC and MPP were stimulated in vitro with 150 ng/mL Flt3L for 7.5 min before being stained for pS6. For starved MPP stimulated with Flt3L, 63.7 ± 9.2% of cells expressed pS6 as compared to a value of 10.5 ± 1.38% for starved and untreated MPP ($p = 0.0079$) (Figure 4C,D). In contrast, Flt3L treatment resulted in only a marginal increase in the percentage of HSC expressing pS6 (from 17.1 ± 5.9% to 20.6 ± 5.4%, Figure 4C,D). In addition, the mean fluorescence intensity (MFI) of pS6 staining in starved and treated MPP was increased 3.9 ± 1.4-fold as compared to the pS6 MFI obtained for starved and untreated MPP, while the corresponding ΔMFI for HSC was 1.2 ± 0.1 (Figure 4E). These data demonstrate that Flt3L stimulates S6 phosphorylation in the majority of MPP. However, the effect of Flt3L on S6 phosphorylation in HSC is much less pronounced and is possibly due to the rarity of Flt3$^+$ cells.

2.4. Flt3 Expression Identifies Distinct Sub-Populations of HSC and MPP

Other hematopoietic growth factor receptors, such as the receptors for Epo (EpoR) and M-CSF (M-CSFR), have been implicated in the pathogenesis of leukemia [64–66]. Both EpoR and M-CSFR are known to be expressed by HSC, and there is evidence to support their role in the development of HSC [67,68]. Therefore, we examined whether Flt3 and EpoR or Flt3 and M-CSFR are co-expressed by cells within the HSC compartment, or if the expression of these receptors identifies distinct sub-populations of HSC.

To determine if HSC and MPP co-express the genes encoding Flt3 and EpoR, we designed a triplex qRT-PCR assay to measure the levels of *Actb*, *Flt3* and *Epor* mRNAs in single cells. The *Epor* qRT-PCR assay detected *Epor* mRNA expression in 0 of 94 CLP and 96 of 98 MEP, suggesting a high degree of specificity of the assay (Figure 5A). As with cells expressing *Actb* and *Flt3* mRNA, cells expressing *Epor* mRNA (*Epor*mRNA$^+$) were readily identified in the HSC and MPP compartments (Figure 5B). One hundred thirty-six LT-HSC, 139 ST-HSC and 148 MPP were analyzed for expression of *Flt3* mRNA and *Epor* mRNA (Figure 5C). For LT-HSC, 12.8 ± 1.46% of cells were *Epor*mRNA$^+$ and none of these cells co-expressed both *Flt3* and *Epor* mRNA. Of the ST-HSC analyzed, 19.3 ± 4.6% of cells expressed *Epor* mRNA. Notably, 2 of the 139 ST-HSC (1.5 ± 2.5%) analyzed for *Flt3* and *Epor* mRNA co-expressed both mRNAs. Within the MPP compartment, 8.2 ± 4.6% were *Epor*mRNA$^+$, and none of them co-expressed *Flt3* mRNA. We could not examine the co-expression of Flt3 and EpoR protein at the cell surface because of the lack of an antibody to EpoR.

We then investigated whether HSC and MPP co-expressed the genes encoding Flt3 and M-CSFR (*Csf1r*) and detected very few cells that expressed *Csf1r* mRNA (*Csf1r*mRNA$^+$) in these compartments. *Csf1r* mRNA was detected in 0.9 ± 0.9% (1 of 122 cells) of LT-HSC, 2.14 ± 1.3% (2 of 118 cells) of ST-HSC, and 4.8 ± 3.8% (7 of 145 cells) of MPP (Figure 6A). As to the few *Csf1r*mRNA$^+$ cells, either the *Csf1r* qRT-PCR assays was not sensitive enough to detect *Csf1r* mRNA expression in HSC and MPP or these cells rarely transcribe the *Csf1r* gene. Mossadegh-Keller et al. have also investigated whether HSC express *Csf1r* mRNA and have reported that very few cells are *Csf1r*mRNA$^+$, suggesting the latter [68]. Surprisingly, a significant proportion of HSC and MPP expressed M-CSFR protein (Figure 6B,D). We therefore examined whether cells within the HSC and MPP compartments co-expressed Flt3 and M-CSFR protein at their surface.

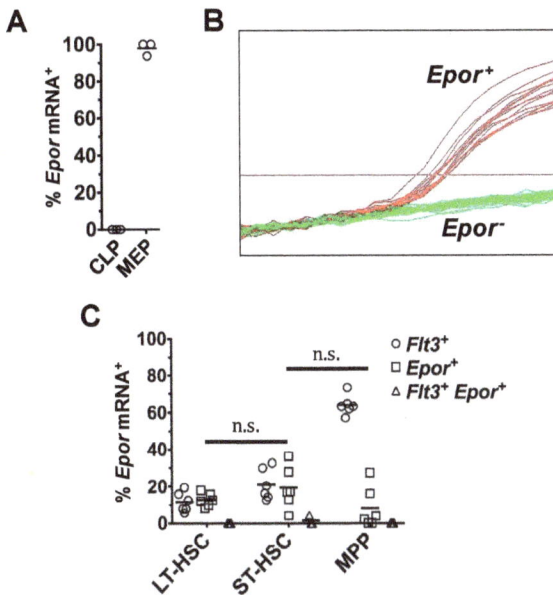

Figure 5. *Flt3* and *Epor* transcripts are rarely co-expressed by LT-HSC, ST-HSC and MPP. Analysis of *Flt3* and *Epor* co-expression by LT-HSC, ST-HSC and MPP was carried out using a triplex qRT-PCR assay specific for *Actb*, *Flt3*, and *Epor* mRNA. Reactions that did not give rise to a detectable amplification of *Actb* were removed from analysis. (**A**) The percentage of CLP and MEP that expressed *Epor* mRNA; (**B**) the amplification of *Epor* mRNA in single ST-HSC where there is a clear distinction between cells that expressed the *Epor* (shown in red) and those that did not (shown in green); and (**C**) the percentage of LT-HSC, ST-HSC and MPP that co-expressed both *Flt3* and *Epor* mRNA. The total fraction of *Flt3*mRNA$^+$ and *Epor*mRNA$^+$ are also shown. Data in (**A**) are the mean of the values obtained from $n = 3$ mice. Data in (**C**) are the mean of the values obtained from $n = 3$–6 mice. p values obtained by two-tailed non-parametric Student's *t*-test, where n.s., non-significant. LT-HSC, long-term reconstituting hematopoietic stem cell; ST-HSC, short-term reconstituting hematopoietic stem cell; MPP, multipotent progenitor; MEP, megakaryocyte-erythrocyte progenitor; CLP, common lymphoid progenitor; Flt3, fms-like tyrosine kinase 3; Epor, erythropoietin receptor.

Within the LT-HSC population, $18.6 \pm 3.8\%$ of cells were M-CSFR$^+$ and only $1.12 \pm 0.48\%$ of these cells expressed both Flt3 and M-CSFR. M-CSFR was detected at the surface of $23.4 \pm 3.6\%$ of ST-HSC, and $2.8 \pm 0.7\%$ co-expressed both Flt3 and M-CSFR protein (Figure 6B). Of the MPP analyzed, $13.4 \pm 2.46\%$ of cells expressed M-CSFR at their surface which was significantly less than the proportion of M-CSFR$^+$ cells in the ST-HSC population ($p = 0.0313$), though the fraction of Flt3$^+$ M-CSFR$^+$ MPP ($9.9 \pm 1.9\%$) was greater than when compared to the ST-HSC population ($p = 0.0313$) (Figure 6B). As Flt3 is only expressed by a small percentage of HSC, we also examined Flt3 and M-CSFR co-expression as a percentage of Flt3$^+$ cells within the HSC and MPP compartments (Figure 6C). Of the Flt3$^+$ LT-HSC and Flt3$^+$ ST-HSC, $20.1 \pm 6.87\%$ and $34.47 \pm 5.35\%$ of cells also expressed M-CSFR, respectively. Of the Flt3$^+$ MPP population, $14.46 \pm 2.62\%$ also expressed M-CSFR at their surface. These data show that Flt3, EpoR and M-CSFR are selectively expressed during early hematopoiesis and that expression of these receptors identifies novel subpopulations of early HSPC.

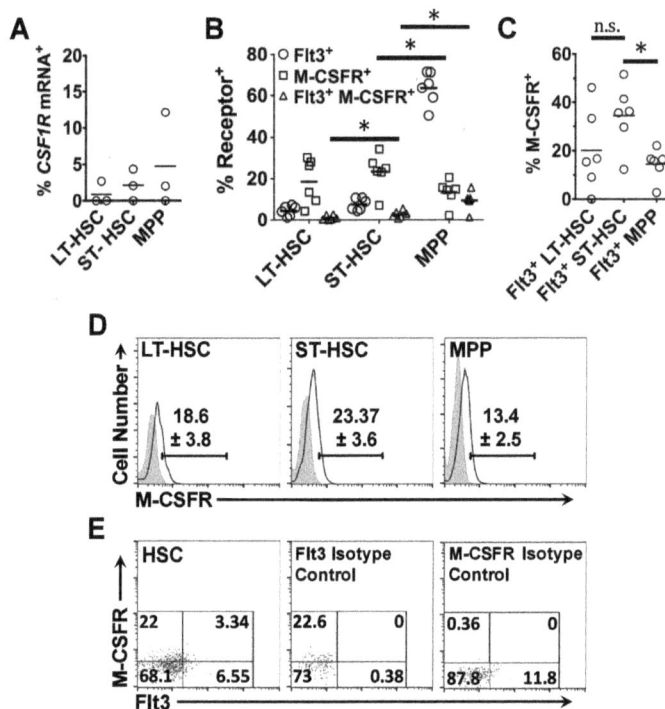

Figure 6. Flt3 and M-CSFR protein are primarily expressed by distinct sub-populations of LT-HSC and ST-HSC: (**A**) analysis of *Csf1r* mRNA expression by single LT-HSC, ST-HSC and MPP. Data are the mean of the values obtained from $n = 3$ mice; (**B**) the percentage LT-HSC, ST-HSC and MPP that expressed Flt3, M-CSFR or both of these receptors at their surface; (**C**) the percentage of Flt3+ LT-HSC, Flt3+ ST-HSC and Flt3+ MPP that expressed M-CSFR at their surface; and (**D**) the gating strategy used to identify M-CSFR+ LT-HSC, M-CSFR+ ST-HSC, and M-CSFR+ MPP (solid black line). Shaded histograms depict the isotype control. Scatter plots in (**E**) represent the gating strategy used to identify HSC that co-expressed Flt3 and M-CSFR at their surface. Isotype controls for both Flt3 and M-CSF staining within the HSC compartment are also shown. Solid arrows in (**D**,**E**) indicate increasing cell number or signal intensity in the designated channel. Data in (**B**–**D**) are the mean of the values obtained from $n = 6$ mice. Values in (**D**,**E**) are the percentages of cells within the corresponding gates. p values were obtained by two-tailed non-parametric Student's t-test, where * $p < 0.05$; n.s., non-significant. HSC, hematopoietic stem cell; LT-HSC, long-term reconstituting hematopoietic stem cell; ST-HSC, short-term reconstituting hematopoietic stem cell; MPP, multipotent progenitor; Flt3, fms-like tyrosine kinase 3; M-CSFR, macrophage colony stimulating factor receptor.

3. Discussion

In this study, the expression of *Flt3* mRNA and Flt3 protein by the various populations of HSPC from the murine bone marrow has been investigated. HSC and MPP were then analyzed to determine whether Flt3 and EpoR or Flt3 and M-CSFR are co-expressed by cells within these compartments. The findings demonstrate a large degree of heterogeneity within the HSPC compartments, and distinct sub-populations of HSC and MPP can be identified by their expression of Flt3, EpoR and M-CSFR.

Flt3 is generally considered to be absent from the surface of HSC in the adult mouse [5,6,69], though recent studies have identified *Flt3* expression within the LSK CD150+ CD48− stem cell compartment [9,70]. Furthermore, Chu et al. have shown that expression of FLT3-ITD under the control of the endogenous Flt3 promoter results in the depletion of the murine stem cell pool [70].

Here, the HSC compartment was divided into LT-HSC (LSK CD150$^+$ CD48$^-$ CD34$^-$) and ST-HSC (LSK CD150$^+$ CD48$^-$ CD34$^+$) and both of these populations contain a small percentage of cells that express *Flt3* mRNA and Flt3 protein. Downstream of HSC, Flt3 expression delineates murine lymphoid/myeloid progenitors from those progressing along the MegE pathway; this is illustrated by Figure 7, which maps HSPC populations and Flt3 expression to the pairwise model (described in [71] and [72]). In humans, Flt3 has been reported to be expressed by HSC [73]. Taking these findings together, the leukemias with Flt3-activating mutations, which include AML and ALL, might well have arisen from the Flt3$^+$ primitive stem cell compartment of the bone marrow. Therefore, the populations identified here are important to understanding the origin of leukemia and their potential lineage affiliation is pertinent to disease progression. This viewpoint is also important to studies that make use of FLT3-ITD mouse models.

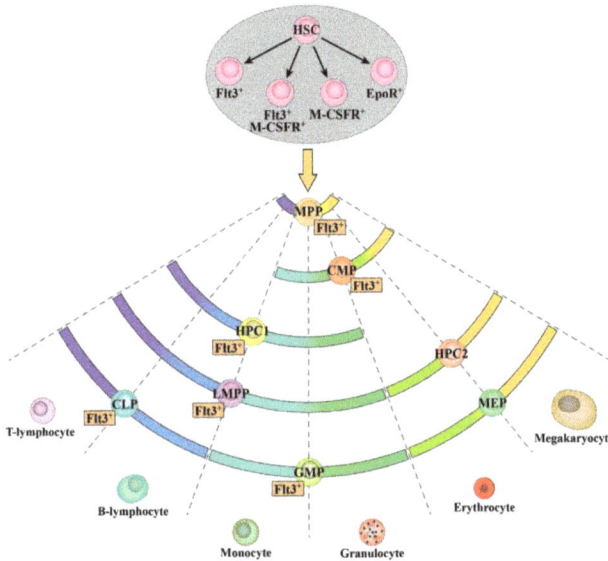

Figure 7. Heterogeneous expression of lineage affiliated receptors Flt3, EpoR and M-CSFR in murine HSPC. Four populations of HSC were identified in this study based on the expression of Flt3, EpoR and M-CSFR (as indicated by the black arrows). These populations are depicted within the shaded area. MPP and HPC are downstream of HSC (as indicated by the yellow arrow) and are heterogeneous as to their expression of Flt3, and this is illustrated using the pairwise model (described in [72] and [71]). The maturation potential of HSPC is depicted by the colored arcs. Expression of Flt3 protein and *Flt3* mRNA was strongly associated with lymphoid-GM potential, as expected. Flt3 expression was rarely observed within the HPC-2 compartment, and was virtually absent from MEP. In this study, HPC-1 and LMPP were defined as LSK CD48$^+$ CD150$^-$ and LSK CD48$^+$ CD150$^-$ Flt3^{hi}, respectively. The percentages of *Flt3mRNA*$^+$/Flt3$^+$ cells within each compartment are as follows: LT-HSC, 11.6 ± 2.2%/4.6 ± 1%; ST-HSC, 21 ± 3.4%/7.7 ± 1.2%; MPP, 64 ± 2.2%/63.7 ± 3%; Flt3$^{-/lo}$ HPC-1, 57.2 ± 14.1%/58.6 ± 3%; HPC-2, 9.9 ± 6.0%/9.3 ± 1.1%; LMPP, 92.8 ± 2.0%/100%; CLP, 66.2 ± 3.8%/80 ± 2.1%; CMP, 24.9 ± 5.4%/36.8 ± 1.4%; GMP, 21.4 ± 2.8%/15.7 ± 0.9%; MEP, 1 ± 1%/0.5 ± 0.06%. HSC, hematopoietic stem cell; LT-HSC, long-term reconstituting hematopoietic stem cell; ST-HSC, short-term reconstituting hematopoietic stem cell; HPC, hematopoietic progenitor cell; MPP, multipotent progenitor; LMPP, lymphoid-primed multipotent progenitor; CLP, common lymphoid progenitor; CMP, common myeloid progenitor; GMP, granulocyte-macrophage progenitor; MEP, megakaryocyte-erythrocyte progenitor; Flt3, fms-like tyrosine kinase 3; M-CSFR, macrophage colony-stimulating factor receptor; EpoR, erythropoietin receptor; LSK, Lineage marker$^-$ Sca1$^+$ c-Kit$^+$.

Previous characterization of Flt3$^+$ cells within the primitive cell compartments of the adult murine bone marrow has indicated that these cells have limited self-renewal capacity [5,7,9]. Indeed, Flt3 expression within the HSC compartment correlates with a low level of CD150 at the cell surface (Figure 3D,E) which is associated with a reduced self-renewal ability [59,60]. However, the reconstitution assays performed in the above studies are limited by their use of bulk Flt3$^+$ LSK cells, which contain very few cells that are Flt3$^+$ LSK CD150$^+$ CD48$^-$, and their assessment of self-renewal capacity by just monitoring myeloid reconstitution. Using their FlkSwitch model, the Forsberg group have recently identified a transient population of fetal Flt3$^+$ HSC that preferentially give rise to lymphoid cells [74]. Like the Flt3$^+$ cells in the adult murine HSC compartment identified here, fetal Flt3$^+$ HSC express a low level of CD150 at their surface. Hence, Flt3 expression in the adult HSC compartment may identify a rare population of lymphoid-biased HSC. Fetal Flt3$^+$ HSC do not persist in the bone marrow past 8 weeks of age, so it is unlikely that the phenotypic Flt3$^+$ HSC identified here are fetal Flt3$^+$ HSC that have persisted into late adulthood [74].

Investigation of the expression of *Epor* mRNA and M-CSFR protein identified subpopulations of cells within the LT-HSC, ST-HSC and MPP compartments (Figure 7). Co-expression of *Flt3* mRNA and *EpoR* mRNA was exceedingly rare, indicating that Flt3 and EpoR expression almost exclusively identify distinct subpopulations of multipotent progenitors. Co-expression of M-CSFR and Flt3 by early HSPC was more common, which was expected since both Flt3 and M-CSFR are associated with the development of myelomonocytic progenitors. Conventionally, HSC are thought to gradually commit to mature cell fates by transitioning through various intermediate progenitor states. However, long-term reconstituting lineage-biased and lineage-restricted progenitors have been identified in the HSC compartment of the mouse bone marrow [59,60,75–83], and megakaryocyte specification occurs primarily within the HSC compartment in humans [84]. These findings suggest that lineage specification occurs at a very early stage of hematopoiesis, without HSC having to transit through a succession of intermediate progenitors. Indeed, a recent study of human HSPC supports this viewpoint [85]. The expression of lineage-associated growth factor receptors by HSC might represent adoption of a fate, providing a selective advantage to cells as to the presence of a particular growth factor. Some hematopoietic growth factor receptors, including Flt3 and M-CSFR, have been used to identify lineage-biased subpopulations within currently defined HPC [83,86–89], and therefore, as already stated in regard to Flt3 expression, EpoR and M-CSFR might be used to enrich for functionally distinct HSC.

As has been previously reported, Flt3 expression dramatically increases during the HSC-to-MPP transition. Approximately 60% of MPP (LSK CD150$^-$ CD48$^-$) express Flt3 and phosphorylate S6 following stimulation with Flt3L [55]. Progenitors downstream of HSC and MPP, other than the MEP, were also found to be heterogeneous as to their expression of Flt3 (Figure 7); the patterns reported here are largely similar to previously published data [55,56,87,90]. As mentioned above, growth factor receptor expression has been used to identify lineage-biased subpopulations of HPC. Further study of the populations identified here might be useful in resolving whether progenitor populations that are viewed as having a number of different lineage options are a uniform population of cells or a mixture of cells with individual lineage potentials.

As expected, Flt3 expression is linked to populations that have robust lymphoid, granulocyte and myelomonocytic potential. Conversely, the frequency of Flt3 expression decreases as cells differentiate towards a MegE fate. This is typified by HPC-1 (LSK CD150$^-$ CD48$^+$), which includes the Flt3$^{-/lo}$ HPC-1 and LMPP populations analyzed in this study, and HPC-2 (LSK CD150$^+$ CD48$^+$). The HPC-1 compartment contains progenitors that preferentially give rise to lymphoid cells and possess limited MegE potential, and Flt3 expression was most commonly detected within this compartment [55]. HPC-2 have significant myeloid potential and a reduced lymphoid potential when compared to HPC-1, and Flt3 expression was infrequently detected in the HPC-2 compartment [55]. The importance of Flt3 expression by lymphoid and GM lineages, as opposed to MegE lineage, is emphasized by the findings from Flt3L-Tg mice [38].

The findings described here, and in a number of other studies, demonstrate the heterogeneity of HSPC. Similarly, compartments that were once thought to comprise of a uniform population of cells have been shown to contain a mixture of cells with varying lineage potentials and cell-intrinsic states [84–86,91]. Moreover, HSC at different stages of the cell cycle display functional heterogeneity [92,93], and hematopoietic progenitors fluctuate between states with differing capacities for particular cell fates [94], highlighting the dynamic nature of cells. These matters need to be considered when using techniques that are reliant on cell surface phenotypes which only provide information about a cell at a single moment in time. In this regard, single cell fate mapping and imaging techniques will be integral to the functional characterization of distinct HSPC subpopulations, such as those identified in this study. Such work is of importance for our understanding of malignant transformation in hematopoiesis and the heterogeneity of leukemia.

4. Materials and Methods

4.1. Animals

Male C57/BL6 mice were purchased from Harlan Laboratories, UK, and housed in the Biomedical Services Unit at The University of Birmingham, UK. All mice were treated in accordance with Home Office guidelines, and were culled by cervical dislocation between 8–14 weeks of age. The use of bone marrow cells from normal mice was conducted under a licence provided to Ciaran Mooney (No. 34942, 18 October 2013) having completed a programme of training approved by the Universities Accreditation Scheme.

4.2. Flow Cytometry

Bone marrow cells were flushed from mouse tibias, femurs and humeri using flow cytometry buffer (Dulbecco's phosphate buffered (DPBS) without Ca^+/Mg^+ (Thermo Fisher Scientific, Waltham, MA, USA) which contained 2% fetal calf serum (FCS) (Thermo Fisher Scientific) and 2 mM ethylenediaminetetraacetic acid (Sigma Aldrich, St. Louis, MO, USA)). Bone marrow cells were strained using 70 µM Cell strainers (BD biosciences, San Diego, CA, USA) and erythrocytes were lysed using ACK Lysing Buffer (Thermo Fisher Scientific). Cells were then stained with fluorescently-labeled antibodies in flow cytometry buffer for 1 h on ice. Other than when CD16/32 expression was examined using fluorescently-labeled antibodies, cells were incubated with purified anti-CD16/CD32 (clone 93, BioLegend, San Diego, CA, USA) for 20 min to block CD16 (FcγIII) and CD32 (FcγII) receptors prior to staining. The following antibodies were used to analyze the phenotype of cells; anti-CD3ε-AF488 (clone 17A2, BioLegend), anti-B220-AF488 (clone RA3-6B2, BioLegend), anti-CD11b-AF488 (clone M1/70, BioLegend), anti-TER-119-AF488 (clone TER-119, BioLegend), anti-Ly-6G-AF488 (clone RB6-8C5, BioLegend), anti-CD16/32-PE (clone 93, BioLegend), anti-CD34-APC (clone RAM34, eBiosciences, San Diego, CA USA), anti-CD48-AF488 (clone HM48-1, BioLegend), anti-CD48-APC-Cy7 (clone HM48-1, BioLegend), anti-CD115(M-CSFR)-BV421 (cloneAFS98, BD Biosciences), anti-CD117-PE-CF594 (clone 2B8, BD Biosciences), anti-CD127-PE-Cy7 (clone A7R34, eBiosciences), anti-CD135-PE (clone A2F10, eBiosciences), anti-CD150-PE (clone TC15-12F12.2, BioLegend), anti-CD150-Pacific Blue (clone TC15-12F12.2, BioLegend), anti-Sca1-PE-Cy7 (clone D7, eBiosciences), anti-pS6(Ser235/Ser236)-PE (clone D57.2.2E, Cell Signalling). Data were acquired using a CyAN FACS Analyser (Beckman Coulter, Fullerton, CA, USA) controlled by Summit v4.3 software.

4.3. Fluorescence Activated Cell Sorting (FACS)

Prior to sorting, stained bone marrow cells were suspended in flow cytometry buffer containing 10% FCS, filtered using Partec CellTrics sterile filters (Sysmex-Partec, Görlitz, Germany) and stored on ice. All of the HSPC populations were sorted twice to ensure high purity. Cells were sorted into LightCycler 480 384-well plates (Roche, Basel, Switzerland) for qRT-PCR analysis or into culture medium for in vitro manipulation. Sorting was carried out using either a MoFlo High Speed Sorter (Beckman Coulter) controlled by Summit v4.3 software or a MoFlo Astrios (Beckman Coulter) controlled by Summit v6.2.3 software.

4.4. Single Cell Quantitative Reverse Transcription Polymerase Chain Reaction

Single cells were sorted into the wells of LightCycler 480 384-well plates (Roche) containing 1 μL of UltraPure DNase/RNase-Free Distilled Water (Thermo Fisher Scientific) for lysis. Gene expression analysis was carried out using a one-step QuantiTect Multiplex RT-PCR Kit (Qiagen, Hilden, Germany) containing limiting concentrations of primers and hydrolysis probes specific for the *Actb* and *Flt3* transcripts. qRT-PCR reactions were carried out and analyzed on a LC480 II instrument (Roche). The program used for qRT-PCR was as follows: reverse transcription, 50 °C for 20 min; polymerase activation, 95 °C for 15 min; and 40–50 cycling steps of denaturation at 95 °C for 45 s, and annealing/extension at 60 °C for 45 s. Oligonucleotides used for qRT-PCR analysis were purchased from Biosearch Technologies (Novato, California, USA) and were as follows; *Actb* forward primer, 5'-CAGCTTCTTTGCAGCTCCTTC-3'; *Actb* reverse primer, 5'-CGACCAGCGCAGCGATAT-3'; *Actb* hydrolysis probe, 5'-CACCAGTTCGCCATGGA-3'; *Flt3* forward primer, 5'-ATCAGCGGGAAAG CCATCATC-3'; *Flt3* reverse primer, 5'-GGGCACACTGGAGGTCTTCT-3'; *Flt3* hydrolysis probe, 5'-TCCTCGCACCATTCGGTA-3'; *Epor* forward primer, 5'- GCAGGAGGGACACAAAGGGT-3'; *Epor* reverse primer, 5'- GGGCTCAGACCAGGCACT-3'; *Epor* hydrolysis probe, 5'-CTCGAACAGCG AAGGTGTAGCGC-3'; *Csf1r* forward primer, 5'-ACCTGTCCTGGTCATCACT-3'; *Csf1r* reverse primer, 5'-AACCTCTTGGGAGCCTGTACTCAC-3'; *Csf1r* hydrolysis probe, 5'-GCATAGCCTCGGCC TTCCTT-3'.

4.5. Phospho-Flow Analysis

Bone marrow cells were stained to identify LT-HSC, ST-HSC and MPP and then washed using serum-free DPBS (without Ca^+/Mg^+) three times before being starved in serum-free Iscove's Modified Dulbecco's Medium (Thermo Fisher Scientific) for 3 h. After the starvation period, each sample was divided in half and one half was stimulated with 150 ng/mL Flt3L for 7.5 min. All samples were then promptly fixed in 1.6% EM-grade paraformaldehyde (PFA) (VWR International, Radnor, PA, USA) for 15 min. Cells were then permeabilized using ice-cold acetone for 15 min at −20 °C, and then stained using an anti-pS6(Ser235/Ser236)-PE (clone D57.2.2E, Cell Signaling Technology, Danvers, MA, USA) and in DPBS containing 2% FCS for 1 h on ice. Cells were washed twice with DPBS containing 2% FCS before data acquisition on a CyAN FACS Analyser (Beckman Coulter) controlled by Summit v4.3 software. During analysis, pS6 staining of stimulated and unstimulated samples of cells from a single mouse was compared.

5. Conclusions

Analyses of single cells within the HSC and HPC bone marrow compartment of C57/BL6 mice, by multiplex RT-PCR for mRNA and flow cytometry for surface protein, has revealed that these cells are heterogeneous in regard to expression of Flt3. Analysis of M-CSFR and EpoR expression within the multipotent compartments of the bone marrow cells revealed further heterogeneity. Flt3 and EpoR are rarely co-expressed whereas co-expression of Flt3 and M-CSFR occurs more commonly. The expression of these receptors for lineage-associated growth factors by HSC indicates their affiliation to or adoption of a fate. The distinct subpopulations of HSPC identified are important to understanding the disease origin and heterogeneity of leukemia.

Acknowledgments: This project has received funding from the European Union's Seventh Framework Programme for research, technological development and demonstration under grant agreement no 315902. Geoffrey Brown is a partner within the Marie Curie Initial Training Network DECIDE (Decision-making within cells and differentiation entity therapies). Ciaran James Mooney and Alan Cunningham gratefully acknowledge receipt of a Marie Curie Research Associate post.

Author Contributions: Ciaran James Mooney, Kai-Michael Toellner and Geoffrey Brown designed the experiments. Ciaran James Mooney performed the experiments. Ciaran James Mooney, Kai-Michael Toellner and Geoffrey Brown analyzed the data. Ciaran James Mooney and Alan Cunningham wrote the manuscript. Panagiotis Tsapogas and Geoffrey Brown revised the manuscript. All authors have read and approved the final manuscript.

Conflicts of Interest: The authors declare no conflict of interest.

References

1. Osawa, M.; Hanada, K.; Hamada, H.; Nakauchi, H. Long-term lymphohematopoietic reconstitution by a single CD34-low/negative hematopoietic stem cell. *Science* **1996**, *273*, 242–245. [CrossRef] [PubMed]

2. Rosnet, O.; Marchetto, S.; deLapeyriere, O.; Birnbaum, D. Murine *Flt3*, a gene encoding a novel tyrosine kinase receptor of the PDGFR/CSF1R family. *Oncogene* **1991**, *6*, 1641–1650. [PubMed]

3. Matthews, W.; Jordan, C.T.; Wiegand, G.W.; Pardoll, D.; Lemischka, I.R. A receptor tyrosine kinase specific to hematopoietic stem and progenitor cell-enriched populations. *Cell* **1991**, *65*, 1143–1152. [CrossRef]

4. Small, D.; Levenstein, M.; Kim, E.; Carow, C.; Amin, S.; Rockwell, P.; Witte, L.; Burrow, C.; Ratajczak, M.Z.; Gewirtz, A.M. Stk-1, the human homolog of Flk-2/Flt-3, is selectively expressed in CD34$^+$ human bone marrow cells and is involved in the proliferation of early progenitor/stem cells. *Proc. Natl. Acad. Sci. USA* **1994**, *91*, 459–463. [CrossRef] [PubMed]

5. Adolfsson, J.; Borge, O.J.; Bryder, D.; Theilgaard-Mönch, K.; Astrand-Grundström, I.; Sitnicka, E.; Sasaki, Y.; Jacobsen, S.E. Upregulation of flt3 expression within the bone marrow Lin$^-$Sca1$^+$c-kit$^+$ stem cell compartment is accompanied by loss of self-renewal capacity. *Immunity* **2001**, *15*, 659–669. [CrossRef]

6. Christensen, J.L.; Weissman, I.L. Flk-2 is a marker in hematopoietic stem cell differentiation: A simple method to isolate long-term stem cells. *Proc. Natl. Acad. Sci. USA* **2001**, *98*, 14541–14546. [CrossRef] [PubMed]

7. Yang, L.; Bryder, D.; Adolfsson, J.; Nygren, J.; Månsson, R.; Sigvardsson, M.; Jacobsen, S.E. Identification of Lin$^-$Sca1$^+$Kit$^+$CD34$^+$Flt3$^-$ short-term hematopoietic stem cells capable of rapidly reconstituting and rescuing myeloablated transplant recipients. *Blood* **2005**, *105*, 2717–2723. [CrossRef] [PubMed]

8. Volpe, G.; Clarke, M.; Garcìa, P.; Walton, D.S.; Vegiopoulos, A.; Del Pozzo, W.; O'Neill, L.P.; Frampton, J.; Dumon, S. Regulation of the *Flt3* gene in haematopoietic stem and early progenitor cells. *PLoS ONE* **2015**, *10*, e0138257. [CrossRef] [PubMed]

9. Buza-Vidas, N.; Woll, P.; Hultquist, A.; Duarte, S.; Lutteropp, M.; Bouriez-Jones, T.; Ferry, H.; Luc, S.; Jacobsen, S.E. Flt3 expression initiates in fully multipotent mouse hematopoietic progenitor cells. *Blood* **2011**, *118*, 1544–1548. [CrossRef] [PubMed]

10. Boyer, S.W.; Schroeder, A.V.; Smith-Berdan, S.; Forsberg, E.C. All hematopoietic cells develop from hematopoietic stem cells through Flk2/Flt3-positive progenitor cells. *Cell Stem Cell* **2011**, *9*, 64–73. [CrossRef] [PubMed]

11. Meshinchi, S.; Appelbaum, F.R. Structural and functional alterations of Flt3 in acute myeloid leukemia. *Clin. Cancer Res.* **2009**, *15*, 4263–4269. [CrossRef] [PubMed]

12. Griffith, J.; Black, J.; Faerman, C.; Swenson, L.; Wynn, M.; Lu, F.; Lippke, J.; Saxena, K. The structural basis for autoinhibition of Flt3 by the juxtamembrane domain. *Mol. Cell* **2004**, *13*, 169–178. [CrossRef]

13. Zhang, S.; Mantel, C.; Broxmeyer, H.E. Flt3 signaling involves tyrosyl-phosphorylation of SHP-2 and SHIP and their association with Grb2 and Shc in Baf3/Flt3 cells. *J. Leukoc. Biol.* **1999**, *65*, 372–380. [PubMed]

14. Marchetto, S.; Fournier, E.; Beslu, N.; Aurran-Schleinitz, T.; Dubreuil, P.; Borg, J.P.; Birnbaum, D.; Rosnet, O. SHC and SHIP phosphorylation and interaction in response to activation of the Flt3 receptor. *Leukemia* **1999**, *13*, 1374–1382. [CrossRef] [PubMed]

15. Lavagna-Sévenier, C.; Marchetto, S.; Birnbaum, D.; Rosnet, O. Flt3 signaling in hematopoietic cells involves CBL, SHC and an unknown P115 as prominent tyrosine-phosphorylated substrates. *Leukemia* **1998**, *12*, 301–310. [CrossRef] [PubMed]

16. Stirewalt, D.L.; Radich, J.P. The role of Flt3 in haematopoietic malignancies. *Nat. Rev. Cancer* **2003**, *3*, 650–665. [CrossRef] [PubMed]

17. Gilliland, D.G.; Griffin, J.D. The roles of Flt3 in hematopoiesis and leukemia. *Blood* **2002**, *100*, 1532–1542. [CrossRef] [PubMed]

18. Veiby, O.P.; Jacobsen, F.W.; Cui, L.; Lyman, S.D.; Jacobsen, S.E. The flt3 ligand promotes the survival of primitive hemopoietic progenitor cells with myeloid as well as B lymphoid potential. Suppression of apoptosis and counteraction by TNF-α and TGF-β. *J. Immunol.* **1996**, *157*, 2953–2960. [PubMed]

19. Broxmeyer, H.E.; Lu, L.; Cooper, S.; Ruggieri, L.; Li, Z.H.; Lyman, S.D. Flt3 ligand stimulates/costimulates the growth of myeloid stem/progenitor cells. *Exp. Hematol.* **1995**, *23*, 1121–1129. [PubMed]

20. von Muenchow, L.; Alberti-Servera, L.; Klein, F.; Capoferri, G.; Finke, D.; Ceredig, R.; Rolink, A.; Tsapogas, P. Permissive roles of cytokines interleukin-7 and Flt3 ligand in mouse B-cell lineage commitment. *Proc. Natl. Acad. Sci. USA* **2016**, *113*, E8122–E8130. [CrossRef] [PubMed]

21. Lyman, S.D.; James, L.; Johnson, L.; Brasel, K.; de Vries, P.; Escobar, S.S.; Downey, H.; Splett, R.R.; Beckmann, M.P.; McKenna, H.J. Cloning of the human homologue of the murine Flt3 ligand: A growth factor for early hematopoietic progenitor cells. *Blood* **1994**, *83*, 2795–2801. [PubMed]

22. Hannum, C.; Culpepper, J.; Campbell, D.; McClanahan, T.; Zurawski, S.; Bazan, J.F.; Kastelein, R.; Hudak, S.; Wagner, J.; Mattson, J. Ligand for Flt3/Flk2 receptor tyrosine kinase regulates growth of haematopoietic stem cells and is encoded by variant rnas. *Nature* **1994**, *368*, 643–648. [CrossRef] [PubMed]

23. Brashem-Stein, C.; Flowers, D.A.; Bernstein, I.D. Regulation of colony forming cell generation by flt-3 ligand. *Br. J. Haematol.* **1996**, *94*, 17–22. [CrossRef] [PubMed]

24. Banu, N.; Deng, B.; Lyman, S.D.; Avraham, H. Modulation of haematopoietic progenitor development by Flt-3 ligand. *Cytokine* **1999**, *11*, 679–688. [CrossRef] [PubMed]

25. Hudak, S.; Hunte, B.; Culpepper, J.; Menon, S.; Hannum, C.; Thompson-Snipes, L.; Rennick, D. Flt3/Flk2 ligand promotes the growth of murine stem cells and the expansion of colony-forming cells and spleen colony-forming units. *Blood* **1995**, *85*, 2747–2755. [PubMed]

26. Gabbianelli, M.; Pelosi, E.; Montesoro, E.; Valtieri, M.; Luchetti, L.; Samoggia, P.; Vitelli, L.; Barberi, T.; Testa, U.; Lyman, S. Multi-level effects of Flt3 ligand on human hematopoiesis: Expansion of putative stem cells and proliferation of granulomonocytic progenitors/monocytic precursors. *Blood* **1995**, *86*, 1661–1670. [PubMed]

27. Jacobsen, S.E.; Okkenhaug, C.; Myklebust, J.; Veiby, O.P.; Lyman, S.D. The Flt3 ligand potently and directly stimulates the growth and expansion of primitive murine bone marrow progenitor cells in vitro: Synergistic interactions with interleukin (IL) 11, IL-12, and other hematopoietic growth factors. *J. Exp. Med.* **1995**, *181*, 1357–1363. [CrossRef] [PubMed]

28. McKenna, H.J.; de Vries, P.; Brasel, K.; Lyman, S.D.; Williams, D.E. Effect of Flt3 ligand on the ex vivo expansion of human CD34$^+$ hematopoietic progenitor cells. *Blood* **1995**, *86*, 3413–3420. [PubMed]

29. Hirayama, F.; Lyman, S.D.; Clark, S.C.; Ogawa, M. The Flt3 ligand supports proliferation of lymphohematopoietic progenitors and early B-lymphoid progenitors. *Blood* **1995**, *85*, 1762–1768. [PubMed]

30. Åhsberg, J.; Tsapogas, P.; Qian, H.; Zetterblad, J.; Zandi, S.; Månsson, R.; Jönsson, J.I.; Sigvardsson, M. Interleukin-7-induced Stat-5 acts in synergy with Flt-3 signaling to stimulate expansion of hematopoietic progenitor cells. *J. Biol. Chem.* **2010**, *285*, 36275–36284. [CrossRef] [PubMed]

31. Turner, A.M.; Lin, N.L.; Issarachai, S.; Lyman, S.D.; Broudy, V.C. Flt3 receptor expression on the surface of normal and malignant human hematopoietic cells. *Blood* **1996**, *88*, 3383–3390. [PubMed]

32. Ratajczak, M.Z.; Ratajczak, J.; Ford, J.; Kregenow, R.; Marlicz, W.; Gewirtz, A.M. Flt3/Flk-2 (Stk-1) ligand does not stimulate human megakaryopoiesis in vitro. *Stem Cells* **1996**, *14*, 146–150. [CrossRef] [PubMed]

33. McKenna, H.J.; Stocking, K.L.; Miller, R.E.; Brasel, K.; de Smedt, T.; Maraskovsky, E.; Maliszewski, C.R.; Lynch, D.H.; Smith, J.; Pulendran, B.; et al. Mice lacking Flt3 ligand have deficient hematopoiesis affecting hematopoietic progenitor cells, dendritic cells, and natural killer cells. *Blood* **2000**, *95*, 3489–3497. [PubMed]

34. Baerenwaldt, A.; von Burg, N.; Kreuzaler, M.; Sitte, S.; Horvath, E.; Peter, A.; Voehringer, D.; Rolink, A.G.; Finke, D. Flt3 ligand regulates the development of innate lymphoid cells in fetal and adult mice. *J. Immunol.* **2016**, *196*, 2561–2571. [CrossRef] [PubMed]

35. Brasel, K.; McKenna, H.J.; Morrissey, P.J.; Charrier, K.; Morris, A.E.; Lee, C.C.; Williams, D.E.; Lyman, S.D. Hematologic effects of Flt3 ligand in vivo in mice. *Blood* **1996**, *88*, 2004–2012. [PubMed]

36. Balciunaite, G.; Ceredig, R.; Massa, S.; Rolink, A.G. A B220$^+$ CD117$^+$ CD19$^-$ hematopoietic progenitor with potent lymphoid and myeloid developmental potential. *Eur. J. Immunol.* **2005**, *35*, 2019–2030. [CrossRef] [PubMed]

37. Ceredig, R.; Rauch, M.; Balciunaite, G.; Rolink, A.G. Increasing Flt3L availability alters composition of a novel bone marrow lymphoid progenitor compartment. *Blood* **2006**, *108*, 1216–1222. [CrossRef] [PubMed]

38. Tsapogas, P.; Swee, L.K.; Nusser, A.; Nuber, N.; Kreuzaler, M.; Capoferri, G.; Rolink, H.; Ceredig, R.; Rolink, A. In vivo evidence for an instructive role of fms-like tyrosine kinase-3 (Flt3) ligand in hematopoietic development. *Haematologica* **2014**, *99*, 638–646. [CrossRef] [PubMed]

39. Carow, C.E.; Levenstein, M.; Kaufmann, S.H.; Chen, J.; Amin, S.; Rockwell, P.; Witte, L.; Borowitz, M.J.; Civin, C.I.; Small, D. Expression of the hematopoietic growth factor receptor Flt3 (STK-1/Flk2) in human leukemias. *Blood* **1996**, *87*, 1089–1096. [PubMed]

40. Rosnet, O.; Bühring, H.J.; Marchetto, S.; Rappold, I.; Lavagna, C.; Sainty, D.; Arnoulet, C.; Chabannon, C.; Kanz, L.; Hannum, C.; et al. Human Flt3/Flk2 receptor tyrosine kinase is expressed at the surface of normal and malignant hematopoietic cells. *Leukemia* **1996**, *10*, 238–248. [PubMed]

41. Piacibello, W.; Fubini, L.; Sanavio, F.; Brizzi, M.F.; Severino, A.; Garetto, L.; Stacchini, A.; Pegoraro, L.; Aglietta, M. Effects of human Flt3 ligand on myeloid leukemia cell growth: Heterogeneity in response and synergy with other hematopoietic growth factors. *Blood* **1995**, *86*, 4105–4114. [PubMed]

42. Nakao, M.; Yokota, S.; Iwai, T.; Kaneko, H.; Horiike, S.; Kashima, K.; Sonoda, Y.; Fujimoto, T.; Misawa, S. Internal tandem duplication of the *Flt3* gene found in acute myeloid leukemia. *Leukemia* **1996**, *10*, 1911–1918. [PubMed]

43. Kuchenbauer, F.; Kern, W.; Schoch, C.; Kohlmann, A.; Hiddemann, W.; Haferlach, T.; Schnittger, S. Detailed analysis of *Flt3* expression levels in acute myeloid leukemia. *Haematologica* **2005**, *90*, 1617–1625. [PubMed]

44. Khaled, S.; Al Malki, M.; Marcucci, G. Acute myeloid leukemia: Biologic, prognostic, and therapeutic insights. *Oncology* **2016**, *30*, 318–329. [PubMed]

45. Horiike, S.; Yokota, S.; Nakao, M.; Iwai, T.; Sasai, Y.; Kaneko, H.; Taniwaki, M.; Kashima, K.; Fujii, H.; Abe, T.; et al. Tandem duplications of the Flt3 receptor gene are associated with leukemic transformation of myelodysplasia. *Leukemia* **1997**, *11*, 1442–1446. [CrossRef] [PubMed]

46. Yamamoto, Y.; Kiyoi, H.; Nakano, Y.; Suzuki, R.; Kodera, Y.; Miyawaki, S.; Asou, N.; Kuriyama, K.; Yagasaki, F.; Shimazaki, C.; et al. Activating mutation of d835 within the activation loop of Flt3 in human hematologic malignancies. *Blood* **2001**, *97*, 2434–2439. [CrossRef] [PubMed]

47. Yokota, S.; Kiyoi, H.; Nakao, M.; Iwai, T.; Misawa, S.; Okuda, T.; Sonoda, Y.; Abe, T.; Kahsima, K.; Matsuo, Y.; et al. Internal tandem duplication of the *Flt3* gene is preferentially seen in acute myeloid leukemia and myelodysplastic syndrome among various hematological malignancies. A study on a large series of patients and cell lines. *Leukemia* **1997**, *11*, 1605–1609. [CrossRef] [PubMed]

48. Marcucci, G.; Maharry, K.; Whitman, S.P.; Vukosavljevic, T.; Paschka, P.; Langer, C.; Mrózek, K.; Baldus, C.D.; Carroll, A.J.; Powell, B.L.; et al. High expression levels of the *ETS*-related gene, *ERG*, predict adverse outcome and improve molecular risk-based classification of cytogenetically normal acute myeloid leukemia: A cancer and leukemia group B study. *J. Clin. Oncol.* **2007**, *25*, 3337–3343. [CrossRef] [PubMed]

49. Fröhling, S.; Schlenk, R.F.; Breitruck, J.; Benner, A.; Kreitmeier, S.; Tobis, K.; Döhner, H.; Döhner, K. Prognostic significance of activating *Flt3* mutations in younger adults (16 to 60 years) with acute myeloid leukemia and normal cytogenetics: A study of the AML study group ulm. *Blood* **2002**, *100*, 4372–4380. [CrossRef] [PubMed]

50. Kottaridis, P.D.; Gale, R.E.; Frew, M.E.; Harrison, G.; Langabeer, S.E.; Belton, A.A.; Walker, H.; Wheatley, K.; Bowen, D.T.; Burnett, A.K.; et al. The presence of a Flt3 internal tandem duplication in patients with acute myeloid leukemia (AML) adds important prognostic information to cytogenetic risk group and response to the first cycle of chemotherapy: Analysis of 854 patients from the united kingdom medical research council AML 10 and 12 trials. *Blood* **2001**, *98*, 1752–1759. [PubMed]

51. Thiede, C.; Steudel, C.; Mohr, B.; Schaich, M.; Schäkel, U.; Platzbecker, U.; Wermke, M.; Bornhäuser, M.; Ritter, M.; Neubauer, A.; et al. Analysis of FLT3-activating mutations in 979 patients with acute myelogenous leukemia: Association with fab subtypes and identification of subgroups with poor prognosis. *Blood* **2002**, *99*, 4326–4335. [CrossRef] [PubMed]

52. Abu-Duhier, F.M.; Goodeve, A.C.; Wilson, G.A.; Care, R.S.; Peake, I.R.; Reilly, J.T. Identification of novel Flt-3 Asp835 mutations in adult acute myeloid leukaemia. *Br. J. Haematol.* **2001**, *113*, 983–988. [CrossRef] [PubMed]

53. De Kouchkovsky, I.; Abdul-Hay, M. Acute myeloid leukemia: A comprehensive review and 2016 update. *Blood Cancer J.* **2016**, *6*, e441. [CrossRef] [PubMed]

54. Kiel, M.J.; Yilmaz, O.H.; Iwashita, T.; Terhorst, C.; Morrison, S.J. Slam family receptors distinguish hematopoietic stem and progenitor cells and reveal endothelial niches for stem cells. *Cell* **2005**, *121*, 1109–1121. [CrossRef] [PubMed]

55. Oguro, H.; Ding, L.; Morrison, S.J. Slam family markers resolve functionally distinct subpopulations of hematopoietic stem cells and multipotent progenitors. *Cell Stem Cell* **2013**, *13*, 102–116. [CrossRef] [PubMed]

56. Adolfsson, J.; Månsson, R.; Buza-Vidas, N.; Hultquist, A.; Liuba, K.; Jensen, C.T.; Bryder, D.; Yang, L.; Borge, O.J.; Thoren, L.A.; et al. Identification of Flt3+ lympho-myeloid stem cells lacking erythro-megakaryocytic potential a revised road map for adult blood lineage commitment. *Cell* **2005**, *121*, 295–306. [CrossRef] [PubMed]

57. Akashi, K.; Traver, D.; Miyamoto, T.; Weissman, I.L. A clonogenic common myeloid progenitor that gives rise to all myeloid lineages. *Nature* **2000**, *404*, 193–197. [CrossRef] [PubMed]

58. Kondo, M.; Weissman, I.L.; Akashi, K. Identification of clonogenic common lymphoid progenitors in mouse bone marrow. *Cell* **1997**, *91*, 661–672. [CrossRef]

59. Beerman, I.; Bhattacharya, D.; Zandi, S.; Sigvardsson, M.; Weissman, I.L.; Bryder, D.; Rossi, D.J. Functionally distinct hematopoietic stem cells modulate hematopoietic lineage potential during aging by a mechanism of clonal expansion. *Proc. Natl. Acad. Sci. USA* **2010**, *107*, 5465–5470. [CrossRef] [PubMed]

60. Morita, Y.; Ema, H.; Nakauchi, H. Heterogeneity and hierarchy within the most primitive hematopoietic stem cell compartment. *J. Exp. Med.* **2010**, *207*, 1173–1182. [CrossRef] [PubMed]

61. Roux, P.P.; Shahbazian, D.; Vu, H.; Holz, M.K.; Cohen, M.S.; Taunton, J.; Sonenberg, N.; Blenis, J. RAS/ERK signaling promotes site-specific ribosomal protein S6 phosphorylation via RSK and stimulates cap-dependent translation. *J. Biol. Chem.* **2007**, *282*, 14056–14064. [CrossRef] [PubMed]

62. Fingar, D.C.; Salama, S.; Tsou, C.; Harlow, E.; Blenis, J. Mammalian cell size is controlled by mtor and its downstream targets s6k1 and 4ebp1/eif4e. *Genes Dev.* **2002**, *16*, 1472–1487. [CrossRef] [PubMed]

63. Kalaitzidis, D.; Neel, B.G. Flow-cytometric phosphoprotein analysis reveals agonist and temporal differences in responses of murine hematopoietic stem/progenitor cells. *PLoS ONE* **2008**, *3*, e3776. [CrossRef] [PubMed]

64. Iacobucci, I.; Li, Y.; Roberts, K.G.; Dobson, S.M.; Kim, J.C.; Payne-Turner, D.; Harvey, R.C.; Valentine, M.; McCastlain, K.; Easton, J.; et al. Truncating erythropoietin receptor rearrangements in acute lymphoblastic leukemia. *Cancer Cell* **2016**, *29*, 186–200. [CrossRef] [PubMed]

65. Specchia, G.; Liso, V.; Capalbo, S.; Fazioli, F.; Bettoni, S.; Bassan, R.; Viero, P.; Barbui, T.; Rambaldi, A. Constitutive expression of IL-1β, M-CSF and c-fms during the myeloid blastic phase of chronic myelogenous leukaemia. *Br. J. Haematol.* **1992**, *80*, 310–316. [CrossRef] [PubMed]

66. Aikawa, Y.; Katsumoto, T.; Zhang, P.; Shima, H.; Shino, M.; Terui, K.; Ito, E.; Ohno, H.; Stanley, E.R.; Singh, H.; et al. 1-mediated upregulation of *CSF1R* is crucial for leukemia stem cell potential induced by MOZ-TIF2. *Nat. Med.* **2010**, *16*, 580–585. [CrossRef] [PubMed]

67. Grover, A.; Mancini, E.; Moore, S.; Mead, A.J.; Atkinson, D.; Rasmussen, K.D.; O'Carroll, D.; Jacobsen, S.E.; Nerlov, C. Erythropoietin guides multipotent hematopoietic progenitor cells toward an erythroid fate. *J. Exp. Med.* **2014**, *211*, 181–188. [CrossRef] [PubMed]

68. Mossadegh-Keller, N.; Sarrazin, S.; Kandalla, P.K.; Espinosa, L.; Stanley, E.R.; Nutt, S.L.; Moore, J.; Sieweke, M.H. M-CSF instructs myeloid lineage fate in single haematopoietic stem cells. *Nature* **2013**, *497*, 239–243. [CrossRef] [PubMed]

69. Sitnicka, E.; Bryder, D.; Theilgaard-Mönch, K.; Buza-Vidas, N.; Adolfsson, J.; Jacobsen, S.E. Key role of Flt3 ligand in regulation of the common lymphoid progenitor but not in maintenance of the hematopoietic stem cell pool. *Immunity* **2002**, *17*, 463–472. [CrossRef]

70. Chu, S.H.; Heiser, D.; Li, L.; Kaplan, I.; Collector, M.; Huso, D.; Sharkis, S.J.; Civin, C.; Small, D. Flt3-itd knockin impairs hematopoietic stem cell quiescence/homeostasis, leading to myeloproliferative neoplasm. *Cell Stem Cell* **2012**, *11*, 346–358. [CrossRef] [PubMed]

71. Ceredig, R.; Rolink, A.G.; Brown, G. Models of haematopoiesis: Seeing the wood for the trees. *Nat. Rev. Immunol.* **2009**, *9*, 293–300. [CrossRef] [PubMed]

72. Brown, G.; Mooney, C.J.; Alberti-Servera, L.; Muenchow, L.; Toellner, K.M.; Ceredig, R.; Rolink, A. Versatility of stem and progenitor cells and the instructive actions of cytokines on hematopoiesis. *Crit. Rev. Clin. Lab. Sci.* **2015**, *52*, 168–179. [PubMed]

73. Kikushige, Y.; Yoshimoto, G.; Miyamoto, T.; Iino, T.; Mori, Y.; Iwasaki, H.; Niiro, H.; Takenaka, K.; Nagafuji, K.; Harada, M.; et al. Human Flt3 is expressed at the hematopoietic stem cell and the granulocyte/macrophage progenitor stages to maintain cell survival. *J. Immunol.* **2008**, *180*, 7358–7367. [CrossRef] [PubMed]

74. Beaudin, A.E.; Boyer, S.W.; Perez-Cunningham, J.; Hernandez, G.E.; Derderian, S.C.; Jujjavarapu, C.; Aaserude, E.; MacKenzie, T.; Forsberg, E.C. A transient developmental hematopoietic stem cell gives rise to innate-like B and T cells. *Cell Stem Cell* **2016**, *19*, 768–783. [CrossRef] [PubMed]

75. Yamamoto, R.; Morita, Y.; Ooehara, J.; Hamanaka, S.; Onodera, M.; Rudolph, K.L.; Ema, H.; Nakauchi, H. Clonal analysis unveils self-renewing lineage-restricted progenitors generated directly from hematopoietic stem cells. *Cell* **2013**, *154*, 1112–1126. [CrossRef] [PubMed]

76. Benz, C.; Copley, M.R.; Kent, D.G.; Wohrer, S.; Cortes, A.; Aghaeepour, N.; Ma, E.; Mader, H.; Rowe, K.; Day, C.; et al. Hematopoietic stem cell subtypes expand differentially during development and display distinct lymphopoietic programs. *Cell Stem Cell.* **2012**, *10*, 273–283. [CrossRef] [PubMed]

77. Naik, S.H.; Perié, L.; Swart, E.; Gerlach, C.; van Rooij, N.; de Boer, R.J.; Schumacher, T.N. Diverse and heritable lineage imprinting of early haematopoietic progenitors. *Nature* **2013**, *496*, 229–232. [CrossRef] [PubMed]
78. Challen, G.A.; Boles, N.C.; Chambers, S.M.; Goodell, M.A. Distinct hematopoietic stem cell subtypes are differentially regulated by tgf-beta1. *Cell Stem Cell* **2010**, *6*, 265–278. [CrossRef] [PubMed]
79. Lu, R.; Neff, N.F.; Quake, S.R.; Weissman, I.L. Tracking single hematopoietic stem cells in vivo using high-throughput sequencing in conjunction with viral genetic barcoding. *Nat. Biotechnol.* **2011**, *29*, 928–933. [CrossRef] [PubMed]
80. Sanjuan-Pla, A.; Macaulay, I.C.; Jensen, C.T.; Woll, P.S.; Luis, T.C.; Mead, A.; Moore, S.; Carella, C.; Matsuoka, S.; Bouriez Jones, T.; et al. Platelet-biased stem cells reside at the apex of the haematopoietic stem-cell hierarchy. *Nature* **2013**, *502*, 232–236. [CrossRef] [PubMed]
81. Gekas, C.; Graf, T. CD41 expression marks myeloid-biased adult hematopoietic stem cells and increases with age. *Blood* **2013**, *121*, 4463–4472. [CrossRef] [PubMed]
82. Shimazu, T.; Iida, R.; Zhang, Q.; Welner, R.S.; Medina, K.L.; Alberola-Lla, J.; Kincade, P.W. CD86 is expressed on murine hematopoietic stem cells and denotes lymphopoietic potential. *Blood* **2012**, *119*, 4889–4897. [CrossRef] [PubMed]
83. Shin, J.Y.; Hu, W.; Naramura, M.; Park, C.Y. High c-kit expression identifies hematopoietic stem cells with impaired self-renewal and megakaryocytic bias. *J. Exp. Med.* **2014**, *211*, 217–231. [CrossRef] [PubMed]
84. Notta, F.; Zandi, S.; Takayama, N.; Dobson, S.; Gan, O.I.; Wilson, G.; Kaufmann, K.B.; McLeod, J.; Laurenti, E.; Dunant, C.F.; et al. Distinct routes of lineage development reshape the human blood hierarchy across ontogeny. *Science* **2016**, *351*, aab2116. [CrossRef] [PubMed]
85. Velten, L.; Haas, S.F.; Raffel, S.; Blaszkiewicz, S.; Islam, S.; Hennig, B.P.; Hirche, C.; Lutz, C.; Buss, E.C.; Nowak, D.; et al. Human haematopoietic stem cell lineage commitment is a continuous process. *Nat. Cell. Biol.* **2017**, *19*, 271–281. [CrossRef] [PubMed]
86. Paul, F.; Arkin, Y.; Giladi, A.; Jaitin, D.A.; Kenigsberg, E.; Keren-Shaul, H.; Winter, D.; Lara-Astiaso, D.; Gury, M.; Weiner, A.; et al. Transcriptional heterogeneity and lineage commitment in myeloid progenitors. *Cell* **2015**, *163*, 1663–1677. [CrossRef] [PubMed]
87. Karsunky, H.; Merad, M.; Cozzio, A.; Weissman, I.L.; Manz, M.G. Flt3 ligand regulates dendritic cell development from Flt3+ lymphoid and myeloid-committed progenitors to Flt3+ dendritic cells in vivo. *J. Exp. Med.* **2003**, *198*, 305–313. [CrossRef] [PubMed]
88. Chi, A.W.; Chavez, A.; Xu, L.; Weber, B.N.; Shestova, O.; Schaffer, A.; Wertheim, G.; Pear, W.S.; Izon, D.; Bhandoola, A. Identification of Flt3 CD150 myeloid progenitors in adult mouse bone marrow that harbor T lymphoid developmental potential. *Blood* **2011**, *118*, 2723–2732. [CrossRef] [PubMed]
89. Luc, S.; Anderson, K.; Kharazi, S.; Buza-Vidas, N.; Böiers, C.; Jensen, C.T.; Ma, Z.; Wittmann, L.; Jacobsen, S.E. Down-regulation of mpl marks the transition to lymphoid-primed multipotent progenitors with gradual loss of granulocyte-monocyte potential. *Blood* **2008**, *111*, 3424–3434. [CrossRef] [PubMed]
90. Böiers, C.; Buza-Vidas, N.; Jensen, C.T.; Pronk, C.J.; Kharazi, S.; Wittmann, L.; Sitnicka, E.; Hultquist, A.; Jacobsen, S.E. Expression and role of Flt3 in regulation of the earliest stage of normal granulocyte-monocyte progenitor development. *Blood* **2010**, *115*, 5061–5068. [CrossRef] [PubMed]
91. Perié, L.; Hodgkin, P.D.; Naik, S.H.; Schumacher, T.N.; de Boer, R.J.; Duffy, K.R. Determining lineage pathways from cellular barcoding experiments. *Cell Rep.* **2014**, *6*, 617–624. [CrossRef] [PubMed]
92. Passegué, E.; Wagers, A.J.; Giuriato, S.; Anderson, W.C.; Weissman, I.L. Global analysis of proliferation and cell cycle gene expression in the regulation of hematopoietic stem and progenitor cell fates. *J. Exp. Med.* **2005**, *202*, 1599–1611. [CrossRef] [PubMed]
93. Fleming, W.H.; Alpern, E.J.; Uchida, N.; Ikuta, K.; Spangrude, G.J.; Weissman, I.L. Functional heterogeneity is associated with the cell cycle status of murine hematopoietic stem cells. *J. Cell Biol.* **1993**, *122*, 897–902. [CrossRef] [PubMed]
94. Chang, H.H.; Hemberg, M.; Barahona, M.; Ingber, D.E.; Huang, S. Transcriptome-wide noise controls lineage choice in mammalian progenitor cells. *Nature* **2008**, *453*, 544–547. [CrossRef] [PubMed]

© 2017 by the authors. Licensee MDPI, Basel, Switzerland. This article is an open access article distributed under the terms and conditions of the Creative Commons Attribution (CC BY) license (http://creativecommons.org/licenses/by/4.0/).

International Journal of
Molecular Sciences

MDPI

Review

The Cytokine Flt3-Ligand in Normal and Malignant Hematopoiesis

Panagiotis Tsapogas [1,*], Ciaran James Mooney [2], Geoffrey Brown [2,3] and Antonius Rolink [1]

[1]   Developmental and Molecular Immunology, Department of Biomedicine, University of Basel, Mattenstrasse 28, Basel 4058, Switzerland; antonius.rolink@unibas.ch

[2]   Institute of Immunology and Immunotherapy, College of Medical and Dental Sciences, University of Birmingham, Edbgaston, Birmingham B15 2TT, UK; c.mooney@smd15.qmul.ac.uk (C.J.M.); g.brown@bham.ac.uk (G.B.)

[3]   Institute of Clinical Sciences, College of Medical and Dental Sciences, University of Birmingham, Edbgaston, Birmingham B15 2TT, UK

*   Correspondence: panagiotis.tsapogas@unibas.ch; Tel.: +41-61-2075-072; Fax: +41-61-2075-070

Academic Editor: Ewa Marcinkowska
Received: 27 April 2017; Accepted: 19 May 2017; Published: 24 May 2017

Abstract: The cytokine Fms-like tyrosine kinase 3 ligand (FL) is an important regulator of hematopoiesis. Its receptor, Flt3, is expressed on myeloid, lymphoid and dendritic cell progenitors and is considered an important growth and differentiation factor for several hematopoietic lineages. Activating mutations of Flt3 are frequently found in acute myeloid leukemia (AML) patients and associated with a poor clinical prognosis. In the present review we provide an overview of our current knowledge on the role of FL in the generation of blood cell lineages. We examine recent studies on Flt3 expression by hematopoietic stem cells and its potential instructive action at early stages of hematopoiesis. In addition, we review current findings on the role of mutated FLT3 in leukemia and the development of FLT3 inhibitors for therapeutic use to treat AML. The importance of mouse models in elucidating the role of Flt3-ligand in normal and malignant hematopoiesis is discussed.

Keywords: Flt3; hematopoiesis; acute myeloid leukemia (AML); cytokines; Flt3 with internal tandem duplications (FLT3-ITD)

1. Introduction

Leukemias occur as a result of the de-regulation of normal hematopoiesis, as evidenced by the significant number of genes important for hematopoietic development that are mutated in leukemias. Therefore, understanding how hematopoiesis is regulated is of utmost importance for the elucidation of the mechanisms that lead to the blood cell malignancies. Cytokines are important regulators of hematopoietic development: they transfer extra-cellular signals to cells to affect their survival, proliferation, differentiation and maturation [1]. The cytokine "Fms-like tyrosine kinase 3 ligand" (hereafter FL) represents a typical example of a hematopoietic cytokine whose receptor is often found to be mutated or over-expressed in leukemias. Therefore, there has been a great deal of attention on the precise role of FL in hematopoiesis.

In this review we will give a general outline of our current knowledge of the role of FL in normal and malignant hematopoiesis and will briefly discuss recent findings from us and others that offer new insights on how FL regulates the generation of blood cells. Furthermore, the potential relevance of these findings to the role of FL in hematopoietic malignancies will be discussed.

2. FL and Its Receptor, Flt3

To date, only one receptor for the cytokine FL has been identified; Fms-like tyrosine kinase 3 (also known as Flk2/CD135; hereafter Flt3). The murine receptor was the first to be identified and cloned by two groups independently [2–4], and soon after its human homologue was cloned [5]. Flt3 belongs to the class III receptor tyrosine kinase group of receptors, which also contains other cytokine receptors, such as the platelet-derived growth factor receptor (PDGF) and c-kit, the receptor for Stem Cell Factor (SCF). Flt3 shows a high sequence and structural homology to these receptors, containing an extra-cellular part with 5 immunoglobulin-like domains, a transmembrane domain and a cytoplasmic region consisting of a juxta-membrane domain followed by two tyrosine kinase domains [6]. The pattern of expression of Flt3 observed in early studies was indicative of its potential importance in hematopoiesis, as it was found to be expressed specifically by early hematopoietic progenitors that contain some degree of stem cell activity, while it was not expressed by more mature blood cell populations [2,3]. In 1993, murine FL was cloned and characterized as a transmembrane protein that can be secreted and, upon binding to Flt3, can stimulate the proliferation of Flt3$^+$ bone marrow and foetal liver progenitors [7,8]. Soon after the human FL was cloned; interestingly, it was shown to be able to bind to murine Flt3 and activate its downstream signaling [9].

Binding of FL to Flt3 leads to homodimerization of the receptor and subsequent conformational changes that result in phosphorylation of the tyrosine kinase domains. This activation of the receptor occurs rapidly and is followed by an equally rapid internalization and degradation of the receptor homodimer [10]. Downstream signaling pathways that are activated by FL-Flt3 binding were studied even before FL was cloned, using a chimeric Flt3 receptor containing the extracellular part of Colony Stimulating Factor 1 receptor (CSF-1R) [11,12]. Subsequent experiments using the native receptor have confirmed some of the initial findings and revealed that Flt3 activation leads to phosphorylation of Src-homology 2 containing proteins (SHC), which directly interact with proteins such as Grb2, Gab2, SHIP, eventually leading to activation of the Ras/MEK/Erk and PI3K pathways [13–16]. Human FLT3 in particular has been shown to bind additionally to tyrosine phosphatase SHP2 and the proto-oncogene CBL, also members of the PI3K pathway [15,17,18]. In addition, Flt3 has been shown to activate Stat3 [19,20] and Stat5a [21] signaling mediators. However, caution should be exercised when interpreting these findings on Flt3 signaling, as most of the above data are derived from work using cell lines, while additional complexity comes from the fact that downstream signaling pathways activated by Flt3 seem to be highly cell-context dependent.

The cloning of FL allowed the investigation of its function and early studies focused mainly on assessing its role in supporting the growth and differentiation of hematopoietic progenitors in vitro. Several of these studies demonstrated that FL has a limited effect on the growth of hematopoietic progenitors when used alone, but it is very potent in synergizing with other hematopoietic cytokines, such as SCF, IL-3, IL-6, Granulocyte-Macrophage Colony stimulating Factor (GM-CSF), Granulocyte Colony Stimulating Factor (G-CSF) and IL-11, to promote the generation of primarily myeloid cell containing colonies [22–28]. There was no significant effect of FL as to promoting the in vitro generation of cells of the erythroid [25] or megakaryocyte lineages [22]. In addition, FL was shown to significantly enhance the in vitro generation of B cells, again mainly in combination with the cytokines SCF and IL-7 [26,29–31]. A similar picture of the in vitro effect of FL emerged from studies using human hematopoietic progenitors [23,32,33]. The synergistic effect of FL with other hematopoietic cytokines was a clear conclusion from these experiments. However, studies have been limited both by the lack of markers that could be used to isolate progenitors at different developmental stages and by the use of an exclusively in vitro approach to determine cell behavior.

Important insights on the role of FL came from the generation of knock-out mouse models. Initial analysis of mice with targeted deletion of the *Flt3* gene showed no significant perturbations in hematopoiesis, apart from a reduction in the numbers of early B cell progenitors and defective repopulation capacity of *Flt3*$^{-/-}$ bone marrow cells upon transplantation into irradiated hosts [34]. In contrast, mice lacking FL showed a more pronounced phenotype, with a reduced overall generation of leukocytes, particularly decreased B cell progenitors, natural-killer cells (NK) and dendritic cells (DC) [35]. The apparent difference in the severity of phenotype between *Flt3*[-/-] and FL[-/-] mice could be indicative of the existence of another receptor for FL or alternatively reflect differences between mouse strains. Subsequent detailed analysis of mice defective in Flt3 signaling showed that apart from committed B cell progenitors, FL is important for the generation and/or maintenance of their uncommitted precursors, CLP (Common Lymphoid Progenitors) [36] and EPLM (Early Progenitors with Lymphoid and Myeloid potential) [37], as well as of early multi-potent progenitors (MPP) within the Lineage[-]kit[+]Sca1[-] (LSK) compartment [38,39]—all of these populations express Flt3 [40,41]. These in vivo studies have shown that active Flt3 signaling is not an absolute requirement for hematopoiesis to occur, but have nevertheless highlighted its importance in regard to several developmental steps in blood cell formation.

3. The Role of FL in Normal Hematopoiesis

3.1. Hematopoietic Stem Cells and Early Progenitors

The most broadly accepted model explaining how the generation of hematopoietic cells occurs from Hematopoietic Stem Cells (HSC) is based on a developmental hierarchy, with HSC residing at the apex as the multi-potent progenitor cell type that gives rise to all of the hematopoietic lineages through the step-wise generation of oligo-potent progenitors with restricted developmental potentials. This model is continuously debated and revised as new findings, often based on new technologies, provide new clues as to how hematopoiesis is regulated. Figure 1 illustrates Flt3 expression by different hematopoietic progenitors and lineages, based on our current knowledge and in the context of a continuum of options and the "pairwise" model for hematopoiesis we have proposed [42,43]. Investigation of Flt3 expression in hematopoietic progenitor stages has greatly contributed in identifying successive developmental stages in the hematopoietic pathway. For example, expression of Flt3 within the HSC-containing LSK compartment has been associated with loss of self-renewal capacity, therefore suggesting that the Flt3[-] fraction of LSK cells is enriched for long-term reconstituting HSC (LT-HSC) [44,45].

The traditional model for hematopoiesis, which is the one most commonly found in textbooks, suggests an early bifurcation in the hematopoietic tree, with progenitors differentiating towards either a lymphoid fate, eventually giving rise to B, T and Innate Lymphoid (ILC) cells, or towards a myeloid fate, which results in the generation of all myeloid cells, platelets and erythrocytes. This model was based on the identification of distinct progenitor types, the CLP and the CMP (Common Myeloid Progenitor), which showed the above developmental potentials, respectively [46,47]. In 2005, the Jacobsen group reported that MPP progenitors with high levels of Flt3 expression (named Lymphoid-primed Multipotent Progenitors, or LMPP) have lost their potential to generate megakaryocytes and erythrocytes while retaining a robust lymphoid and myeloid potential (shown in Figure 1), thereby suggesting that the earliest branching point in hematopoiesis occurs between the megakaryocyte/erythrocyte and lymphoid/myeloid lineages [48]. Whether Flt3[+] MPP progenitors can indeed give rise to cells of the megakaryocyte and erythrocyte lineages has been debated for some time [49–52]. Lineage tracing experiments have shown that all hematopoietic lineages, including megakaryocyte/erythrocyte cells, are derived from progenitors that at some point expressed *Flt3* mRNA [49,53]. However, these results could be explained by significant expression of *Flt3* mRNA prior to expression of the Flt3 protein (or a low, therefore FACS-undetectable, protein expression) on the surface of progenitors that give rise to megakaryocytes and erythrocytes.

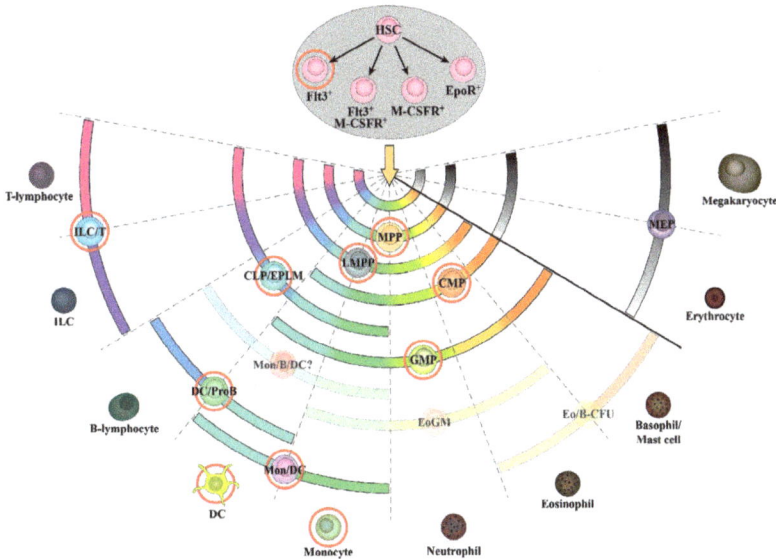

Figure 1. Flt3 expression in murine hematopoietic cells. Flt3 expression in progenitor and mature hematopoietic cells. The fate choices that are available to HSC are a continuum as shown by the short central arc below the yellow arrow. The fates choices of each of the indicated progenitors are shown as a shorter arc that spans the end cell types each progenitor cell population can give rise to. Red circles indicate Flt3 expression by the corresponding cell type. The grey section of the spectrum and grey shading of the MEP and mature cells indicates that these cells do not express Flt3. Progenitor cells that have not been investigated for expression of Flt3 are shown in a faded color. Expression is confined to myeloid and lymphoid progenitors as opposed to megakaryocyte/erythroid progenitors. HSC: Hematopoietic Stem Cell; MPP: Multi-Potent Progenitor; LMPP: Lymphoid-primed Multi-potent Progenitor; MEP: Megakaryocyte-Erythrocyte Progenitor; CMP: Common Myeloid Progenitor; GMP: Granulocyte-Macrophage Progenitor; CLP: Common Lymphoid Progenitor; EPLM: Early Progenitors with Lymphoid and Myeloid potential; ILC: Innate Lymphoid Cell; DC: Dendritic Cell; Eo: Eosinophil; CFU: Colony Forming Unit; Mon: Monocyte; M-CSFR: Macrophage–Colony Stimulating Factor Receptor; EpoR: Erythropoietin Receptor; GM: Granulocyte-Macrophage; ProB: progenitor B-lymphocyte; B: B-lymphocyte; T: T-lymphocyte.

The above data raise the issue of potential Flt3 expression by HSC. We have addressed this question by examining *Flt3* mRNA and protein at the single-cell level within HSC [54] phenotypically defined using the CD150/CD48 staining strategy first reported by the Morrison group [55,56]. Our findings have revealed that a small fraction of both LT- and short-term (ST)-HSC express Flt3 (Figure 1). These results are in agreement with a currently emerging view of heterogeneity within HSC [57,58] and have implications for our understanding of hematopoietic malignancies and the heterogeneity of leukemias, as discussed below. Interestingly, our single-cell analysis finds co-expression of Flt3 and the receptor for macrophage-colony stimulating factor (M-CSFR) by some HSC but virtually no co-expression of Flt3 with the receptor for the erythrocyte-lineage promoting cytokine erythropoietin (EpoR). In light of recent suggestions that lineage commitment might occur at a much earlier developmental stage than previously thought [59], these data indicate that some level of lineage skewing occurs already within the HSC population, with some cells being "primed" to respond to lineage-instructing cytokines and differentiate towards the corresponding cell fates.

The observed heterogeneity in Flt3 expression amongst early hematopoietic progenitors and its clear association specifically with the lymphoid and myeloid pathways (Figure 1) raises the question of whether FL has a functional role in promoting the differentiation of a particular cohort of hematopoietic progenitors. As mentioned previously, knock-out mouse models for both FL and Flt3 have shown a reduction in the numbers of early lymphoid and myeloid progenitors. This could be due to either a proliferative and/or survival role of FL on Flt3+ progenitors or it could reflect a function of the cytokine as a differentiating factor, since hematopoietic cytokines can promote the generation of different lineages by acting either in a permissive way (selective expansion of receptor-positive lineages through promotion of their proliferation and/or survival), or in an instructive way (activation of a lineage-specific genetic program) [60,61].

Prompted by the significant changes observed in mice repeatedly injected with FL [62,63], we have recently followed a reverse to the loss-of-function approach, by generating transgenic mice (hereafter FLtg mice) that express human FL under the control of the actin promoter, resulting in a sustained and high level expression of FL in vivo. As expected, FLtg mice exhibit a tremendous expansion of Flt3+ cells, including myeloid cells, DC, MPP, CMP, Granulocyte-Macrophage Progenitors (GMP), CLP and EPLM progenitors, resulting in leukocytosis in the blood and splenomegaly [64]. One of the most prominent phenotypes in these mice is a severe anemia that they develop quite early in life, due to reduced erythropoiesis, as demonstrated by the profound reduction in numbers of both Ter119+ enucleated erythroid progenitors as well as Megakaryocyte-Erythrocyte Progenitors (MEP) in the bone marrow. This negative effect of over-expression of FL on erythropoiesis could be due to an instructive action of FL, with the cytokine actively promoting the differentiation of early, multi-potent progenitors towards a lympho-myeloid fate. Alternatively, it could just be the result of a selective expansion of Flt3+ CMP and GMP progenitors at the expense of Flt3− MEP, thereby leading to a growth disadvantage for megakaryocyte-erythrocyte progenitors and thus to their subsequent reduction. However, a kinetic analysis of these progenitors post injecting FL into wild-type mice showed that by day 3 Ter119+ erythrocyte progenitors are significantly reduced [64]. Considering the turnover of these progenitors and the fact that, at day 3 after FL injection, no other cell type shows any significant increase, this reduction in erythrocyte progenitors is more likely to be the result of a negative effect of FL on their generation, rather than a disadvantage in their expansion. Moreover, analysis of the MEP, CMP and GMP progenitors revealed that Flt3− MEP are significantly reduced by day 3 of treatment and at a time when Flt3+ GMP and CMP are only slightly increased (Figure 2A). Even though absolute proof of an instructive or permissive action of a cytokine can only be derived from experiments at the single-cell level [65–67], careful analysis of our FLtg mouse model and of mice injected with FL provides strong in vivo evidence for an instructive role of FL in driving hematopoiesis towards the lympho-myeloid and away from the megakaryocyte/erythrocyte lineages. We suggest a working model for the role of FL in early hematopoiesis (Figure 2B) whereby activation of Flt3 signaling by FL above a certain level leads to up-regulation of lymphoid and myeloid lineage associated genes and a "priming" of progenitors to these cell fates. As a result of this priming Flt3 expression is further increased and the cells acquire the LMPP phenotype. In contrast, progenitors that do not receive a strong enough FL signal, either due to low Flt3 expression or due to low FL availability in their immediate microenvironment, will further down-regulate Flt3 and become "primed" to differentiate along the megakaryocyte/erythrocyte pathways, possibly under the influence of other cytokine signals, such as Erythropoietin (Epo) and Thrombopoietin (Tpo). This hypothesis could explain the discrepancy as to whether platelets and erythrocytes originate from Flt3+ progenitors or not, since it postulates that the level of Flt3 expression in vivo is a continuum and possibly under the influence of FL itself.

Interestingly, and considering the question of Flt3 expression on HSC, CD48−CD150+ HSC are significantly reduced in FLtg mice compared to wild-type [64]. However, it remains unknown whether this reduction is a direct effect of FL on HSC or a secondary effect due to aberrant expression of other cytokines in the FLtg bone marrow. In light of the recent data on Flt3 expression by HSC [54], identifying the precise stage where FL might exert its instructive action will be integral to elucidating the precise mechanism by which it does so.

A

B

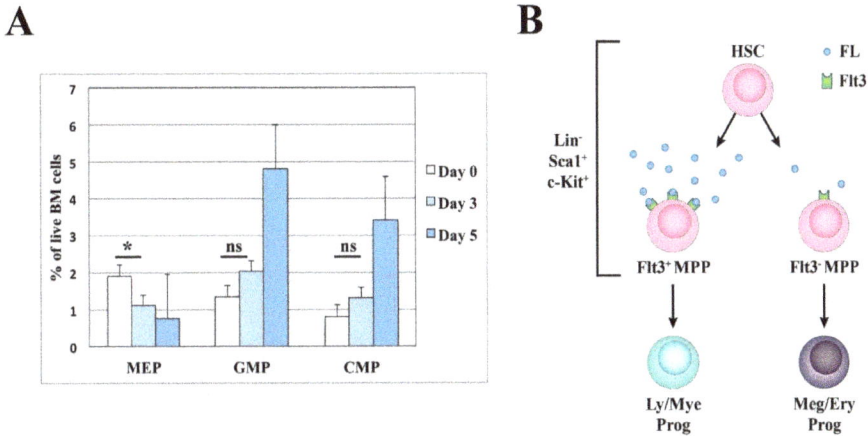

Figure 2. Potential instructive role of FL in early hematopoiesis. (**A**) Relative percentages of MEP, GMP and CMP populations in wild-type mice ($n = 4$) injected with 10 μg of FL daily, for a period of 5 days. Cells were pre-gated as Live, Lin⁻kit⁺Sca1/CD127⁻ and identified as: MEP: CD16^{low}CD34⁻, GMP: CD16⁺CD34⁺, CMP: CD16^{low}CD34⁺. *: significant ($p = 0.017$), ns: not significant. FL: Fms-like tyrosine kinase 3 ligand, MEP: Megakaryocyte-Erythrocyte Progenitor, GMP: Granulocyte-Macrophage Progenitor, CMP: Common Lymphoid Progenitor. BM: Bone Marrow. (**B**) Schematic representation of the proposed model for the instructive action of FL. HSC: Hematopoietic Stem Cell; MPP: Multipotent Progenitor; Ly/Mye Prog: Lymphoid/Myeloid Progenitor; Meg/Ery Prog: Megakaryocyte/Erythrocyte Progenitor; FL: Fms-like tyrosine kinase 3 ligand.

3.2. Dendritic Cells

FL is essential to the generation of DC, as manifested by their severe reduction in the absence of active Flt3 signaling [35,68]. FL has a unique role as a factor that promotes the generation of both human and mouse DC, of all types, in bone marrow cultures [69–73]. Also, in vivo administration of FL in mice leads to a dramatic increase in numbers of both plasmacytoid and conventional DC [74–76]. The same effect is observed in transgenic mice expressing high levels of FL in either an inducible [77] or sustained manner [64,78]. The developmental origin of DC cell types is still not entirely resolved, but putative progenitor populations identified express the Flt3 receptor [79,80]. However, mature DC are also Flt3⁺ [70] and as a result it remains unclear whether FL regulates their generation from precursors, or expands them, or both.

Considering the central role of DC as antigen-presenting cells of the immune system, efforts have been made to utilize FL treatment in order to enhance immune response against malignancies. Accordingly, studies have shown a beneficial effect of FL administration against solid tumors [81–83], but the results from Phase I clinical trials have proved inconclusive [84,85]. Care should be taken in such therapeutic approaches, as FL would most probably have a growth-promoting effect on leukemic cells. In addition, FL-mediated DC expansion can be accompanied by increased proliferation of regulatory T cells (Treg) [63] and, therefore, a suppression of the desired anti-tumor immune response.

3.3. B Cells

As discussed before, one of the phenotypes of both FL⁻ᐟ⁻ and *Flt3*⁻ᐟ⁻ mice is a significant reduction in the numbers of B cell progenitors in the bone marrow, while early in vitro experiments showed a positive effect of FL in B cell generation. Furthermore, FL enhances reconstitution of the B cell compartment after irradiation or chemically-induced myeloablation [86] and appears to be a critical factor for foetal B lymphopoiesis as well [87]. It has been shown that the B cell commitment transcription factor Pax5, which is responsible for the irreversible commitment of pro-B cells to the B cell lineage [88,89], downregulates Flt3 expression [90]. Therefore, FL cannot promote the expansion of committed B cell progenitors that express CD19 (another Pax5 target) but its positive effect on B cell generation should rather be attributed to a role in early, un-committed precursors that generate the B cell pool. Indeed, absence of FL signaling leads to a decrease in the numbers of CLP and EPLM progenitors [36–39], the Ly6D$^+$ fraction of which is Flt3$^+$ and represents the latest un-committed B cell progenitors prior to Pax5 expression [91,92].

FL has been shown to promote the survival of Flt3$^+$CD19$^-$ progenitors, as over-expressing the pro-survival gene *Bcl2* in FL$^{-/-}$ mice can significantly restore their numbers [37,39]. However, this rescue is only partial and even though numbers are increased, B cell priming is not restored in these progenitors. It should be noted that due to the anti-proliferative effect of Bcl2 [37,93] the potential role of FL as a survival factor for Flt3$^+$CD19$^-$ progenitors might not have been fully assessed in these studies. A more clear role of FL in promoting the proliferation of these progenitors has been revealed, as FL over-expression increased the percentage of cycling Ly6D$^+$ EPLM, while FL deficiency had exactly the opposite effect [37]. Therefore, FL seems to be pivotal for promoting the expansion and possibly the survival of Flt3$^+$CD19$^-$ un-committed progenitors, which will further differentiate to committed B cell progenitors. But does FL also act as an instructive cytokine for this commitment? The findings that neither Bcl2 over-expression in FL$^{-/-}$ mice [39], nor increased availability of FL itself [37] lead to increased B cell priming in early CLP/EPLM progenitors argue against such an instructive role. On the contrary, FLtg mice exhibit a reduction in the percentage of B-lineage skewed Ly6D$^+$ EPLM cells expressing Pax5 [37]. Thus, FL seems to exert only a permissive role in B cell development, by promoting the proliferation and survival of early progenitors, therefore facilitating the generation of a significant number of CLP/EPLM precursors, some of which will eventually commit to the B cell fate through Pax5 up-regulation. Pax5 will in turn shut down Flt3 expression in CD19$^+$ committed pro-B cells. Considering the above discussed evidence for an instructive role of FL in early hematopoietic lineage decisions, this permissive role in B cell development highlights the functional versatility of hematopoietic cytokines, since their action can be very much cell context dependent.

3.4. T Cells

Thymopoiesis seems to be largely unaffected in FL$^{-/-}$ or Flt3$^{-/-}$ mice. However, and in keeping with the notion of the synergistic action of FL with other cytokines, ablation of Flt3 signaling further exacerbates the defect in T cell development observed in Interleukin-7 (IL-7) knock-out mice [94]. This is probably due to the fact that the bone marrow progenitors that seed the thymus are Flt3$^+$ and the earliest identified thymic T cell progenitors also express Flt3 [95]. Indeed, FL treatment of immunodeficient mice after bone marrow transplantation accelerates T cell recovery because it expands bone marrow Flt3$^+$ LSK progenitors [96]. Further, this Flt3 expression by the thymus seeding progenitors seems to be functionally important, since FL production by the thymic microenvironment has been shown to promote the maintenance of these progenitors, both in steady-state conditions and after irradiation [97,98]. Interestingly, high levels of FL in vivo lead to an increase in the numbers of Treg [63,64], but this seems to be an indirect effect, through an IL-2 dependent activation of Treg proliferation following interactions with DC, a cell type that is vastly increased under conditions of increased FL availability [63].

3.5. Overview

Overall, our current knowledge points towards a critical role of FL in the generation of lymphoid cells. However, this role is mainly exerted on early, un-committed lymphoid progenitors and it seems to be of a permissive nature, i.e., promoting their proliferation and survival. Commitment and further maturation of lymphoid cells coincides with downregulation of Flt3, thereby rendering them unresponsive to FL. Apart from B and T cells, this seems to hold true for Innate Lymphoid cells (ILC) as well, since FL regulates their numbers through the maintenance of their early progenitors [99,100]. Even though the intracellular response to FL might be different for different lymphoid lineages, this common pattern of Flt3 downregulation upon commitment and maturation to functional lymphocytes seems to indicate that active Flt3 signaling possibly contributes to the maintenance of an "immature" lymphoid phenotype and therefore it needs to be silenced in order for mature lymphoid cells to become functional. There have been reports of Flt3 re-expression on mature, activated B cells [101,102] and activated human T cells [103]. For B cells, this could be the result of Pax5 downregulation, which is necessary for further differentiation to antibody secreting plasma cells. Some evidence for a potential functional significance of Flt3 expression on activated lymphocytes has been reported [102] but further investigations are required to elucidate the exact role of FL in this context.

4. FL and Flt3 in Hematopoietic Malignancies

Aberrant expression of FLT3 is very commonly found in hematopoietic malignancies. In most cases, this is due to activating mutations in the *FLT3* gene but a significant number of leukemias are also characterized by a higher than normal expression level of un-mutated, wild-type FLT3, thus underscoring the importance of FLT3 signaling perturbations in malignant transformation. In most cases FLT3 mutations are associated with a poor clinical prognosis [104]. FLT3 mutations are most commonly found in AML patients, almost one third of which harbor such a mutation. In addition, 5–10% of patients with Myelodysplasia (MDS) and 1–3% of Acute Lymphoblastic Leukemia (ALL) patients have mutations in the *FLT3* gene. In pediatric leukemias FLT3 mutations are somewhat more rare, but they also clearly associate with poor clinical prognosis [105].

There are two types of FLT3 mutations found in AML: internal tandem duplications of the juxta-membrane domain (the mutated receptor thus termed FLT3-ITD) and point mutations in the tyrosine kinase domains (collectively named FLT3-TKD).

4.1. FLT3-ITD

In 1996, Nakao et al. described for the first time the presence of a mutated FLT3 receptor in AML patients, which exhibited tandem duplications in the juxta-membrane domain [106]. Since then this particular type of mutation has been studied extensively and identified as one of the most common mutations in AML [107–109]. The mutation consists of a head-to-tail replication of sequences coding part of the juxta-membrane domain of the receptor (Figure 3). These sequences can be variable in length (from 3 to >400 base pairs) and, as they are always found to be in-frame, they result in the transcription and translation of a receptor with an elongated juxta-membrane domain. The consequence of this elongation is that the mutated receptor can dimerize, phosphorylate and activate the kinase domains without the need to bind FL, therefore resulting in ligand-independent, constitutive activation of FLT3 [110–112]. This probably occurs by eliminating an auto-inhibitory function of the wild-type receptor, which ensures that, without ligand binding and dimerization, the kinase domains cannot be activated [110,113]. It has been proposed that these tandem duplications occur as a result of DNA replication mistakes and they provide a growth advantage to the cells harboring them [111].

Studies on the signaling events occurring downstream of the constitutively activated FLT3-ITD have initially been carried out using cell lines transduced with the mutated receptor and have indicated that there are qualitative differences between FLT3-ITD and wild-type FLT3 signaling. In addition to activating the Ras/MEK/Erk and PI3K pathways [111,112], which are also activated by wild-type FLT3, FLT3-ITD has been shown to promote STAT5 phosphorylation and subsequent DNA binding [114]. As a consequence, FLT3-ITD can activate an array of STAT5 target genes, which would normally not be expressed upon binding of FL to its wild-type receptor [115]. Interestingly, amongst them are not only cell-cycle regulating genes, which would account for a growth advantage of FLT3-ITD harboring cells, but also myeloid differentiation transcription factors, such as PU1 and C/EBPα, which seem to be suppressed by FLT3-ITD [115,116]. Another outcome of the aberrant STAT5 activation triggered by FLT3-ITD seems to be an increase in reactive oxygen species production and in the frequency of double-strand DNA breaks, therefore resulting in genomic instability and providing a potential mechanism for the apparent poor clinical prognosis of patients harboring FLT3-ITD mutations [117,118]. In addition, a potential mechanism by which FLT3-ITD might promote survival and proliferation of AML cells is through phosphorylation and subsequent suppression of the Forkhead family of transcription factors member FOXO3a, an important pro-apoptotic regulator [119].

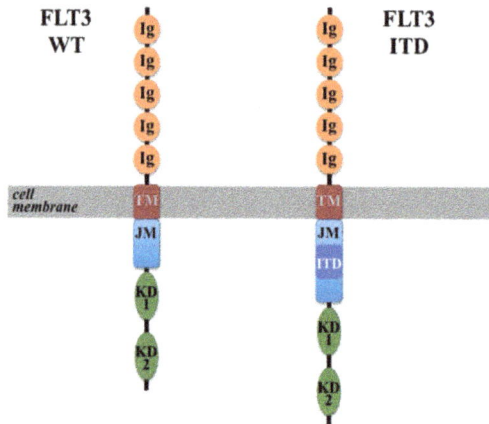

Figure 3. Schematic diagram of the structure of wild-type FLT3 (**left**) and FLT3-ITD (**right**) receptors. Ig: Immunoglobulin-like domain; TM: transmembrane domain; JM: juxta-membrane domain; KD: kinase domain; ITD: internal tandem duplications.

Important insights into the mechanisms by which FLT3-ITD can promote leukemogenesis have come from in vivo studies. Apart from injecting FLT3-ITD transduced cell lines into mice [114,120], early attempts to create mouse models where the mechanism of FLT3-ITD leukemogenesis could be studied involved retroviral transduction of bone marrow cells and subsequent transplantation into recipient mice. These experiments showed that FLT3-ITD expressing bone marrow cells caused a myeloproliferative disease in the recipient mice, characterized by splenomegaly, leukocytosis and expansion of myeloid lineages, but without developing an AML phenotype similar to human patients [121,122]. Similar results were obtained from a transgenic mouse model expressing FLT3-ITD under the control of the *vav* promoter, with the difference that few transgenic lines in this system developed a lymphoid disease as well [123]. Mouse models that resemble more closely the leukemogenic effect of FLT3-ITD in humans have been generated by two groups, through a knock-in approach, whereby the human FLT3-ITD gene was inserted into the endogenous *Flt3* locus [124,125]. In both studies, mice developed symptoms of myeloproliferative disease, with splenomegaly, leukocytosis, expansion of myeloid progenitors and dendritic cells, as well as a decrease in the numbers of B cell progenitors. Interestingly, while both mouse models exhibited a significant increase in early LSK progenitors, further analysis of LSK subpopulations in one of them revealed a reduction in the number of CD48$^-$CD150$^+$ HSC [126]. This reduction was shown to be the result of increased proliferation and cell-cycle entry of HSC, eventually leading to their exhaustion. Considering that FLT3-ITD in this mouse model is expressed under the control of the endogenous *Flt3* regulatory elements, these results would argue in favor of Flt3 expression within the HSC compartment, as discussed previously (Figure 1) [54].

4.2. FLT3-TKD

Point mutations in the tyrosine kinase domain of FLT3 are the second most common type of FLT3 mutations found in AML. They have also been associated with an unfavorable clinical outcome in patients although, due to their somewhat lower frequency, very large studies are needed to precisely evaluate their impact on clinical outcome [113]. FLT3-TKD point mutations also promote ligand-independent phosphorylation of the receptor and cell growth [127,128]. In wild-type FLT3, in the absence of ligand binding, the activation loop of the kinase domain remains in a closed conformation, therefore preventing ATP and protein binding [129]. It is believed that TKD mutations result in opening of this region and subsequent activation without the need for ligand binding. Even though both FLT3-ITD and FLT3-TKD mutations confer ligand-independent receptor activation, the mechanisms by which they contribute to the development of leukemia might be different. Indeed, signaling downstream of both mutated receptors seems to differ as to which signal transduction pathways get activated [130], and a study comparing gene expression profiles of the two types of mutations in childhood AML patients showed significant differences in the genetic program they induce [131]. FLT3-ITD has been shown to localize to a large extent intra-cellularly [132], resulting in aberrant interactions with signaling molecules [133], and this has been hypothesized to be one mechanism responsible for differential signaling by FLT3-ITD and FLT3-TKD [134]. Furthermore, in a mouse bone marrow transplantation model, FLT3-TKD not only manifested a malignancy with longer latency compared to FLT3-ITD, but it was also found to cause an oligoclonal lymphoid disease, in contrast to FLT3-ITD, which led to the development of a myeloproliferative disorder [121]. The reason for the difference in lineage outcome of the disease was differential activation of STAT5 from the two types of mutations, since the FLT3-TKD mutation expressed in this model did not result in STAT5 phosphorylation, as was the case with FLT3-ITD. Intriguingly, STAT5 phosphorylation seemed to be the decisive factor as to the lineage phenotype of the disorder, since deletion of STAT5 in FLT3-ITD-mediated malignancy significantly increased survival and switched the immunophenotype of the disease from a myeloid to a lymphoid one [135]. The differences in the in vivo effects of FLT3-ITD and FLT3-TKD were also demonstrated in a knock-in mouse model expressing the FLT3-TKD most commonly found in AML [136]. In agreement with the previous study, this mouse model showed that FLT3-TKD manifested a less aggressive malignancy than FLT3-ITD, and even though myeloid progenitors were increased, a significant expansion of B cell progenitors was also observed.

Apart from activating mutations, increased expression levels of the wild-type FLT3 receptor have also been observed in cases of leukemia [137–140]. Consequently, high FLT3 expression and/or constitutively active mutated FLT3 are found in 70–100% of AML cases, as well as in a high percentage of ALL cases [104]. The importance or potential mechanism of high FLT3 expression in leukemias remains unknown. Some receptor tyrosine kinases can exhibit dimerization even in the absence of ligand binding [141]. Therefore, a significant increase in the amount of FLT3 on the cell surface might facilitate some degree of ligand-independent dimerization and activation of the receptor [142]. Alternatively, it could be hypothesized that in steady-state conditions the amount of FLT3 on the cell surface is the limiting factor for FLT3 signaling, with FL being abundant in the bone marrow. Therefore, higher FLT3 expression could result in stronger downstream signaling. In that context, it should be noted that there is evidence for autocrine FL signaling from studies of AML patients' cells and leukemic cell lines [143].

4.3. FLT3 Inhibitors

The poor response of FLT3-ITD AML patients to conventional therapies has prompted investigation of the use of FLT3 tyrosine kinase inhibitors (TKIs) to treat relapsed and refractory AML. Interestingly, blast cells from FLT3-ITD AML patients at the time of relapse are more sensitive to FLT3 TKIs when compared to presentation blasts [144]. This suggests that chemotherapy selects for cells that are dependent on FLT3-ITD signalling, underscoring the potential of FLT3 TKIs in treating relapsed AML.

First generation FLT3 inhibitors were non-specific TKIs and inhibited other tyrosine kinase receptors, such as c-KIT and vascular endothelial growth factor receptor [145–148]. These include two TKIs, sunitinib and sorfenib, which have been approved for the treatment of solid tumours [149–153]. Phase I and Phase I/II clinical trials have demonstrated that sunitinib inhibits FLT3 signalling in AML patients' cells [154] and is well tolerated as both a monotherapy and with intensive chemotherapy [155,156]. As a monotherapy, sunitinib induces short-term (4–16 weeks) and partial remission [156]. Sunitinib with conventional chemotherapy seems promising, but this combination has only been investigated in a single-arm Phase I/II study [155]. Sorefenib has been more intensely studied as to its use in AML. Early studies of sorefenib as a monotherapy to treat AML generated conflicting outcomes [157–159]. However, when combined with chemotherapy, sorefenib has been shown to increase event- and relapse-free survival in patients with previously untreated AML who are above, but not below, the age of 60 [160], though toxicity is increased [161,162]. Sorafenib also sustains remission in AML patients harbouring FLT3-ITD mutations following allogenic stem cell transplant [163,164] and a Phase IV trial is currently underway to investigate this further (NCT02474290). Midostaurin is another FLT3 TKI that has shown promise in young AML patients. Stone et al. have shown that midostaurin with intense chemotherapy increases both event-free and overall survival in patients with newly diagnosed AML ($\leq$60 years) [165,166]. Lestaurnib has been studied as a potential treatment for AML but has shown limited clinical benefit and achieving sustained FLT3 inhibition with lestaurnib has proven challenging [167,168]. Second generation FLT3 TKIs are highly selective for FLT3 and have shown significant promise in treating relapsed and refractory AML. Quizartnib is one of the most effective monotherapies for the treatment of FLT3-ITD AML and in one Phase II study complete remission was seen in approximately half of the patients and was sustained for an average of 11–13 weeks [169,170]. A pilot study has also shown that quizartnib is an effective treatment of newly diagnosed AML when combined with chemotherapy (a complete remission seen in 79% of patients) [171], and a phase III clinical trial is currently investigating the use of quizartnib in combination with chemotherapy in newly diagnosed FLT3-ITD AML (NCT02668653). Two other second generation FLT3 TKIs, gilteritinib and crenolanib, have also been shown to have activity against AML cells with FLT3-ITD and FLT3-TKD mutations. A phase I/II clinical trial has demonstrated that gilteritinib is effective as a monotherapy in the treatment of relapsed and refractory AML [172]. Complete remission was observed in 49% and 29% of FLT3-ITD and FLT3-TKD AML patients, respectively. Crenolanib is effective in treating relapsed and refractory AML [173] and has been shown to be particularly useful in the treatment of cases of AML that are resistant to previous TKI therapy [174–176]. In a recent trial of crenolanib in patients with relapsed or refractory AML harbouring *FLT3* mutations, the overall response rate was 31% in patients that had received prior FLT3 TKI treatment and 39% in FLT3 TKI naïve patients [177].

Mechanisms of FLT3-TKI resistance in AML have been identified, such as the FLT3-TKD mutations that commonly confer resistance to quizartnib [178,179]. One proposed mechanism is that elevated plasma FL levels that are induced following chemotherapy reduce the activity of FLT3-TKIs, and this is supported by in vitro studies [180]. Therefore, several studies have investigated the use of FLT3-TKIs with azacytidine, a hypomethylating agent that reduces plasma FL concentrations, in the treatment of AML. An interim report of a study examining the combination of quizartinib with azacitidine has shown promising results, with a response observed in 69% of patients with relapsed or refractory MDS, chronic myelomonocytic leukemia or AML, including 4 of 7 patients that had received prior FLT3 TKI treatment [181]. Another approach to overcoming resistance is the use of antibodies to FL in combination with FLT3-TKIs ([182] and NCT02789254).

5. Open Questions and Future Challenges

The large amount of data briefly discussed above have significantly increased our knowledge as to the role of the FL-FLT3 axis, both in normal and malignant hematopoeisis. However, several questions with clinical importance remain open. It is clear that activating mutations of FLT3 are an important factor for disease progression in AML. Nevertheless, mouse models indicate that the presence of a FLT3 mutation alone might not be sufficient to cause leukemia. It has been hypothesized, in accordance with the "two hit" model for neoplastic transformation [183], that FLT3 mutations have to act in concert with at least one other mutation, in order for hematopoietic cells to become leukemic. This hypothesis

Int. J. Mol. Sci. **2017**, *18*, 1115

postulates that aberrant FLT3 signaling would provide the increased survival and growth signal to cells, while the additional mutation would confer a differentiation block, therefore resulting in the transformation of the cell to a leukemic blast. In support of this hypothesis, it has been shown that FLT3-ITD can cause a leukemia-like phenotype in a mouse model only when present in combination with the t(15;17) translocation [184] and FLT3-ITD is frequently found in leukemias together with genetic rearrangements and point mutations [104].

Several lines of evidence point towards a very important role of FL in lymphoid development. Absence of Flt3 signaling seems to affect lymphoid cells more than myeloid [34,35], while high expression of Flt3 in MPP has identified a progenitor population with more robust differentiation potential for lymphoid rather than myeloid cells [48]. However, FLT3 mutations are found mainly in myeloid leukemias and mouse models have shown that such mutations predominantly confer a myeloproliferative rather than lymphoproliferative disease. This remains a paradox, though it can be hypothesized that the key to answering this question lies in the specific signaling pathways and genes that the mutated receptor activates in comparison to the wild-type one. Therefore, further investigations are needed in this direction—not only at the gene expression level, since a lot of important differences between wild-type and FLT3-ITD might occur at a post-translational level. Recent evidence would suggest that the activation of STAT5 specifically by FLT3-ITD might be very important in this context [135]. Further research into the mechanistic aspects of this lineage-specific leukemogenic transformation by mutated FLT3 is required, with the final aim to assist the development of specific inhibitors for therapeutic treatment.

It is interesting that the phenotype caused by expression of mutated FLT3 in mouse models highly resembles the one observed in mice with high levels of FL; i.e., leukocytosis, splenomegaly, anemia, expansion of DC and myeloid progenitors and reduced B cell generation [64,78]. It is tempting to hypothesize that constitutive, ligand-independent Flt3 signaling might elicit a cellular response that is similar to the one caused by activation of the receptor due to sustained FL exposure. To this end, it will be interesting to assess whether continuous wild-type Flt3 engagement might trigger downstream signals that are significantly different from the ones generated by transient or low-level signaling. This could be of particular interest for potential therapeutic applications in leukemias with wild-type FLT3 over-expression.

Further investigation of the role of FL in normal hematopoiesis might provide clues as to the manner in which FLT3-related leukemias can be treated. However, the reverse is also true, since studies on mutated FLT3 biology can provide insights on the physiological role of the cytokine. A recent study showed that FLT3-ITD can instruct myeloid differentiation in lympho-myeloid oligo-potent progenitors in a mouse model [185]. Apart from providing further evidence and a potential mechanism for the presence of FLT3-ITD mainly in myeloid leukemias, this study also provides some support for the hypothesis of an instructive role of growth factor receptor signaling in early hematopoiesis. Another example is the finding that FLT3-ITD expressed under the control of the endogenous *Flt3* promoter directly affects HSC proliferation in vivo, thus supporting the notion that Flt3 expression is initiated already in HSC (Figure 1).

The studies that have been briefly outlined herein have greatly advanced our understanding not only of how the cytokine FL exerts its function in hematopoiesis, but also of how normal blood cell generation is regulated. The clinical importance of Flt3 signaling requires further investigations into the role of FL and its receptor in leukemia.

Acknowledgments: This project has received funding from the European Union's Seventh Framework Programme for research, technological development and demonstration under grant agreement no. 315902. Geoffrey Brown and Antonius Rolink are partners within the Marie Curie Initial Training Network DECIDE (Decision-making within cells and differentiation entity therapies). Ciaran Mooney gratefully acknowledges receipt of a Marie Curie Research Associate post. Panagiotis Tsapogas and Antonius Rolink are supported by the Swiss National Science Foundation and the Novartis Foundation for Medical and Biological Research. Antonius Rolink is holder of the chair in Immunology endowed by Fritz Hoffman—La Roche Ltd.

Conflicts of Interest: The authors declare no conflict of interest.

References

1. Metcalf, D. Hematopoietic cytokines. *Blood* **2008**, *111*, 485–491. [CrossRef] [PubMed]
2. Matthews, W.; Jordan, C.T.; Wiegand, G.W.; Pardoll, D.; Lemischka, I.R. A receptor tyrosine kinase specific to hematopoietic stem and progenitor cell-enriched populations. *Cell* **1991**, *65*, 1143–1152. [CrossRef]
3. Rosnet, O.; Marchetto, S.; deLapeyriere, O.; Birnbaum, D. Murine Flt3, a gene encoding a novel tyrosine kinase receptor of the PDGFR/CSF1R family. *Oncogene* **1991**, *6*, 1641–1650. [PubMed]
4. Rosnet, O.; Mattei, M.G.; Marchetto, S.; Birnbaum, D. Isolation and chromosomal localization of a novel FMS-like tyrosine kinase gene. *Genomics* **1991**, *9*, 380–385. [CrossRef]
5. Small, D.; Levenstein, M.; Kim, E.; Carow, C.; Amin, S.; Rockwell, P.; Witte, L.; Burrow, C.; Ratajczak, M.Z.; Gewirtz, A.M.; et al. STK-1, the human homolog of Flk-2/Flt-3, is selectively expressed in CD34+ human bone marrow cells and is involved in the proliferation of early progenitor/stem cells. *Proc. Natl. Acad. Sci. USA* **1994**, *91*, 459–463. [CrossRef] [PubMed]
6. Rosnet, O.; Birnbaum, D. Hematopoietic receptors of class III receptor-type tyrosine kinases. *Crit. Rev. Oncog.* **1993**, *4*, 595–613. [PubMed]
7. Lyman, S.D.; James, L.; Vanden Bos, T.; de Vries, P.; Brasel, K.; Gliniak, B.; Hollingsworth, L.T.; Picha, K.S.; McKenna, H.J.; Splett, R.R.; et al. Molecular cloning of a ligand for the flt3flk-2 tyrosine kinase receptor: A proliferative factor for primitive hematopoietic cells. *Cell* **1993**, *75*, 1157–1167. [CrossRef]
8. Lyman, S.D.; James, L.; Escobar, S.; Downey, H.; de Vries, P.; Brasel, K.; Stocking, K.; Beckmann, M.P.; Copeland, N.G.; Cleveland, L.S.; et al. Identification of soluble and membrane-bound isoforms of the murine flt3 ligand generated by alternative splicing of mRNAs. *Oncogene* **1995**, *10*, 149–157. [PubMed]
9. Lyman, S.D.; James, L.; Johnson, L.; Brasel, K.; de Vries, P.; Escobar, S.S.; Downey, H.; Splett, R.R.; Beckmann, M.P.; McKenna, H.J. Cloning of the human homologue of the murine flt3 ligand: A growth factor for early hematopoietic progenitor cells. *Blood* **1994**, *83*, 2795–2801. [PubMed]
10. Turner, A.M.; Lin, N.L.; Issarachai, S.; Lyman, S.D.; Broudy, V.C. Flt3 receptor expression on the surface of normal and malignant human hematopoietic cells. *Blood* **1996**, *88*, 3383–3390. [PubMed]
11. Dosil, M.; Wang, S.; Lemischka, I.R. Mitogenic signalling and substrate specificity of the Flk2/Flt3 receptor tyrosine kinase in fibroblasts and interleukin 3-dependent hematopoietic cells. *Mol. Cell. Biol.* **1993**, *13*, 6572–6585. [CrossRef] [PubMed]
12. Rottapel, R.; Turck, C.W.; Casteran, N.; Liu, X.; Birnbaum, D.; Pawson, T.; Dubreuil, P. Substrate specificities and identification of a putative binding site for PI3K in the carboxy tail of the murine Flt3 receptor tyrosine kinase. *Oncogene* **1994**, *9*, 1755–1765. [PubMed]
13. Lavagna-Sevenier, C.; Marchetto, S.; Birnbaum, D.; Rosnet, O. FLT3 signaling in hematopoietic cells involves CBL, SHC and an unknown P115 as prominent tyrosine-phosphorylated substrates. *Leukemia* **1998**, *12*, 301–310. [CrossRef] [PubMed]
14. Marchetto, S.; Fournier, E.; Beslu, N.; Aurran-Schleinitz, T.; Dubreuil, P.; Borg, J.P.; Birnbaum, D.; Rosnet, O. SHC and SHIP phosphorylation and interaction in response to activation of the FLT3 receptor. *Leukemia* **1999**, *13*, 1374–1382. [CrossRef] [PubMed]
15. Zhang, S.; Broxmeyer, H.E. p85 subunit of pi3 kinase does not bind to human Flt3 receptor, but associates with SHP2, SHIP, and a tyrosine-phosphorylated 100-kDa protein in Flt3 ligand-stimulated hematopoietic cells. *Biochem. Biophys. Res. Commun.* **1999**, *254*, 440–445. [CrossRef] [PubMed]
16. Ahsberg, J.; Tsapogas, P.; Qian, H.; Zetterblad, J.; Zandi, S.; Mansson, R.; Jonsson, J.I.; Sigvardsson, M. Interleukin-7-induced Stat-5 acts in synergy with Flt-3 signaling to stimulate expansion of hematopoietic progenitor cells. *J. Biol. Chem.* **2010**, *285*, 36275–36284. [CrossRef] [PubMed]
17. Zhang, S.; Broxmeyer, H.E. Flt3 ligand induces tyrosine phosphorylation of gab1 and gab2 and their association with shp-2, grb2, and PI3 kinase. *Biochem. Biophys. Res. Commun.* **2000**, *277*, 195–199. [CrossRef] [PubMed]
18. Zhang, S.; Mantel, C.; Broxmeyer, H.E. Flt3 signaling involves tyrosyl-phosphorylation of SHP-2 and SHIP and their association with Grb2 and Shc in Baf3/Flt3 cells. *J. Leukoc. Biol.* **1999**, *65*, 372–380. [PubMed]
19. Laouar, Y.; Welte, T.; Fu, X.Y.; Flavell, R.A. STAT3 is required for Flt3L-dependent dendritic cell differentiation. *Immunity* **2003**, *19*, 903–912. [CrossRef]

20. Onai, N.; Obata-Onai, A.; Tussiwand, R.; Lanzavecchia, A.; Manz, M.G. Activation of the Flt3 signal transduction cascade rescues and enhances type I interferon-producing and dendritic cell development. *J. Exp. Med.* **2006**, *203*, 227–238. [CrossRef] [PubMed]

21. Zhang, S.; Fukuda, S.; Lee, Y.; Hangoc, G.; Cooper, S.; Spolski, R.; Leonard, W.J.; Broxmeyer, H.E. Essential role of signal transducer and activator of transcription (Stat)5a but not Stat5b for Flt3-dependent signaling. *J. Exp. Med.* **2000**, *192*, 719–728. [CrossRef] [PubMed]

22. Banu, N.; Deng, B.; Lyman, S.D.; Avraham, H. Modulation of haematopoietic progenitor development by FLT-3 ligand. *Cytokine* **1999**, *11*, 679–688. [CrossRef] [PubMed]

23. Brashem-Stein, C.; Flowers, D.A.; Bernstein, I.D. Regulation of colony forming cell generation by flt-3 ligand. *Br. J. Haematol.* **1996**, *94*, 17–22. [CrossRef] [PubMed]

24. Broxmeyer, H.E.; Lu, L.; Cooper, S.; Ruggieri, L.; Li, Z.H.; Lyman, S.D. Flt3 ligand stimulates/costimulates the growth of myeloid stem/progenitor cells. *Exp. Hematol.* **1995**, *23*, 1121–1129. [PubMed]

25. Gabbianelli, M.; Pelosi, E.; Montesoro, E.; Valtieri, M.; Luchetti, L.; Samoggia, P.; Vitelli, L.; Barberi, T.; Testa, U.; Lyman, S.; et al. Multi-level effects of flt3 ligand on human hematopoiesis: Expansion of putative stem cells and proliferation of granulomonocytic progenitors/monocytic precursors. *Blood* **1995**, *86*, 1661–1670. [PubMed]

26. Hirayama, F.; Lyman, S.D.; Clark, S.C.; Ogawa, M. The flt3 ligand supports proliferation of lymphohematopoietic progenitors and early B-lymphoid progenitors. *Blood* **1995**, *85*, 1762–1768. [PubMed]

27. Hudak, S.; Hunte, B.; Culpepper, J.; Menon, S.; Hannum, C.; Thompson-Snipes, L.; Rennick, D. FLT3/FLK2 ligand promotes the growth of murine stem cells and the expansion of colony-forming cells and spleen colony-forming units. *Blood* **1995**, *85*, 2747–2755. [PubMed]

28. Jacobsen, S.E.; Okkenhaug, C.; Myklebust, J.; Veiby, O.P.; Lyman, S.D. The FLT3 ligand potently and directly stimulates the growth and expansion of primitive murine bone marrow progenitor cells in vitro: Synergistic interactions with interleukin (IL) 11, IL-12, and other hematopoietic growth factors. *J. Exp. Med.* **1995**, *181*, 1357–1363. [CrossRef] [PubMed]

29. Namikawa, R.; Muench, M.O.; de Vries, J.E.; Roncarolo, M.G. The FLK2/FLT3 ligand synergizes with interleukin-7 in promoting stromal-cell-independent expansion and differentiation of human fetal pro-B cells in vitro. *Blood* **1996**, *87*, 1881–1890. [PubMed]

30. Ray, R.J.; Paige, C.J.; Furlonger, C.; Lyman, S.D.; Rottapel, R. Flt3 ligand supports the differentiation of early B cell progenitors in the presence of interleukin-11 and interleukin-7. *Eur. J. Immunol.* **1996**, *26*, 1504–1510. [CrossRef] [PubMed]

31. Veiby, O.P.; Lyman, S.D.; Jacobsen, S.E. Combined signaling through interleukin-7 receptors and flt3 but not c-kit potently and selectively promotes B-cell commitment and differentiation from uncommitted murine bone marrow progenitor cells. *Blood* **1996**, *88*, 1256–1265. [PubMed]

32. McKenna, H.J.; de Vries, P.; Brasel, K.; Lyman, S.D.; Williams, D.E. Effect of flt3 ligand on the ex vivo expansion of human CD34+ hematopoietic progenitor cells. *Blood* **1995**, *86*, 3413–3420. [PubMed]

33. Rusten, L.S.; Lyman, S.D.; Veiby, O.P.; Jacobsen, S.E. The FLT3 ligand is a direct and potent stimulator of the growth of primitive and committed human CD34+ bone marrow progenitor cells in vitro. *Blood* **1996**, *87*, 1317–1325. [PubMed]

34. Mackarehtschian, K.; Hardin, J.D.; Moore, K.A.; Boast, S.; Goff, S.P.; Lemischka, I.R. Targeted disruption of the flk2/flt3 gene leads to deficiencies in primitive hematopoietic progenitors. *Immunity* **1995**, *3*, 147–161. [CrossRef]

35. McKenna, H.J.; Stocking, K.L.; Miller, R.E.; Brasel, K.; De Smedt, T.; Maraskovsky, E.; Maliszewski, C.R.; Lynch, D.H.; Smith, J.; Pulendran, B.; et al. Mice lacking flt3 ligand have deficient hematopoiesis affecting hematopoietic progenitor cells, dendritic cells, and natural killer cells. *Blood* **2000**, *95*, 3489–3497. [PubMed]

36. Sitnicka, E.; Bryder, D.; Theilgaard-Monch, K.; Buza-Vidas, N.; Adolfsson, J.; Jacobsen, S.E. Key role of flt3 ligand in regulation of the common lymphoid progenitor but not in maintenance of the hematopoietic stem cell pool. *Immunity* **2002**, *17*, 463–472. [CrossRef]

37. Von Muenchow, L.; Alberti-Servera, L.; Klein, F.; Capoferri, G.; Finke, D.; Ceredig, R.; Rolink, A.; Tsapogas, P. Permissive roles of cytokines interleukin-7 and Flt3 ligand in mouse B-cell lineage commitment. *Proc. Natl. Acad. Sci. USA* **2016**, *113*, E8122–E8130. [CrossRef] [PubMed]

38. Beaudin, A.E.; Boyer, S.W.; Forsberg, E.C. Flk2/Flt3 promotes both myeloid and lymphoid development by expanding non-self-renewing multipotent hematopoietic progenitor cells. *Exp. Hematol.* **2014**, *42*, 218–229. [CrossRef] [PubMed]

39. Dolence, J.J.; Gwin, K.A.; Shapiro, M.B.; Medina, K.L. Flt3 signaling regulates the proliferation, survival, and maintenance of multipotent hematopoietic progenitors that generate B cell precursors. *Exp. Hematol.* **2014**, *42*, 380–393. [CrossRef] [PubMed]

40. Balciunaite, G.; Ceredig, R.; Massa, S.; Rolink, A.G. A B220+ CD117+ CD19− hematopoietic progenitor with potent lymphoid and myeloid developmental potential. *Eur. J. Immunol.* **2005**, *35*, 2019–2030. [CrossRef] [PubMed]

41. Karsunky, H.; Inlay, M.A.; Serwold, T.; Bhattacharya, D.; Weissman, I.L. Flk2+ common lymphoid progenitors possess equivalent differentiation potential for the B and T lineages. *Blood* **2008**, *111*, 5562–5570. [CrossRef] [PubMed]

42. Brown, G.; Mooney, C.J.; Alberti-Servera, L.; Muenchow, L.; Toellner, K.M.; Ceredig, R.; Rolink, A. Versatility of stem and progenitor cells and the instructive actions of cytokines on hematopoiesis. *Crit. Rev. Clin. Lab. Sci.* **2015**, *52*, 168–179. [PubMed]

43. Ceredig, R.; Rolink, A.G.; Brown, G. Models of haematopoiesis: Seeing the wood for the trees. *Nat. Rev. Immunol.* **2009**, *9*, 293–300. [CrossRef] [PubMed]

44. Adolfsson, J.; Borge, O.J.; Bryder, D.; Theilgaard-Monch, K.; Astrand-Grundstrom, I.; Sitnicka, E.; Sasaki, Y.; Jacobsen, S.E. Upregulation of Flt3 expression within the bone marrow Lin−Sca1+c-kit+ stem cell compartment is accompanied by loss of self-renewal capacity. *Immunity* **2001**, *15*, 659–669. [CrossRef]

45. Christensen, J.L.; Weissman, I.L. Flk-2 is a marker in hematopoietic stem cell differentiation: A simple method to isolate long-term stem cells. *Proc. Natl. Acad. Sci. USA* **2001**, *98*, 14541–14546. [CrossRef] [PubMed]

46. Akashi, K.; Traver, D.; Miyamoto, T.; Weissman, I.L. A clonogenic common myeloid progenitor that gives rise to all myeloid lineages. *Nature* **2000**, *404*, 193–197. [CrossRef] [PubMed]

47. Kondo, M.; Weissman, I.L.; Akashi, K. Identification of clonogenic common lymphoid progenitors in mouse bone marrow. *Cell* **1997**, *91*, 661–672. [CrossRef]

48. Adolfsson, J.; Mansson, R.; Buza-Vidas, N.; Hultquist, A.; Liuba, K.; Jensen, C.T.; Bryder, D.; Yang, L.; Borge, O.J.; Thoren, L.A.; et al. Identification of Flt3+ lympho-myeloid stem cells lacking erythro-megakaryocytic potential a revised road map for adult blood lineage commitment. *Cell* **2005**, *121*, 295–306. [CrossRef] [PubMed]

49. Boyer, S.W.; Schroeder, A.V.; Smith-Berdan, S.; Forsberg, E.C. All hematopoietic cells develop from hematopoietic stem cells through Flk2/Flt3-positive progenitor cells. *Cell Stem Cell* **2011**, *9*, 64–73. [CrossRef] [PubMed]

50. Forsberg, E.C.; Serwold, T.; Kogan, S.; Weissman, I.L.; Passegue, E. New evidence supporting megakaryocyte-erythrocyte potential of flk2/flt3+ multipotent hematopoietic progenitors. *Cell* **2006**, *126*, 415–426. [CrossRef] [PubMed]

51. Luc, S.; Buza-Vidas, N.; Jacobsen, S.E. Biological and molecular evidence for existence of lymphoid-primed multipotent progenitors. *Ann. N. Y. Acad. Sci.* **2007**, *1106*, 89–94. [CrossRef] [PubMed]

52. Luc, S.; Anderson, K.; Kharazi, S.; Buza-Vidas, N.; Boiers, C.; Jensen, C.T.; Ma, Z.; Wittmann, L.; Jacobsen, S.E. Down-regulation of Mpl marks the transition to lymphoid-primed multipotent progenitors with gradual loss of granulocyte-monocyte potential. *Blood* **2008**, *111*, 3424–3434. [CrossRef] [PubMed]

53. Buza-Vidas, N.; Woll, P.; Hultquist, A.; Duarte, S.; Lutteropp, M.; Bouriez-Jones, T.; Ferry, H.; Luc, S.; Jacobsen, S.E. Flt3 expression initiates in fully multipotent mouse hematopoietic progenitor cells. *Blood* **2011**, *118*, 1544–1548. [CrossRef] [PubMed]

54. Buza-Vidas, N.; Woll, P.; Hultquist, A.; Duarte, S.; Lutteropp, M.; Bouriez-Jones, T.; Ferry, H.; Luc, S.; Jacobsen, S.E. Selective expression of Flt3 within the mouse hematopoietic stem cell compartment. *Int. J. Mol. Sci.* **2017**, *18*, 1037.

55. Kiel, M.J.; Yilmaz, O.H.; Iwashita, T.; Yilmaz, O.H.; Terhorst, C.; Morrison, S.J. Slam family receptors distinguish hematopoietic stem and progenitor cells and reveal endothelial niches for stem cells. *Cell* **2005**, *121*, 1109–1121. [CrossRef] [PubMed]

56. Oguro, H.; Ding, L.; Morrison, S.J. SLAM family markers resolve functionally distinct subpopulations of hematopoietic stem cells and multipotent progenitors. *Cell Stem Cell* **2013**, *13*, 102–116. [CrossRef] [PubMed]

57. Crisan, M.; Dzierzak, E. The many faces of hematopoietic stem cell heterogeneity. *Development* **2016**, *143*, 4571–4581. [CrossRef] [PubMed]

58. Eaves, C.J. Hematopoietic stem cells: Concepts, definitions, and the new reality. *Blood* **2015**, *125*, 2605–2613. [CrossRef] [PubMed]

59. Notta, F.; Zandi, S.; Takayama, N.; Dobson, S.; Gan, O.I.; Wilson, G.; Kaufmann, K.B.; McLeod, J.; Laurenti, E.; Dunant, C.F.; et al. Distinct routes of lineage development reshape the human blood hierarchy across ontogeny. *Science* **2016**, *351*, aab2116. [CrossRef] [PubMed]

60. Endele, M.; Etzrodt, M.; Schroeder, T. Instruction of hematopoietic lineage choice by cytokine signaling. *Exp. Cell Res.* **2014**, *329*, 207–213. [CrossRef] [PubMed]

61. Sarrazin, S.; Sieweke, M. Integration of cytokine and transcription factor signals in hematopoietic stem cell commitment. *Semin Immunol.* **2011**, *23*, 326–334. [CrossRef] [PubMed]

62. Ceredig, R.; Rauch, M.; Balciunaite, G.; Rolink, A.G. Increasing Flt3L availability alters composition of a novel bone marrow lymphoid progenitor compartment. *Blood* **2006**, *108*, 1216–1222. [CrossRef] [PubMed]

63. Swee, L.K.; Bosco, N.; Malissen, B.; Ceredig, R.; Rolink, A. Expansion of peripheral naturally occurring T regulatory cells by Fms-like tyrosine kinase 3 ligand treatment. *Blood* **2009**, *113*, 6277–6287. [CrossRef] [PubMed]

64. Tsapogas, P.; Swee, L.K.; Nusser, A.; Nuber, N.; Kreuzaler, M.; Capoferri, G.; Rolink, H.; Ceredig, R.; Rolink, A. In vivo evidence for an instructive role of fms-like tyrosine kinase-3 (FLT3) ligand in hematopoietic development. *Haematologica* **2014**, *99*, 638–646. [CrossRef] [PubMed]

65. Grover, A.; Mancini, E.; Moore, S.; Mead, A.J.; Atkinson, D.; Rasmussen, K.D.; O'Carroll, D.; Jacobsen, S.E.; Nerlov, C. Erythropoietin guides multipotent hematopoietic progenitor cells toward an erythroid fate. *J. Exp. Med.* **2014**, *211*, 181–188. [CrossRef] [PubMed]

66. Mossadegh-Keller, N.; Sarrazin, S.; Kandalla, P.K.; Espinosa, L.; Stanley, E.R.; Nutt, S.L.; Moore, J.; Sieweke, M.H. M-CSF instructs myeloid lineage fate in single haematopoietic stem cells. *Nature* **2013**, *497*, 239–243. [CrossRef] [PubMed]

67. Rieger, M.A.; Hoppe, P.S.; Smejkal, B.M.; Eitelhuber, A.C.; Schroeder, T. Hematopoietic cytokines can instruct lineage choice. *Science* **2009**, *325*, 217–218. [CrossRef] [PubMed]

68. Waskow, C.; Liu, K.; Darrasse-Jeze, G.; Guermonprez, P.; Ginhoux, F.; Merad, M.; Shengelia, T.; Yao, K.; Nussenzweig, M. The receptor tyrosine kinase Flt3 is required for dendritic cell development in peripheral lymphoid tissues. *Nat. Immunol.* **2008**, *9*, 676–683. [CrossRef] [PubMed]

69. Gilliet, M.; Boonstra, A.; Paturel, C.; Antonenko, S.; Xu, X.L.; Trinchieri, G.; O'Garra, A.; Liu, Y.J. The development of murine plasmacytoid dendritic cell precursors is differentially regulated by FLT3-ligand and granulocyte/macrophage colony-stimulating factor. *J. Exp. Med.* **2002**, *195*, 953–958. [CrossRef] [PubMed]

70. Karsunky, H.; Merad, M.; Cozzio, A.; Weissman, I.L.; Manz, M.G. Flt3 ligand regulates dendritic cell development from Flt3+ lymphoid and myeloid-committed progenitors to Flt3+ dendritic cells in vivo. *J. Exp. Med.* **2003**, *198*, 305–313. [CrossRef] [PubMed]

71. Brasel, K.; De Smedt, T.; Smith, J.L.; Maliszewski, C.R. Generation of murine dendritic cells from flt3-ligand-supplemented bone marrow cultures. *Blood* **2000**, *96*, 3029–3039. [PubMed]

72. Brawand, P.; Fitzpatrick, D.R.; Greenfield, B.W.; Brasel, K.; Maliszewski, C.R.; De Smedt, T. Murine plasmacytoid pre-dendritic cells generated from Flt3 ligand-supplemented bone marrow cultures are immature APCs. *J. Immunol.* **2002**, *169*, 6711–6719. [CrossRef] [PubMed]

73. Naik, S.H.; Proietto, A.I.; Wilson, N.S.; Dakic, A.; Schnorrer, P.; Fuchsberger, M.; Lahoud, M.H.; O'Keeffe, M.; Shao, Q.X.; Chen, W.F.; et al. Cutting edge: Generation of splenic CD8+ and CD8− dendritic cell equivalents in Fms-like tyrosine kinase 3 ligand bone marrow cultures. *J. Immunol.* **2005**, *174*, 6592–6597. [CrossRef] [PubMed]

74. Daro, E.; Pulendran, B.; Brasel, K.; Teepe, M.; Pettit, D.; Lynch, D.H.; Vremec, D.; Robb, L.; Shortman, K.; McKenna, H.J.; et al. Polyethylene glycol-modified GM-CSF expands CD11b^{high}CD11c^{high} but notCD11b^{low}CD11c^{high} murine dendritic cells in vivo: A comparative analysis with Flt3 ligand. *J. Immunol.* **2000**, *165*, 49–58. [CrossRef] [PubMed]

75. Maraskovsky, E.; Brasel, K.; Teepe, M.; Roux, E.R.; Lyman, S.D.; Shortman, K.; McKenna, H.J. Dramatic increase in the numbers of functionally mature dendritic cells in Flt3 ligand-treated mice: Multiple dendritic cell subpopulations identified. *J. Exp. Med.* **1996**, *184*, 1953–1962. [CrossRef] [PubMed]

76. O'Keeffe, M.; Hochrein, H.; Vremec, D.; Pooley, J.; Evans, R.; Woulfe, S.; Shortman, K. Effects of administration of progenipoietin 1, Flt-3 ligand, granulocyte colony-stimulating factor, and pegylated granulocyte-macrophage colony-stimulating factor on dendritic cell subsets in mice. *Blood* **2002**, *99*, 2122–2130. [CrossRef] [PubMed]

77. Manfra, D.J.; Chen, S.C.; Jensen, K.K.; Fine, J.S.; Wiekowski, M.T.; Lira, S.A. Conditional expression of murine Flt3 ligand leads to expansion of multiple dendritic cell subsets in peripheral blood and tissues of transgenic mice. *J. Immunol.* **2003**, *170*, 2843–2852. [CrossRef] [PubMed]

78. Juan, T.S.; McNiece, I.K.; Van, G.; Lacey, D.; Hartley, C.; McElroy, P.; Sun, Y.; Argento, J.; Hill, D.; Yan, X.Q.; et al. Chronic expression of murine flt3 ligand in mice results in increased circulating white blood cell levels and abnormal cellular infiltrates associated with splenic fibrosis. *Blood* **1997**, *90*, 76–84. [PubMed]

79. Liu, K.; Victora, G.D.; Schwickert, T.A.; Guermonprez, P.; Meredith, M.M.; Yao, K.; Chu, F.F.; Randolph, G.J.; Rudensky, A.Y.; Nussenzweig, M. In vivo analysis of dendritic cell development and homeostasis. *Science* **2009**, *324*, 392–397. [CrossRef] [PubMed]

80. Onai, N.; Kurabayashi, K.; Hosoi-Amaike, M.; Toyama-Sorimachi, N.; Matsushima, K.; Inaba, K.; Ohteki, T. A clonogenic progenitor with prominent plasmacytoid dendritic cell developmental potential. *Immunity* **2013**, *38*, 943–957. [CrossRef] [PubMed]

81. Chakravarty, P.K.; Alfieri, A.; Thomas, E.K.; Beri, V.; Tanaka, K.E.; Vikram, B.; Guha, C. Flt3-ligand administration after radiation therapy prolongs survival in a murine model of metastatic lung cancer. *Cancer Res.* **1999**, *59*, 6028–6032. [PubMed]

82. Chen, K.; Braun, S.; Lyman, S.; Fan, Y.; Traycoff, C.M.; Wiebke, E.A.; Gaddy, J.; Sledge, G.; Broxmeyer, H.E.; Cornetta, K. Antitumor activity and immunotherapeutic properties of Flt3-ligand in a murine breast cancer model. *Cancer Res.* **1997**, *57*, 3511–3516. [PubMed]

83. Lynch, D.H.; Andreasen, A.; Maraskovsky, E.; Whitmore, J.; Miller, R.E.; Schuh, J.C. Flt3 ligand induces tumor regression and antitumor immune responses in vivo. *Nat. Med.* **1997**, *3*, 625–631. [CrossRef] [PubMed]

84. Disis, M.L.; Rinn, K.; Knutson, K.L.; Davis, D.; Caron, D.; dela Rosa, C.; Schiffman, K. Flt3 ligand as a vaccine adjuvant in association with HER-2/neu peptide-based vaccines in patients with HER-2/neu-overexpressing cancers. *Blood* **2002**, *99*, 2845–2850. [CrossRef] [PubMed]

85. Fong, L.; Hou, Y.; Rivas, A.; Benike, C.; Yuen, A.; Fisher, G.A.; Davis, M.M.; Engleman, E.G. Altered peptide ligand vaccination with Flt3 ligand expanded dendritic cells for tumor immunotherapy. *Proc. Natl. Acad. Sci. USA* **2001**, *98*, 8809–8814. [CrossRef] [PubMed]

86. Buza-Vidas, N.; Cheng, M.; Duarte, S.; Nozad, H.; Jacobsen, S.E.; Sitnicka, E. Crucial role of FLT3 ligand in immune reconstitution after bone marrow transplantation and high-dose chemotherapy. *Blood* **2007**, *110*, 424–432. [CrossRef] [PubMed]

87. Jensen, C.T.; Kharazi, S.; Boiers, C.; Cheng, M.; Lubking, A.; Sitnicka, E.; Jacobsen, S.E. FLT3 ligand and not TSLP is the key regulator of IL-7-independent B-1 and B-2 B lymphopoiesis. *Blood* **2008**, *112*, 2297–2304. [CrossRef] [PubMed]

88. Nutt, S.L.; Heavey, B.; Rolink, A.G.; Busslinger, M. Commitment to the B-lymphoid lineage depends on the transcription factor Pax5. *Nature* **1999**, *401*, 556–562. [PubMed]

89. Rolink, A.G.; Nutt, S.L.; Melchers, F.; Busslinger, M. Long-term in vivo reconstitution of T-cell development by Pax5-deficient B-cell progenitors. *Nature* **1999**, *401*, 603–606. [CrossRef] [PubMed]

90. Holmes, M.L.; Carotta, S.; Corcoran, L.M.; Nutt, S.L. Repression of Flt3 by Pax5 is crucial for B-cell lineage commitment. *Genes Dev.* **2006**, *20*, 933–938. [CrossRef] [PubMed]

91. Inlay, M.A.; Bhattacharya, D.; Sahoo, D.; Serwold, T.; Seita, J.; Karsunky, H.; Plevritis, S.K.; Dill, D.L.; Weissman, I.L. Ly6d marks the earliest stage of B-cell specification and identifies the branchpoint between B-cell and T-cell development. *Genes Dev.* **2009**, *23*, 2376–2381. [CrossRef] [PubMed]

92. Mansson, R.; Zandi, S.; Welinder, E.; Tsapogas, P.; Sakaguchi, N.; Bryder, D.; Sigvardsson, M. Single-cell analysis of the common lymphoid progenitor compartment reveals functional and molecular heterogeneity. *Blood* **2010**, *115*, 2601–2609. [CrossRef] [PubMed]

93. O'Reilly, L.A.; Huang, D.C.; Strasser, A. The cell death inhibitor Bcl-2 and its homologues influence control of cell cycle entry. *EMBO J.* **1996**, *15*, 6979–6990. [PubMed]

94. Sitnicka, E.; Buza-Vidas, N.; Ahlenius, H.; Cilio, C.M.; Gekas, C.; Nygren, J.M.; Mansson, R.; Cheng, M.; Jensen, C.T.; Svensson, M.; et al. Critical role of FLT3 ligand in IL-7 receptor independent T lymphopoiesis and regulation of lymphoid-primed multipotent progenitors. *Blood* **2007**, *110*, 2955–2964. [CrossRef] [PubMed]

95. Luc, S.; Luis, T.C.; Boukarabila, H.; Macaulay, I.C.; Buza-Vidas, N.; Bouriez-Jones, T.; Lutteropp, M.; Woll, P.S.; Loughran, S.J.; Mead, A.J.; et al. The earliest thymic T cell progenitors sustain B cell and myeloid lineage potential. *Nat. Immunol.* **2012**, *13*, 412–419. [CrossRef] [PubMed]

96. Wils, E.J.; Braakman, E.; Verjans, G.M.; Rombouts, E.J.; Broers, A.E.; Niesters, H.G.; Wagemaker, G.; Staal, F.J.; Lowenberg, B.; Spits, H.; et al. Flt3 ligand expands lymphoid progenitors prior to recovery of thymopoiesis and accelerates T cell reconstitution after bone marrow transplantation. *J. Immunol.* **2007**, *178*, 3551–3557. [CrossRef] [PubMed]

97. Kenins, L.; Gill, J.W.; Boyd, R.L.; Hollander, G.A.; Wodnar-Filipowicz, A. Intrathymic expression of Flt3 ligand enhances thymic recovery after irradiation. *J. Exp. Med.* **2008**, *205*, 523–531. [CrossRef] [PubMed]

98. Kenins, L.; Gill, J.W.; Hollander, G.A.; Wodnar-Filipowicz, A. Flt3 ligand-receptor interaction is important for maintenance of early thymic progenitor numbers in steady-state thymopoiesis. *Eur. J. Immunol.* **2010**, *40*, 81–90. [CrossRef] [PubMed]

99. Baerenwaldt, A.; von Burg, N.; Kreuzaler, M.; Sitte, S.; Horvath, E.; Peter, A.; Voehringer, D.; Rolink, A.G.; Finke, D. Flt3 ligand regulates the development of innate lymphoid cells in fetal and adult mice. *J. Immunol.* **2016**, *196*, 2561–2571. [CrossRef] [PubMed]

100. Shaw, S.G.; Maung, A.A.; Steptoe, R.J.; Thomson, A.W.; Vujanovic, N.L. Expansion of functional NK cells in multiple tissue compartments of mice treated with Flt3-ligand: Implications for anti-cancer and anti-viral therapy. *J. Immunol.* **1998**, *161*, 2817–2824. [PubMed]

101. Kallies, A.; Hasbold, J.; Fairfax, K.; Pridans, C.; Emslie, D.; McKenzie, B.S.; Lew, A.M.; Corcoran, L.M.; Hodgkin, P.D.; Tarlinton, D.M.; et al. Initiation of plasma-cell differentiation is independent of the transcription factor Blimp-1. *Immunity* **2007**, *26*, 555–566. [CrossRef] [PubMed]

102. Svensson, M.N.; Andersson, K.M.; Wasen, C.; Erlandsson, M.C.; Nurkkala-Karlsson, M.; Jonsson, I.M.; Brisslert, M.; Bemark, M.; Bokarewa, M.I. Murine germinal center B cells require functional Fms-like tyrosine kinase 3 signaling for IgG1 class-switch recombination. *Proc. Natl. Acad. Sci. USA* **2015**, *112*, E6644–E6653. [CrossRef] [PubMed]

103. Astier, A.L.; Beriou, G.; Eisenhaure, T.M.; Anderton, S.M.; Hafler, D.A.; Hacohen, N. RNA interference screen in primary human T cells reveals FLT3 as a modulator of IL-10 levels. *J. Immunol.* **2010**, *184*, 685–693. [CrossRef] [PubMed]

104. Gilliland, D.G.; Griffin, J.D. The roles of FLT3 in hematopoiesis and leukemia. *Blood* **2002**, *100*, 1532–1542. [CrossRef] [PubMed]

105. Annesley, C.E.; Brown, P. The biology and targeting of FLT3 in pediatric leukemia. *Front. Oncol.* **2014**, *4*, 263. [CrossRef] [PubMed]

106. Nakao, M.; Yokota, S.; Iwai, T.; Kaneko, H.; Horiike, S.; Kashima, K.; Sonoda, Y.; Fujimoto, T.; Misawa, S. Internal tandem duplication of the flt3 gene found in acute myeloid leukemia. *Leukemia* **1996**, *10*, 1911–1918. [PubMed]

107. Meshinchi, S.; Woods, W.G.; Stirewalt, D.L.; Sweetser, D.A.; Buckley, J.D.; Tjoa, T.K.; Bernstein, I.D.; Radich, J.P. Prevalence and prognostic significance of Flt3 internal tandem duplication in pediatric acute myeloid leukemia. *Blood* **2001**, *97*, 89–94. [CrossRef] [PubMed]

108. Schnittger, S.; Schoch, C.; Dugas, M.; Kern, W.; Staib, P.; Wuchter, C.; Loffler, H.; Sauerland, C.M.; Serve, H.; Buchner, T.; et al. Analysis of FLT3 length mutations in 1003 patients with acute myeloid leukemia: Correlation to cytogenetics, FAB subtype, and prognosis in the AMLCG study and usefulness as a marker for the detection of minimal residual disease. *Blood* **2002**, *100*, 59–66. [CrossRef] [PubMed]

109. Thiede, C.; Steudel, C.; Mohr, B.; Schaich, M.; Schakel, U.; Platzbecker, U.; Wermke, M.; Bornhauser, M.; Ritter, M.; Neubauer, A.; et al. Analysis of FLT3-activating mutations in 979 patients with acute myelogenous leukemia: Association with FAB subtypes and identification of subgroups with poor prognosis. *Blood* **2002**, *99*, 4326–4335. [CrossRef] [PubMed]

110. Kiyoi, H.; Ohno, R.; Ueda, R.; Saito, H.; Naoe, T. Mechanism of constitutive activation of FLT3 with internal tandem duplication in the juxtamembrane domain. *Oncogene* **2002**, *21*, 2555–2563. [CrossRef] [PubMed]

111. Kiyoi, H.; Towatari, M.; Yokota, S.; Hamaguchi, M.; Ohno, R.; Saito, H.; Naoe, T. Internal tandem duplication of the FLT3 gene is a novel modality of elongation mutation which causes constitutive activation of the product. *Leukemia* **1998**, *12*, 1333–1337. [CrossRef] [PubMed]

112. Hayakawa, F.; Towatari, M.; Kiyoi, H.; Tanimoto, M.; Kitamura, T.; Saito, H.; Naoe, T. Tandem-duplicated Flt3 constitutively activates STAT5 and MAP kinase and introduces autonomous cell growth in IL-3-dependent cell lines. *Oncogene* **2000**, *19*, 624–631. [CrossRef] [PubMed]

113. Stirewalt, D.L.; Radich, J. The role of FLT3 in haematopoietic malignancies. *Nat. Rev. Cancer* **2003**, *3*, 650–665. [CrossRef] [PubMed]

114. Mizuki, M.; Fenski, R.; Halfter, H.; Matsumura, I.; Schmidt, R.; Muller, C.; Gruning, W.; Kratz-Albers, K.; Serve, S.; Steur, C.; et al. Flt3 mutations from patients with acute myeloid leukemia induce transformation of 32D cells mediated by the Ras and STAT5 pathways. *Blood* **2000**, *96*, 3907–3914. [PubMed]

115. Mizuki, M.; Schwable, J.; Steur, C.; Choudhary, C.; Agrawal, S.; Sargin, B.; Steffen, B.; Matsumura, I.; Kanakura, Y.; Bohmer, F.D.; et al. Suppression of myeloid transcription factors and induction of STAT response genes by AML-specific Flt3 mutations. *Blood* **2003**, *101*, 3164–3173. [CrossRef] [PubMed]

116. Zheng, R.; Friedman, A.D.; Levis, M.; Li, L.; Weir, E.G.; Small, D. Internal tandem duplication mutation of FLT3 blocks myeloid differentiation through suppression of C/EBPα expression. *Blood* **2004**, *103*, 1883–1890. [CrossRef] [PubMed]

117. Fan, J.; Li, L.; Small, D.; Rassool, F. Cells expressing FLT3/ITD mutations exhibit elevated repair errors generated through alternative NHEJ pathways: Implications for genomic instability and therapy. *Blood* **2010**, *116*, 5298–5305. [CrossRef] [PubMed]

118. Sallmyr, A.; Fan, J.; Datta, K.; Kim, K.T.; Grosu, D.; Shapiro, P.; Small, D.; Rassool, F. Internal tandem duplication of FLT3 (FLT3/ITD) induces increased ROS production, DNA damage, and misrepair: Implications for poor prognosis in AML. *Blood* **2008**, *111*, 3173–3182. [CrossRef] [PubMed]

119. Scheijen, B.; Ngo, H.T.; Kang, H.; Griffin, J.D. FLT3 receptors with internal tandem duplications promote cell viability and proliferation by signaling through Foxo proteins. *Oncogene* **2004**, *23*, 3338–3349. [CrossRef] [PubMed]

120. Levis, M.; Allebach, J.; Tse, K.F.; Zheng, R.; Baldwin, B.R.; Smith, B.D.; Jones-Bolin, S.; Ruggeri, B.; Dionne, C.; Small, D. A FLT3-targeted tyrosine kinase inhibitor is cytotoxic to leukemia cells in vitro and in vivo. *Blood* **2002**, *99*, 3885–3891. [CrossRef] [PubMed]

121. Grundler, R.; Miething, C.; Thiede, C.; Peschel, C.; Duyster, J. FLT3-ITD and tyrosine kinase domain mutants induce 2 distinct phenotypes in a murine bone marrow transplantation model. *Blood* **2005**, *105*, 4792–4799. [CrossRef] [PubMed]

122. Kelly, L.M.; Liu, Q.; Kutok, J.L.; Williams, I.R.; Boulton, C.L.; Gilliland, D.G. FLT3 internal tandem duplication mutations associated with human acute myeloid leukemias induce myeloproliferative disease in a murine bone marrow transplant model. *Blood* **2002**, *99*, 310–318. [CrossRef] [PubMed]

123. Lee, B.H.; Williams, I.R.; Anastasiadou, E.; Boulton, C.L.; Joseph, S.W.; Amaral, S.M.; Curley, D.P.; Duclos, N.; Huntly, B.J.; Fabbro, D.; et al. FLT3 internal tandem duplication mutations induce myeloproliferative or lymphoid disease in a transgenic mouse model. *Oncogene* **2005**, *24*, 7882–7892. [CrossRef] [PubMed]

124. Lee, B.H.; Tothova, Z.; Levine, R.L.; Anderson, K.; Buza-Vidas, N.; Cullen, D.E.; McDowell, E.P.; Adelsperger, J.; Frohling, S.; Huntly, B.J.; et al. FLT3 mutations confer enhanced proliferation and survival properties to multipotent progenitors in a murine model of chronic myelomonocytic leukemia. *Cancer Cell* **2007**, *12*, 367–380. [CrossRef] [PubMed]

125. Li, L.; Piloto, O.; Nguyen, H.B.; Greenberg, K.; Takamiya, K.; Racke, F.; Huso, D.; Small, D. Knock-in of an internal tandem duplication mutation into murine FLT3 confers myeloproliferative disease in a mouse model. *Blood* **2008**, *111*, 3849–3858. [CrossRef] [PubMed]

126. Chu, S.H.; Heiser, D.; Li, L.; Kaplan, I.; Collector, M.; Huso, D.; Sharkis, S.J.; Civin, C.; Small, D. FLT3-ITD knockin impairs hematopoietic stem cell quiescence/homeostasis, leading to myeloproliferative neoplasm. *Cell Stem Cell* **2012**, *11*, 346–358. [CrossRef] [PubMed]

127. Abu-Duhier, F.M.; Goodeve, A.C.; Wilson, G.A.; Care, R.S.; Peake, I.R.; Reilly, J.T. Identification of novel FLT-3 Asp835 mutations in adult acute myeloid leukaemia. *Br. J. Haematol.* **2001**, *113*, 983–988. [CrossRef] [PubMed]

128. Yamamoto, Y.; Kiyoi, H.; Nakano, Y.; Suzuki, R.; Kodera, Y.; Miyawaki, S.; Asou, N.; Kuriyama, K.; Yagasaki, F.; Shimazaki, C.; et al. Activating mutation of D835 within the activation loop of FLT3 in human hematologic malignancies. *Blood* **2001**, *97*, 2434–2439. [CrossRef] [PubMed]

129. Griffith, J.; Black, J.; Faerman, C.; Swenson, L.; Wynn, M.; Lu, F.; Lippke, J.; Saxena, K. The structural basis for autoinhibition of FLT3 by the juxtamembrane domain. *Mol. Cell* **2004**, *13*, 169–178. [CrossRef]

130. Choudhary, C.; Schwable, J.; Brandts, C.; Tickenbrock, L.; Sargin, B.; Kindler, T.; Fischer, T.; Berdel, W.E.; Muller-Tidow, C.; Serve, H. AML-associated Flt3 kinase domain mutations show signal transduction differences compared with Flt3 ITD mutations. *Blood* **2005**, *106*, 265–273. [CrossRef] [PubMed]

131. Lacayo, N.J.; Meshinchi, S.; Kinnunen, P.; Yu, R.; Wang, Y.; Stuber, C.M.; Douglas, L.; Wahab, R.; Becton, D.L.; Weinstein, H.; et al. Gene expression profiles at diagnosis in de novo childhood AML patients identify FLT3 mutations with good clinical outcomes. *Blood* **2004**, *104*, 2646–2654. [CrossRef] [PubMed]

132. Koch, S.; Jacobi, A.; Ryser, M.; Ehninger, G.; Thiede, C. Abnormal localization and accumulation of FLT3-ITD, a mutant receptor tyrosine kinase involved in leukemogenesis. *Cells Tissues Organs* **2008**, *188*, 225–235. [CrossRef] [PubMed]

133. Choudhary, C.; Olsen, J.V.; Brandts, C.; Cox, J.; Reddy, P.N.; Bohmer, F.D.; Gerke, V.; Schmidt-Arras, D.E.; Berdel, W.E.; Muller-Tidow, C.; et al. Mislocalized activation of oncogenic RTKs switches downstream signaling outcomes. *Mol. Cell* **2009**, *36*, 326–339. [CrossRef] [PubMed]

134. Chan, P.M. Differential signaling of Flt3 activating mutations in acute myeloid leukemia: A working model. *Protein Cell* **2011**, *2*, 108–115. [CrossRef] [PubMed]

135. Muller, T.A.; Grundler, R.; Istvanffy, R.; Rudelius, M.; Hennighausen, L.; Illert, A.L.; Duyster, J. Lineage-specific STAT5 target gene activation in hematopoietic progenitor cells predicts the FLT3$^+$-mediated leukemic phenotype. *Leukemia* **2016**, *30*, 1725–1733. [CrossRef] [PubMed]

136. Bailey, E.; Li, L.; Duffield, A.S.; Ma, H.S.; Huso, D.L.; Small, D. FLT3/D835Y mutation knock-in mice display less aggressive disease compared with FLT3/internal tandem duplication (ITD) mice. *Proc. Natl. Acad. Sci. USA* **2013**, *110*, 21113–21118. [CrossRef] [PubMed]

137. Birg, F.; Courcoul, M.; Rosnet, O.; Bardin, F.; Pebusque, M.J.; Marchetto, S.; Tabilio, A.; Mannoni, P.; Birnbaum, D. Expression of the FMS/KIT-like gene FLT3 in human acute leukemias of the myeloid and lymphoid lineages. *Blood* **1992**, *80*, 2584–2593. [PubMed]

138. Carow, C.E.; Levenstein, M.; Kaufmann, S.H.; Chen, J.; Amin, S.; Rockwell, P.; Witte, L.; Borowitz, M.J.; Civin, C.I.; Small, D. Expression of the hematopoietic growth factor receptor FLT3 (STK-1/Flk2) in human leukemias. *Blood* **1996**, *87*, 1089–1096. [PubMed]

139. Rosnet, O.; Buhring, H.J.; Marchetto, S.; Rappold, I.; Lavagna, C.; Sainty, D.; Arnoulet, C.; Chabannon, C.; Kanz, L.; Hannum, C.; et al. Human FLT3/FLK2 receptor tyrosine kinase is expressed at the surface of normal and malignant hematopoietic cells. *Leukemia* **1996**, *10*, 238–248. [PubMed]

140. Stacchini, A.; Fubini, L.; Severino, A.; Sanavio, F.; Aglietta, M.; Piacibello, W. Expression of type III receptor tyrosine kinases FLT3 and KIT and responses to their ligands by acute myeloid leukemia blasts. *Leukemia* **1996**, *10*, 1584–1591. [PubMed]

141. Lemmon, M.A.; Schlessinger, J. Cell signaling by receptor tyrosine kinases. *Cell* **2010**, *141*, 1117–1134. [CrossRef] [PubMed]

142. Ozeki, K.; Kiyoi, H.; Hirose, Y.; Iwai, M.; Ninomiya, M.; Kodera, Y.; Miyawaki, S.; Kuriyama, K.; Shimazaki, C.; Akiyama, H.; et al. Biologic and clinical significance of the FLT3 transcript level in acute myeloid leukemia. *Blood* **2004**, *103*, 1901–1908. [CrossRef] [PubMed]

143. Zheng, R.; Levis, M.; Piloto, O.; Brown, P.; Baldwin, B.R.; Gorin, N.C.; Beran, M.; Zhu, Z.; Ludwig, D.; Hicklin, D.; et al. FLT3 ligand causes autocrine signaling in acute myeloid leukemia cells. *Blood* **2004**, *103*, 267–274. [CrossRef] [PubMed]

144. Pratz, K.W.; Sato, T.; Murphy, K.M.; Stine, A.; Rajkhowa, T.; Levis, M. FLT3-mutant allelic burden and clinical status are predictive of response to FLT3 inhibitors in AML. *Blood* **2010**, *115*, 1425–1432. [CrossRef] [PubMed]

145. Fabbro, D.; Buchdunger, E.; Wood, J.; Mestan, J.; Hofmann, F.; Ferrari, S.; Mett, H.; O'Reilly, T.; Meyer, T. Inhibitors of protein kinases: CGP 41251, a protein kinase inhibitor with potential as an anticancer agent. *Pharmacol. Ther.* **1999**, *82*, 293–301. [CrossRef]

146. Hexner, E.O.; Serdikoff, C.; Jan, M.; Swider, C.R.; Robinson, C.; Yang, S.; Angeles, T.; Emerson, S.G.; Carroll, M.; Ruggeri, B.; et al. Lestaurtinib (CEP701) is a JAK2 inhibitor that suppresses JAK2/STAT5 signaling and the proliferation of primary erythroid cells from patients with myeloproliferative disorders. *Blood* **2008**, *111*, 5663–5671. [CrossRef] [PubMed]

147. Mendel, D.B.; Laird, A.D.; Xin, X.; Louie, S.G.; Christensen, J.G.; Li, G.; Schreck, R.E.; Abrams, T.J.; Ngai, T.J.; Lee, L.B.; et al. In vivo antitumor activity of SU11248, a novel tyrosine kinase inhibitor targeting vascular endothelial growth factor and platelet-derived growth factor receptors: Determination of a pharmacokinetic/pharmacodynamic relationship. *Clin. Cancer Res.* **2003**, *9*, 327–337. [PubMed]

148. O'Farrell, A.M.; Abrams, T.J.; Yuen, H.A.; Ngai, T.J.; Louie, S.G.; Yee, K.W.; Wong, L.M.; Hong, W.; Lee, L.B.; Town, A.; et al. SU11248 is a novel FLT3 tyrosine kinase inhibitor with potent activity in vitro and in vivo. *Blood* **2003**, *101*, 3597–3605. [CrossRef] [PubMed]

149. Demetri, G.D.; van Oosterom, A.T.; Garrett, C.R.; Blackstein, M.E.; Shah, M.H.; Verweij, J.; McArthur, G.; Judson, I.R.; Heinrich, M.C.; Morgan, J.A.; et al. Efficacy and safety of sunitinib in patients with advanced gastrointestinal stromal tumour after failure of imatinib: A randomised controlled trial. *Lancet* **2006**, *368*, 1329–1338. [CrossRef]

150. Escudier, B.; Eisen, T.; Stadler, W.M.; Szczylik, C.; Oudard, S.; Siebels, M.; Negrier, S.; Chevreau, C.; Solska, E.; Desai, A.A.; et al. Sorafenib in advanced clear-cell renal-cell carcinoma. *N. Engl. J. Med.* **2007**, *356*, 125–134. [CrossRef] [PubMed]

151. Hsieh, J.J.; Purdue, M.P.; Signoretti, S.; Swanton, C.; Albiges, L.; Schmidinger, M.; Heng, D.Y.; Larkin, J.; Ficarra, V. Renal cell carcinoma. *Nat. Rev. Dis. Primers* **2017**, *3*, 17009. [CrossRef] [PubMed]

152. Llovet, J.M.; Ricci, S.; Mazzaferro, V.; Hilgard, P.; Gane, E.; Blanc, J.F.; de Oliveira, A.C.; Santoro, A.; Raoul, J.L.; Forner, A.; et al. Sorafenib in advanced hepatocellular carcinoma. *N. Engl. J. Med.* **2008**, *359*, 378–390. [CrossRef] [PubMed]

153. Motzer, R.J.; Hutson, T.E.; Tomczak, P.; Michaelson, M.D.; Bukowski, R.M.; Rixe, O.; Oudard, S.; Negrier, S.; Szczylik, C.; Kim, S.T.; et al. Sunitinib versus interferon alfa in metastatic renal-cell carcinoma. *N. Engl. J. Med.* **2007**, *356*, 115–124. [CrossRef] [PubMed]

154. O'Farrell, A.M.; Foran, J.M.; Fiedler, W.; Serve, H.; Paquette, R.L.; Cooper, M.A.; Yuen, H.A.; Louie, S.G.; Kim, H.; Nicholas, S.; et al. An innovative phase I clinical study demonstrates inhibition of FLT3 phosphorylation by SU11248 in acute myeloid leukemia patients. *Clin. Cancer Res.* **2003**, *9*, 5465–5476. [PubMed]

155. Fiedler, W.; Kayser, S.; Kebenko, M.; Janning, M.; Krauter, J.; Schittenhelm, M.; Gotze, K.; Weber, D.; Gohring, G.; Teleanu, V.; et al. A phase I/II study of sunitinib and intensive chemotherapy in patients over 60 years of age with acute myeloid leukaemia and activating FLT3 mutations. *Br. J. Haematol.* **2015**, *169*, 694–700. [CrossRef] [PubMed]

156. Fiedler, W.; Serve, H.; Dohner, H.; Schwittay, M.; Ottmann, O.G.; O'Farrell, A.M.; Bello, C.L.; Allred, R.; Manning, W.C.; Cherrington, J.M.; et al. A phase 1 study of SU11248 in the treatment of patients with refractory or resistant acute myeloid leukemia (AML) or not amenable to conventional therapy for the disease. *Blood* **2005**, *105*, 986–993. [CrossRef] [PubMed]

157. Metzelder, S.; Wang, Y.; Wollmer, E.; Wanzel, M.; Teichler, S.; Chaturvedi, A.; Eilers, M.; Enghofer, E.; Neubauer, A.; Burchert, A. Compassionate use of sorafenib in FLT3-ITD-positive acute myeloid leukemia: Sustained regression before and after allogeneic stem cell transplantation. *Blood* **2009**, *113*, 6567–6571. [CrossRef] [PubMed]

158. Sharma, M.; Ravandi, F.; Bayraktar, U.D.; Chiattone, A.; Bashir, Q.; Giralt, S.; Chen, J.; Qazilbash, M.; Kebriaei, P.; Konopleva, M.; et al. Treatment of FLT3-ITD-positive acute myeloid leukemia relapsing after allogeneic stem cell transplantation with sorafenib. *Biol. Blood Marrow Transplant.* **2011**, *17*, 1874–1877. [CrossRef] [PubMed]

159. Zhang, W.; Konopleva, M.; Shi, Y.X.; McQueen, T.; Harris, D.; Ling, X.; Estrov, Z.; Quintas-Cardama, A.; Small, D.; Cortes, J.; et al. Mutant FLT3: A direct target of sorafenib in acute myelogenous leukemia. *J. Natl. Cancer Inst.* **2008**, *100*, 184–198. [CrossRef] [PubMed]

160. Serve, H.; Krug, U.; Wagner, R.; Sauerland, M.C.; Heinecke, A.; Brunnberg, U.; Schaich, M.; Ottmann, O.; Duyster, J.; Wandt, H.; et al. Sorafenib in combination with intensive chemotherapy in elderly patients with acute myeloid leukemia: Results from a randomized, placebo-controlled trial. *J. Clin. Oncol.* **2013**, *31*, 3110–3118. [CrossRef] [PubMed]

161. Ravandi, F.; Cortes, J.E.; Jones, D.; Faderl, S.; Garcia-Manero, G.; Konopleva, M.Y.; O'Brien, S.; Estrov, Z.; Borthakur, G.; Thomas, D.; et al. Phase I/II study of combination therapy with sorafenib, idarubicin, and cytarabine in younger patients with acute myeloid leukemia. *J. Clin. Oncol.* **2010**, *28*, 1856–1862. [CrossRef] [PubMed]

162. Rollig, C.; Serve, H.; Huttmann, A.; Noppeney, R.; Muller-Tidow, C.; Krug, U.; Baldus, C.D.; Brandts, C.H.; Kunzmann, V.; Einsele, H.; et al. Addition of sorafenib versus placebo to standard therapy in patients aged 60 years or younger with newly diagnosed acute myeloid leukaemia (SORAML): A multicentre, phase 2, randomised controlled trial. *Lancet Oncol.* **2015**, *16*, 1691–1699. [CrossRef]

163. Antar, A.; Kharfan-Dabaja, M.A.; Mahfouz, R.; Bazarbachi, A. Sorafenib Maintenance Appears Safe and Improves Clinical Outcomes in FLT3-ITD Acute Myeloid Leukemia After Allogeneic Hematopoietic Cell Transplantation. *Clin. Lymphoma Myeloma Leuk.* **2015**, *15*, 298–302. [CrossRef] [PubMed]

164. Metzelder, S.K.; Schroeder, T.; Finck, A.; Scholl, S.; Fey, M.; Gotze, K.; Linn, Y.C.; Kroger, M.; Reiter, A.; Salih, H.R.; et al. High activity of sorafenib in FLT3-ITD-positive acute myeloid leukemia synergizes with allo-immune effects to induce sustained responses. *Leukemia* **2012**, *26*, 2353–2359. [CrossRef] [PubMed]

165. Stone, R.M.; DeAngelo, D.J.; Klimek, V.; Galinsky, I.; Estey, E.; Nimer, S.D.; Grandin, W.; Lebwohl, D.; Wang, Y.; Cohen, P.; et al. Patients with acute myeloid leukemia and an activating mutation in FLT3 respond to a small-molecule FLT3 tyrosine kinase inhibitor, PKC412. *Blood* **2005**, *105*, 54–60. [CrossRef] [PubMed]

166. Stone, R.M.; Mandrekar, S.; Sanford, B.L.; Geyer, S.; Bloomfield, C.D.; Dohner, K.; Thiede, C.; Marcucci, G.; Lo-Coco, F.; Klisovic, R.B.; et al. The multi-kinase inhibitor midostaurin (M) prolongs survival compared with placebo (P) in combination with daunorubicin (D)/cytarabine (C) induction (ind), high-dose c consolidation (consol), and as maintenance (maint) therapy in newly diagnosed acute myeloid leukemia (AML) patients (pts) age 18–60 with FLT3 mutations (muts): An international prospective randomized (rand) P-controlled double-blind trial (CALGB 10603/RATIFY [Alliance]). *Blood* **2015**, *126*, 6.

167. Knapper, S.; Russell, N.; Gilkes, A.; Hills, R.K.; Gale, R.E.; Cavenagh, J.D.; Jones, G.; Kjeldsen, L.; Grunwald, M.R.; Thomas, I.; et al. A randomized assessment of adding the kinase inhibitor lestaurtinib to first-line chemotherapy for FLT3-mutated AML. *Blood* **2017**, *129*, 1143–1154. [CrossRef] [PubMed]

168. Levis, M.; Ravandi, F.; Wang, E.S.; Baer, M.R.; Perl, A.; Coutre, S.; Erba, H.; Stuart, R.K.; Baccarani, M.; Cripe, L.D.; et al. Results from a randomized trial of salvage chemotherapy followed by lestaurtinib for patients with FLT3 mutant AML in first relapse. *Blood* **2011**, *117*, 3294–3301. [CrossRef] [PubMed]

169. Cortes, J.E.; Perl, A.E.; Dombret, H.; Kayser, S.; Steffen, B.; Rousselot, P.; Martinelli, G.; Estey, E.H.; Burnett, A.K.; Gammon, G.; et al. Final results of a phase 2 open-label, monotherapy efficacy and safety study of quizartinib (AC220) in patients ≥ 60 years of age with FLT3 ITD positive or negative relapsed/refractory acute myeloid leukemia. *Blood* **2012**, *120*, 48.

170. Levis, M.J.; Perl, A.E.; Dombret, H.; Döhner, H.; Steffen, B.; Rousselot, P.; Martinelli, G.; Estey, E.H.; Burnett, A.K.; Gammon, G.; et al. Final results of a phase 2 open-label, monotherapy efficacy and safety study of quizartinib (AC220) in patients with FLT3-ITD positive or negative relapsed/refractory acute myeloid leukemia after second-line chemotherapy or hematopoietic stem cell transplantation. *Blood* **2012**, *120*, 673.

171. Burnett, A.K.; Bowen, D.; Russell, N.; Knapper, S.; Milligan, D.; Hunter, A.E.; Khwaja, A.; Clark, R.E.; Culligan, D.; Clark, H.; et al. AC220 (quizartinib) can be safely combined with conventional chemotherapy in older patients with newly diagnosed acute myeloid leukaemia: Experience from the AML 18 pilot trial. *Blood* **2013**, *122*, 622.

172. Altman, J.K.; Perl, A.E.; Cortes, J.E.; Levis, M.J.; Smith, C.C.; Litzow, M.R.; Baer, M.R.; Claxton, D.F.; Erba, H.P.; Gill, S.C.; et al. Antileukemic activity and tolerability of ASP2215 80mg and greater in FLT3 mutation-positive subjects with relapsed or refractory acute myeloid leukemia: Results from a phase 1/2, open-label, dose-escalation/dose-response study. *Blood* **2015**, *126*, 321.

173. Randhawa, J.K.; Kantarjian, H.M.; Borthakur, G.; Thompson, P.A.; Konopleva, M.; Daver, N.; Pemmaraju, N.; Jabbour, E.; Kadia, T.M.; Estrov, Z.; et al. Results of a phase II study of crenolanib in relapsed/refractory acute myeloid leukemia patients (Pts) with activating FLT3 mutations. *Blood* **2014**, *124*, 389.

174. Galanis, A.; Ma, H.; Rajkhowa, T.; Ramachandran, A.; Small, D.; Cortes, J.; Levis, M. Crenolanib is a potent inhibitor of FLT3 with activity against resistance-conferring point mutants. *Blood* **2014**, *123*, 94–100. [CrossRef] [PubMed]

175. Smith, C.C.; Lasater, E.A.; Lin, K.C.; Wang, Q.; McCreery, M.Q.; Stewart, W.K.; Damon, L.E.; Perl, A.E.; Jeschke, G.R.; Sugita, M.; et al. Crenolanib is a selective type I pan-FLT3 inhibitor. *Proc. Natl. Acad. Sci. USA* **2014**, *111*, 5319–5324. [CrossRef] [PubMed]

176. Zimmerman, E.I.; Turner, D.C.; Buaboonnam, J.; Hu, S.; Orwick, S.; Roberts, M.S.; Janke, L.J.; Ramachandran, A.; Stewart, C.F.; Inaba, H.; et al. Crenolanib is active against models of drug-resistant FLT3-ITD-positive acute myeloid leukemia. *Blood* **2013**, *122*, 3607–3615. [CrossRef] [PubMed]

177. Wang, E.S.; Stone, R.M.; Tallman, M.S.; Walter, R.B.; Eckardt, J.R.; Collins, R. Crenolanib, a type I FLT3 TKI, can be safely combined with cytarabine and anthracycline induction chemotherapy and results in high response rates in patients with newly diagnosed FLT3 mutant acute myeloid leukemia (AML). *Blood* **2016**, *128*, 1071.

178. Smith, C.C.; Lin, K.; Stecula, A.; Sali, A.; Shah, N.P. FLT3 D835 mutations confer differential resistance to type II FLT3 inhibitors. *Leukemia* **2015**, *29*, 2390–2392. [CrossRef] [PubMed]

179. Smith, C.C.; Zhang, C.; Lin, K.C.; Lasater, E.A.; Zhang, Y.; Massi, E.; Damon, L.E.; Pendleton, M.; Bashir, A.; Sebra, R.; et al. Characterizing and overriding the structural mechanism of the Quizartinib-Resistant FLT3 "Gatekeeper" F691L mutation with PLX3397. *Cancer Discov.* **2015**, *5*, 668–679. [CrossRef] [PubMed]

180. Sato, T.; Yang, X.; Knapper, S.; White, P.; Smith, B.D.; Galkin, S.; Small, D.; Burnett, A.; Levis, M. FLT3 ligand impedes the efficacy of FLT3 inhibitors in vitro and in vivo. *Blood* **2011**, *117*, 3286–3293. [CrossRef] [PubMed]

181. Borthakur, G.; Kantarjian, H.M.; O'Brien, S.; Garcia-Manero, G.; Jabbour, E.; Daver, N.; Kadia, T.M.; Gborogen, R.; Konopleva, M.; Andreeff, M.; et al. The combination of quizartinib with azacitidine or low dose cytarabine is highly active in patients (Pts) with FLT3-ITD mutated myeloid leukemias: Interim report of a phase I/II trial. *Blood* **2014**, *124*, 388.

182. Hofmann, M.; Grosse-Hovest, L.; Nubling, T.; Pyz, E.; Bamberg, M.L.; Aulwurm, S.; Buhring, H.J.; Schwartz, K.; Haen, S.P.; Schilbach, K.; et al. Generation, selection and preclinical characterization of an Fc-optimized FLT3 antibody for the treatment of myeloid leukemia. *Leukemia* **2012**, *26*, 1228–1237. [CrossRef] [PubMed]

183. Knudson, A.G., Jr.; Hethcote, H.W.; Brown, B.W. Mutation and childhood cancer: A probabilistic model for the incidence of retinoblastoma. *Proc. Natl. Acad. Sci. USA* **1975**, *72*, 5116–5120. [CrossRef] [PubMed]

184. Kelly, L.M.; Kutok, J.L.; Williams, I.R.; Boulton, C.L.; Amaral, S.M.; Curley, D.P.; Ley, T.J.; Gilliland, D.G. PML/RARα and FLT3-ITD induce an APL-like disease in a mouse model. *Proc. Natl. Acad. Sci. USA* **2002**, *99*, 8283–8288. [CrossRef] [PubMed]

185. Mead, A.J.; Kharazi, S.; Atkinson, D.; Macaulay, I.; Pecquet, C.; Loughran, S.; Lutteropp, M.; Woll, P.; Chowdhury, O.; Luc, S.; et al. FLT3-ITDs instruct a myeloid differentiation and transformation bias in lymphomyeloid multipotent progenitors. *Cell Rep.* **2013**, *3*, 1766–1776. [CrossRef] [PubMed]

© 2017 by the authors. Licensee MDPI, Basel, Switzerland. This article is an open access article distributed under the terms and conditions of the Creative Commons Attribution (CC BY) license (http://creativecommons.org/licenses/by/4.0/).

International Journal of
Molecular Sciences

MDPI

Article

Apoptosis Induced by the Curcumin Analogue EF-24 Is Neither Mediated by Oxidative Stress-Related Mechanisms nor Affected by Expression of Main Drug Transporters ABCB1 and ABCG2 in Human Leukemia Cells

Nikola Skoupa, Petr Dolezel, Eliska Ruzickova and Petr Mlejnek *

Department of Anatomy, Faculty of Medicine and Dentistry, Palacky University Olomouc, Hnevotinska 3, Olomouc 77515, Czech Republic; NikolaSkoupa@seznam.cz (N.S.); p.dolezel@atlas.cz (P.D.); elis.ruzickova@gmail.com (E.R.)
* Correspondence: mlejnek_petr@volny.cz; Tel.: +420-585-632-203

Received: 9 October 2017; Accepted: 24 October 2017; Published: 31 October 2017

Abstract: The synthetic curcumin analogue, 3,5-bis[(2-fluorophenyl)methylene]-4-piperidinone (EF-24), suppresses NF-κB activity and exhibits antiproliferative effects against a variety of cancer cells in vitro. Recently, it was reported that EF-24-induced apoptosis was mediated by a redox-dependent mechanism. Here, we studied the effects of *N*-acetylcysteine (NAC) on EF-24-induced cell death. We also addressed the question of whether the main drug transporters, ABCB1 and ABCG2, affect the cytotoxic of EF-24. We observed that EF-24 induced cell death with apoptotic hallmarks in human leukemia K562 cells. Importantly, the loss of cell viability was preceded by production of reactive oxygen species (ROS), and by a decrease of reduced glutathione (GSH). However, neither ROS production nor the decrease in GSH predominantly contributed to the EF-24-induced cell death. We found that EF-24 formed an adduct with GSH, which is likely the mechanism contributing to the decrease of GSH. Although NAC abrogated ROS production, decreased GSH and prevented cell death, its protective effect was mainly due to a rapid conversion of intra- and extra-cellular EF-24 into the EF-24-NAC adduct without cytotoxic effects. Furthermore, we found that neither overexpression of ABCB1 nor ABCG2 reduced the antiproliferative effects of EF-24. In conclusion, a redox-dependent-mediated mechanism only marginally contributes to the EF-24-induced apoptosis in K562 cells. The main mechanism of NAC protection against EF-24-induced apoptosis is conversion of cytotoxic EF-24 into the noncytotoxic EF-24-NAC adduct. Neither ABCB1 nor ABCG2 mediated resistance to EF-24.

Keywords: NF-κB; Nrf2; EF-24-GSH adduct; EF-24-NAC adduct; K562 cells

1. Introduction

Many polyphenolic compounds extracted from plants have been demonstrated to have cancer-preventing activities in laboratory studies. Curcumin, a component of turmeric (*Curcuma longa*), is a typical example. This agent was reported to inhibit proliferation and survival of cancer cells in vitro, while retaining a pharmacologically safety profile in vivo. However, clinical studies show that curcumin has low efficacy due to rapid excretion in vivo [1,2]. This prompted the development of analogues, which are more potent inducers of apoptosis in in vitro assays and also more efficient in vivo.

Adams and co-workers synthesized 3,5-bis[(2-fluorophenyl)methylene]-4-piperidinone (EF-24), a fluorinated curcumin analog which exhibits antiprolifative effects against a variety of cancer cells in vitro [3–9] and also in animal models in vivo [10–12]. Unlike curcumin, EF-24 is bioavailable with a greater potency to induce cell death in many tumors and shows relatively low toxicity [13]. It does this through direct inhibition of IκB kinase (IKK), resulting in suppression of NF-κB activity [5,12]. In addition, the antiproliferative effects of EF-24 may also be mediated by deregulation of important signaling pathways that include upregulation of PTEN, downregulation of Akt or inhibition HIF-1α in certain cancer cell lines in vitro [7–10]. Several reports also suggest that the anticancer activity of EF-24 may be mediated, in part, by a redox-dependent induction of apoptosis [14,15] or due to increased reactive oxygen species (ROS) production [9,16,17].

Here we address the question of how much oxidative stress contributes to the cell death induction in EF-24-treated human leukemia K562 cells. We also studied whether ABCB1 (P-glycoprotein) and ABCG2 (BCRP, breast cancer resistance protein), which are usually considered the main drug transporters underlying the multidrug resistance (MDR) phenotype in cancer cells, compromise the antiproliferative effects of EF-24 in K562 cells. We observed that ROS production and glutathione (GSH) depletion contributed to EF-24-induced cell death only slightly. Importantly, neither ABCB1 nor ABCG2 expression reduced the cytotoxic effects of EF-24.

2. Results

2.1. N-Acetylcysteine (NAC) Prevented 3,5-bis[(2-Fluorophenyl)Methylene]-4-Piperidinone (EF-24)-Induced Cell Death with Apoptotic Hallmarks in K562 Cells

In this study we used human leukemia K562 cells, since they represent a type of malignancy with impaired NF-κB activity [18]. We observed that EF-24 induced cell death with apoptotic hallmarks, including chromatin condensation, DNA fragmentation and caspase-3 activation (Figure 1, Table 1). Importantly, EF-24-induced cell death was preceded by ROS production, which was significant but moderate (Figure 2a,b), and by the reduction of GSH levels within K562 cells (Figure 2c). Despite the fact that enhanced ROS production and the drop in GSH levels might indicate oxidative stress, an increased level of oxidized glutathione (GSSG) was not found as the GSH/GSSG ratio was not dramatically changed (Figure 2d). Addition of N-acetylcysteine (NAC) at millimolar concentrations into the growth medium abrogated ROS production (Figure 2a,b), GSH depletion (Figure 2c) and prevented apoptosis induction in EF-24-treated K562 cells (Figure 1, Table 1). Interestingly, pretreatment of cells with NAC at sub-milimolar concentrations diminished ROS production, restored GSH levels but, it prevented cell death only partially (not shown). In contrast, addition of catalase (CAT) into the growth medium abrogated ROS production (Figure 3a), but it did not prevent GSH depletion (Figure 3b), and failed to reduced cell death (Figure 3c, Table 1).

Table 1. Effect of N-acetylcysteine (NAC) and catalase (CAT) on cell proliferation and viability in EF-24-treated K562 cells.

Cell Treatment	IC_{50}
EF-24	0.73 ± 0.11 μM
EF-24 + 2 mM NAC	>20 μM
EF-24 + CAT (50 U/mL)	0.84 ± 0.15 μM

2.2. EF-24 Did Not Activate Nuclear Factor-Erythroid 2 Related Factor 2 (Nrf2) in K562 Cells

To evaluate the significance of ROS production in the fate of the cell, we addressed the question of whether EF-24 is capable of activating the nuclear factor-erythroid 2 related factor 2 (Nrf2). It is well documented that cells respond to oxidative stress by activation of cellular defense mechanisms that comprise several antioxidant pathways regulated by Nrf2 [19,20]. However, our results clearly indicated that EF-24 did not activate the Nrf2 signaling pathway. Indeed, neither increased levels of Nrf2 nor elevated expression levels of stress-response proteins, including heme oxygenase-1 (HO-1) and NAD(P)H:quinon oxidoreductase 1 (NQO1) were observed in EF-24-treated K562 cells (Figure 4).

These results, together with those described in the previous paragraph, collectively suggest that increased production of ROS played only a marginal role in the induction of the death of EF-24-treated K562 cells.

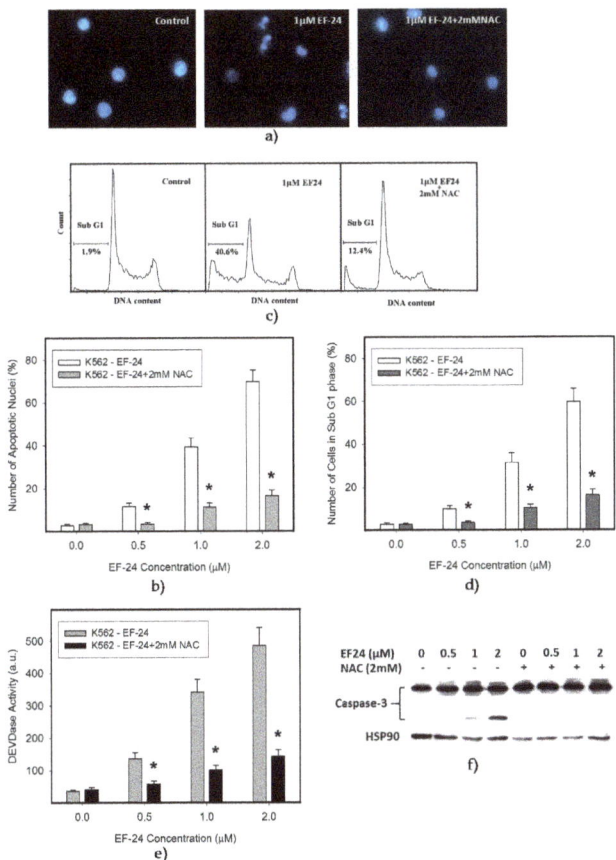

Figure 1. Effect of *N*-acetylcysteine (NAC) on cell death in 3,5-bis[(2-fluorophenyl)methylene]-4-piperidinone (EF-24)-treated cells. Cells were treated with EF-24 or with EF-24 + 2 mM NAC, as indicated. Incubation proceeded for 24 h prior to appropriate analysis. The experimental points represent mean values from three replicate experiments, with standard deviations. (**a**) Effect of NAC on nuclear morphology in EF-24 treated cells. Pictures represent typical examples. (**b**) A quantitative analysis of the effect of NAC on nuclear morphology in EF-24 treated cells. At least 300 cells were examined in each experiment; * denotes significant change in the number of cells with apoptotic nuclei ($p < 0.05$) between K562 cells treated with EF-24 and cells treated with EF-24 + NAC. (**c**) Effect of NAC on occurrence of cells in sub G1 phase in response to EF-24 treatment. A representative analysis. (**d**) A quantitative analysis of the effect of NAC on occurrence of cells in sub G1 phase in response to EF-24 treatment; * denotes significant change in the number of cells sub G1 phase ($p < 0.05$) between K562 cells treated with EF-24 and cells treated with EF-24 + NAC. (**e**) Effect of NAC on caspase-3 activation in EF-24-treated cells. After 16 h, cells DEVDase enzymatic activity was determined in cell lysates using Ac-DEVD-AMC; * denotes significant change in DEVDase enzymatic activity ($p < 0.05$) between K562 cells treated with EF-24 and cells treated with EF-24 + NAC. (**f**) Effect of NAC on caspase-3 processing in EF-24 treated cells. After 16 h, caspase-3 processing was monitored using western blot analysis. Picture represents a typical example.

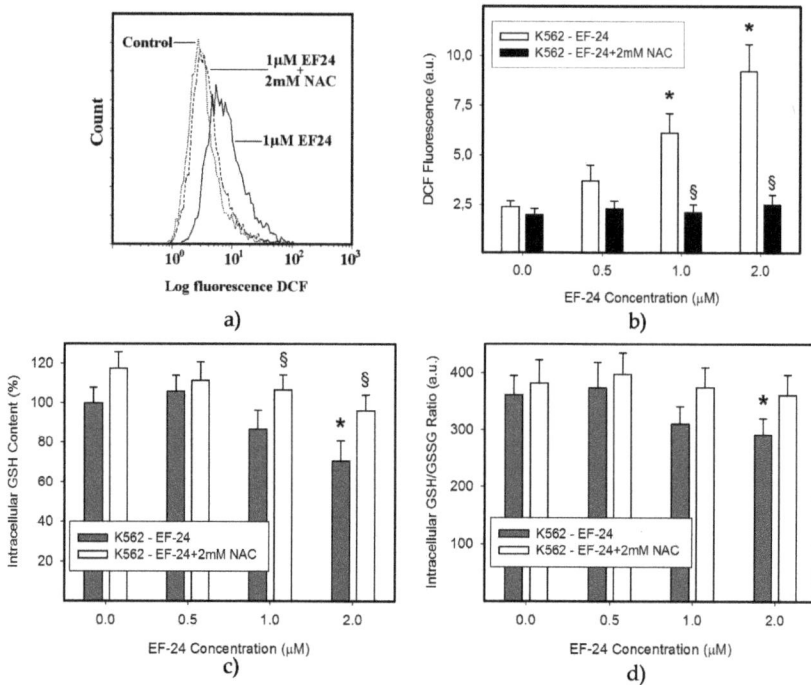

Figure 2. Effect of NAC on EF-24-induced oxidative stress. Cells were treated with EF-24 or with EF-24 + 2 mM NAC, as indicated. The experimental points represent mean values from three replicate experiments, with standard deviations. (**a**) Effect of NAC on reactive oxygen species (ROS) production in EF-24-treated cells. ROS production was determined using flow cytometry after 3 h incubation. A representative analysis. (**b**) A quantitative analysis of the effect of NAC on ROS production in response to EF-24 treatment. ROS production was determined using flow cytometry after 3 h incubation; * denotes significant change in ROS production ($p < 0.05$) between control (untreated) K562 cells and cells treated with EF-24; § denotes significant change in ROS production ($p < 0.05$) between EF-24 treated K562 cells and cells treated with EF-24 + NAC. (**c**) Effect of NAC on intracellular glutathione (GSH) level in EF-24 treated cells. After 3 h incubation, the intracellular content of GSH was determined using LC/MS/MS analysis; * denotes significant change in the intracellular level of GSH ($p < 0.05$) between control (untreated) K562 cells and cells treated with EF-24; § denotes significant change in intracellular level of GSH ($p < 0.05$) between K562 cells treated with EF-24 and cells treated with EF-24 + NAC. (**d**) Effect of NAC on GSH/oxidized glutathione (GSSG) ratio in EF-24-treated cells. After 3 h incubation, the intracellular content of GSH and GSSG were determined using LC/MS/MS analysis; * denotes significant change in GSH/GSSG ratio ($p < 0.05$) between K562 cells treated with EF-24 and cells treated with EF-24 + NAC.

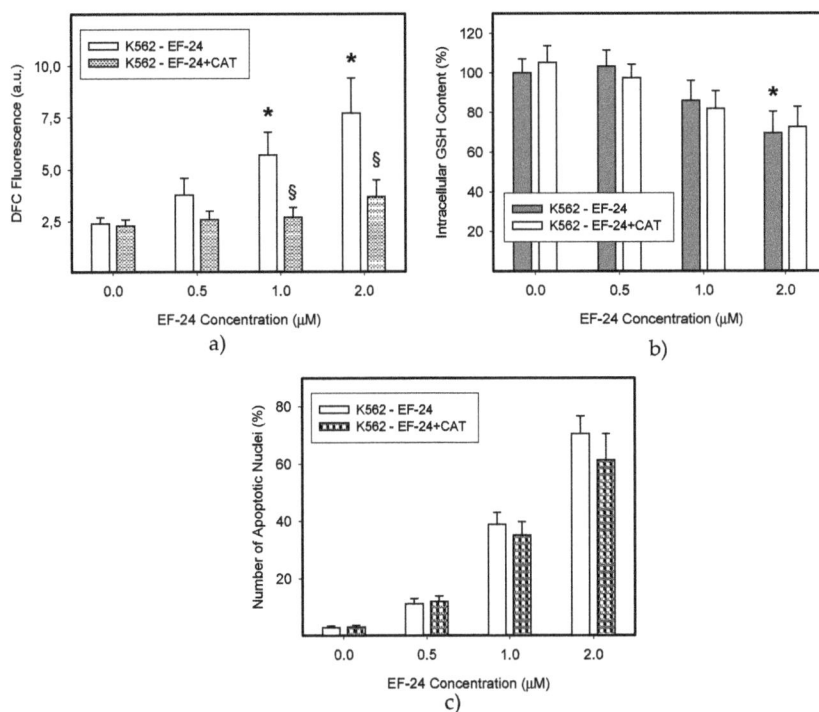

Figure 3. Effect of catalase (CAT) on EF-24-induced oxidative stress. Cells were treated with EF-24 or with EF-24 + CAT (50 U/mL), as indicated. The experimental points represent mean values from three replicate experiments, with standard deviations. (**a**) Effect of CAT on ROS production in EF-24 treated cells. After 3 h incubation, ROS production was determined using flow cytometry; * denotes significant change in ROS production ($p < 0.05$) between control (untreated) K562 cells and cells treated with EF-24; § denotes significant change in ROS production ($p < 0.05$) between EF-24-treated K562 cells and cells treated with EF-24 + CAT. (**b**) Effect of CAT on intracellular GSH level in EF-24 treated cells. After 6 h incubation, the intracellular content of GSH was determined using LC/MS/MS analysis; * denotes significant change in intracellular level of GSH ($p < 0.05$) between control (untreated) K562 cells and cells treated with EF-24. (**c**) Effect of CAT on nuclear morphology in EF-24-treated cells. After 24 h, cells were stained using Hoechst33342, and nuclear morphology was examined using fluorescence microscopy.

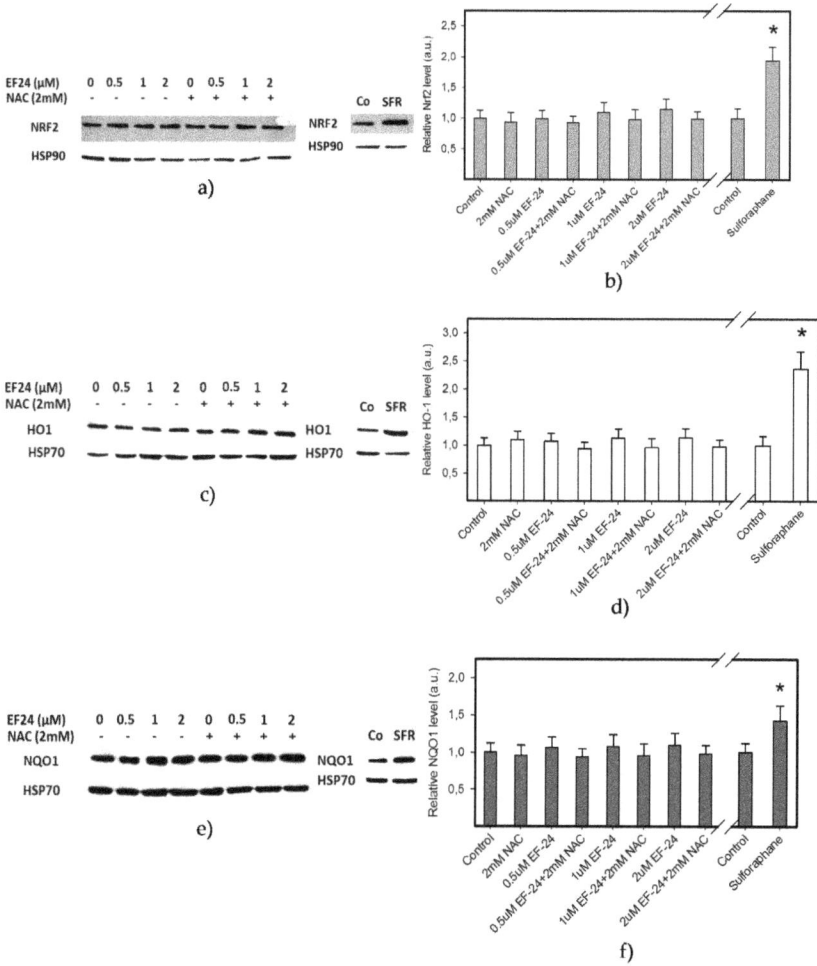

Figure 4. Effect of EF-24 on the activation of Nrf2 regulated signaling pathway. Cells were treated with EF-24 or with EF-24 + 2 mM NAC, as indicated. Incubation proceeded for 6 h prior to Western blot analysis. Treatment of cells with sulforaphane was used as a positive control [21]. (**a**) Effect of EF-24 on the activation of Nrf2. A representative example. (**b**) A quantitative analysis of the effect of EF-24 on the activation of Nrf2; * denotes significant change in the Nrf2 level ($p < 0.05$) between untreated K562 cells and cells treated with sulforaphane. (**c**) Effect of EF-24 on the activation of heme oxygenase-1 (HO-1). A representative example. (**d**) A quantitative analysis of the effect of EF-24 on the activation of HO-1; * denotes significant change in the HO-1 level ($p < 0.05$) between untreated K562 cells and cells treated with sulforaphane. (**e**) Effect of EF-24 on the activation of NAD(P)H:quinon oxidoreductase 1 (NQO1). A representative example. (**f**) A quantitative analysis of the effect of EF-24 on the activation of NQO1; * denotes significant change in NQO1 levels ($p < 0.05$) between untreated K562 cells and cells treated with sulforaphane.

2.3. Conversion of Cytotoxic EF-24 Into the Non-Cytotoxic EF-24-NAC Adduct is the Main Mechanism of the NAC Protection Against the Cytotoxic Effects of EF-24

Excessive oxidation of GSH as a result of oxidative stress, or its conjugation with xenobiotics represent important mechanisms that may lead to the depletion of cellular GSH [22,23]. Given that no increased level of GSSG was found in cells (Figure 2d) or culture medium (not shown), we studied the interaction of EF-24 and GSH in more detail. As expected, we found intracellular formation of the EF-24-GSH adduct in K562 cells (Figure 5a,b). The amount of EF-24-GSH adduct was proportional to the concentration of EF-24 (not shown). Formation of the EF-24-GSH adduct was significantly reduced by the addition of NAC (Figure 5b). While GSH depletion seemed to be related to the EF-24-GSH adduct formation, the restoration of GSH synthesis using NAC could not fully explain the mechanism of NAC protection against EF-24-induced apoptosis. Indeed, we found that adding NAC to the culture medium induced a rapid intra- and extra-cellular clearance of EF-24 (Figure 6a,b). This was accompanied by intra- and extra-cellular formation of the EF-24 adduct with NAC (Figure 6c–e). These results clearly indicated that conversion of cytotoxic EF-24 into the non-cytotoxic EF-24-NAC adduct was the main mechanism of NAC protection. It is important to point out that EF-24 formed mono- and di-adducts with NAC and/or GSH, however, the mono-adducts were predominant (not shown). For this reason, the data presented here refer to the appropriate mono-adducts.

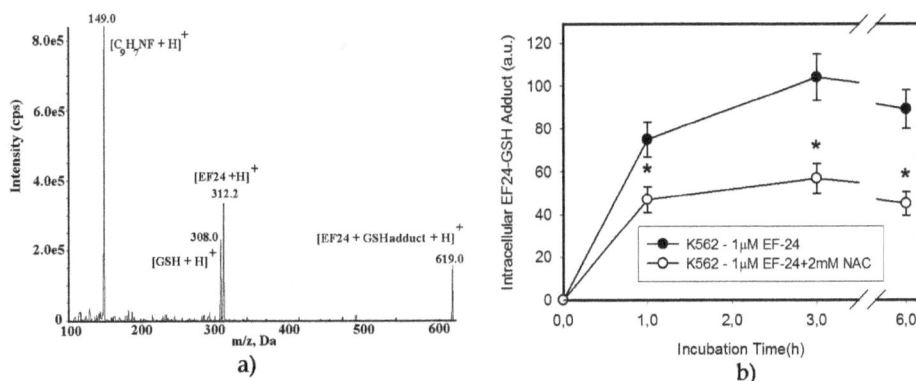

Figure 5. EF-24 forms adduct with GSH. (**a**) MS/MS analysis of the EF-24-GSH adduct: Mass spectrum of the molecular ion of EF-24-GSH adduct (m/z 619.0), result fragment ion [EF-24 + H]+, m/z 312.2, [GSH + H]+, m/z 308.0, and [C9H7NF+H]+, m/z 149.0. (**b**) Effect of NAC on EF-24-GSH adduct formation. Cells were treated with EF-24 or in the combination with 2 mM NAC for given incubation times. The amount of the EF-24-GSH adduct in cells was determined using LC/MS/MS. The experimental points represent mean values from three replicate experiments, with standard deviations; * denotes significant change in EF-24-GSH adduct content between EF-24 and EF-24 + NAC-treated K562 cells.

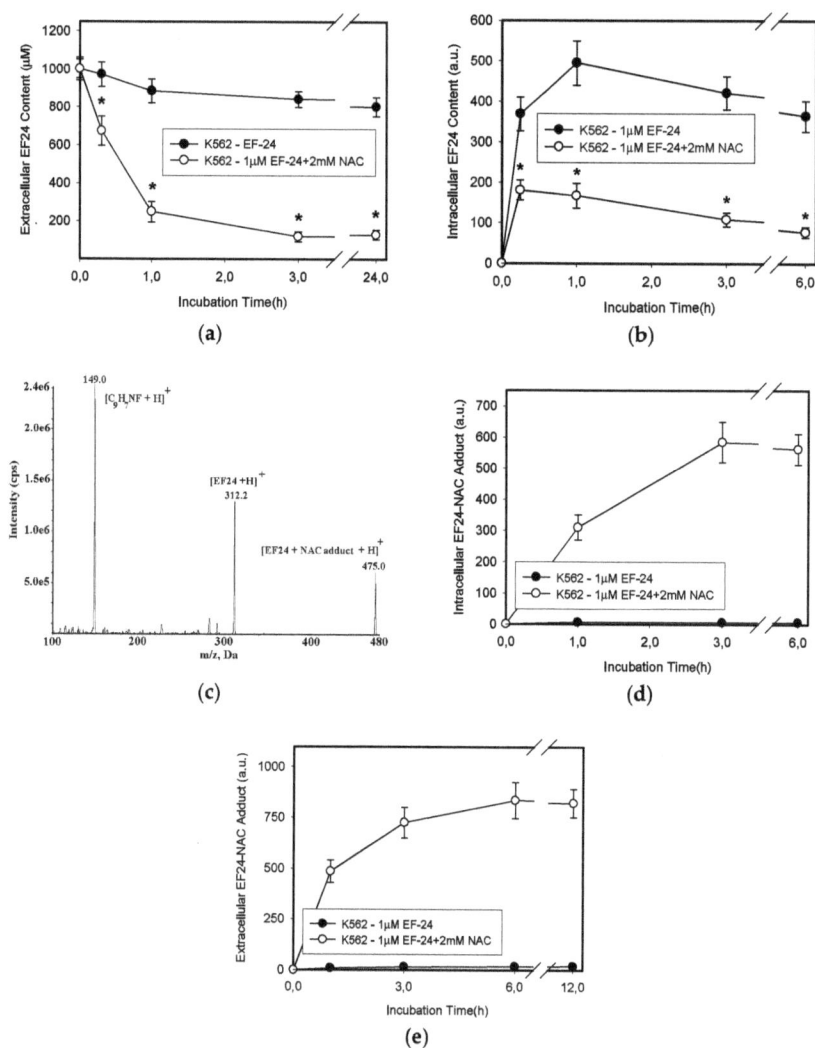

Figure 6. Effect of NAC on extracellular and intracellular levels of EF-24. Cells were treated with EF-24 or with EF-24 + 2 mM NAC. At indicated time intervals, extracellular and intracellular levels of EF-24 were determined using LC/MS/MS analysis. (**a**) Extracellular EF-24 level; * denotes significant change in EF-24 content in growth medium ($p < 0.05$) between EF-24 and EF-24 + NAC-treated K562 cells. (**b**) Intracellular EF-24 level; * denotes significant change in intracellular EF-24 content ($p < 0.05$) between EF-24 and EF-24 + NAC-treated K562 cells. The experimental points represent mean values from three replicate experiments, with standard deviations. (**c**) MS/MS analysis of EF-24-NAC adduct: Mass spectrum of the molecular ion of [EF-24-NAC-adduct]$^+$, m/z 475.0, result fragment ion [EF-24+H]$^+$, m/z 312.2, and [C9H7NF + H]$^+$, m/z 149.0. (**d**) Time course of intracellular EF-24-NAC adduct formation. Cells were treated with EF-24 or with EF-24 + NAC for given incubation time. The amount of EF-24-NAC adduct in cells was determined using LC/MS/MS. (**e**) Time–course of extracellular EF-24-NAC adduct formation. Cells were treated with EF-24 or with EF-24 + NAC for given incubation times. The amount of EF-24-NAC adduct in growth medium was determined using LC/MS/MS. The experimental points represent mean values from three replicate experiments, with standard deviations.

To directly demonstrate that the EF-24-NAC adduct was not cytotoxic, we used a mixture of EF-24 and NAC in distilled water, where the reaction was allowed to proceed until 50% of EF-24 was converted into the EF-24-NAC adduct (not shown). K562 cells were then treated with the diluted reaction mixture. We observed that the proapoptotic effect of the reaction mixture, containing approximately one half of remnant free EF-24 and one half of EF-24-NAC adduct, was equivalent to the effect of remnant free EF-24 (Figure 7). For example, a diluted reaction mixture containing approximately 1 µM EF-24 and 1 µM EF-24-NAC adduct exhibited a proapoptotic effect, which was similar to that of 1 µM EF-24 itself rather than to the effect of 2 µM EF-24 (Figure 7).

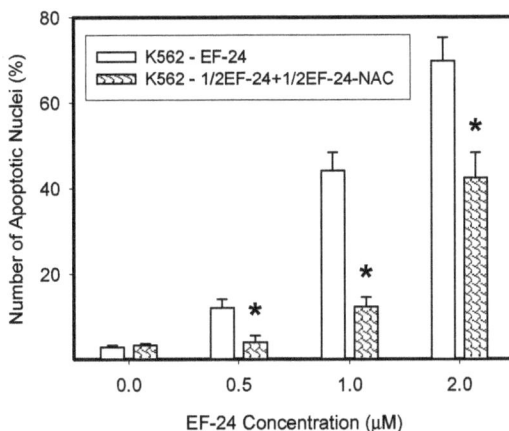

Figure 7. Proapoptotic effects of the EF-24-NAC adduct. EF-24 (50 µM) was mixed with 2 mM NAC in distilled water at ambient temperature and reaction was monitored using LC/MS/MS analysis. When 50% of EF-24 was converted into the EF-24-NAC adduct, reaction was stopped by dilution. Then K562 cells were treated with EF-24 alone or with diluted reaction mixture containing the same amount of EF-24, however, approximately 50% of it was converted into EF-24-NAC adduct. The experimental points represent mean values from three replicate experiments, with standard deviations; * denotes significant change in the number of cells with apoptotic nuclei ($p < 0.05$) between K562 cells treated with "free" EF-24 and cells treated with a diluted mixture containing the same amount of EF-24 (50% "free" EF-24 + 50% EF-24-NAC adduct).

2.4. Overexpression of ABCB1 and ABCG2 Did Not Compromise the Antiproliferative Effects of EF-24 in K562 Cells

Decreased intracellular drug levels, which prevent effective interaction between the drug and its cellular target, is a generally accepted mechanism of resistance mediated by ABC transporters [24,25]. Therefore, we analyzed extracts from cells expressing ABCB1 and ABCG2, the main drug transporters (Figures 8 and 9). We found that neither the overexpression of ABCB1 nor ABCG2 significantly reduced the intracellular levels of EF-24 (Figure 9). Accordingly, neither ABCB1 nor ABCG2 expression mediated resistance to EF-24 (Table 2). Importantly, EF-24 exhibited a proapoptotic effect in ABCB1- and ABCG2-expressing cells similar to that found in parental K562 cells (not shown). Our results clearly indicate that EF-24 would exhibit antiproliferative effects, irrespective of ABCB1 or ABCG2 expression.

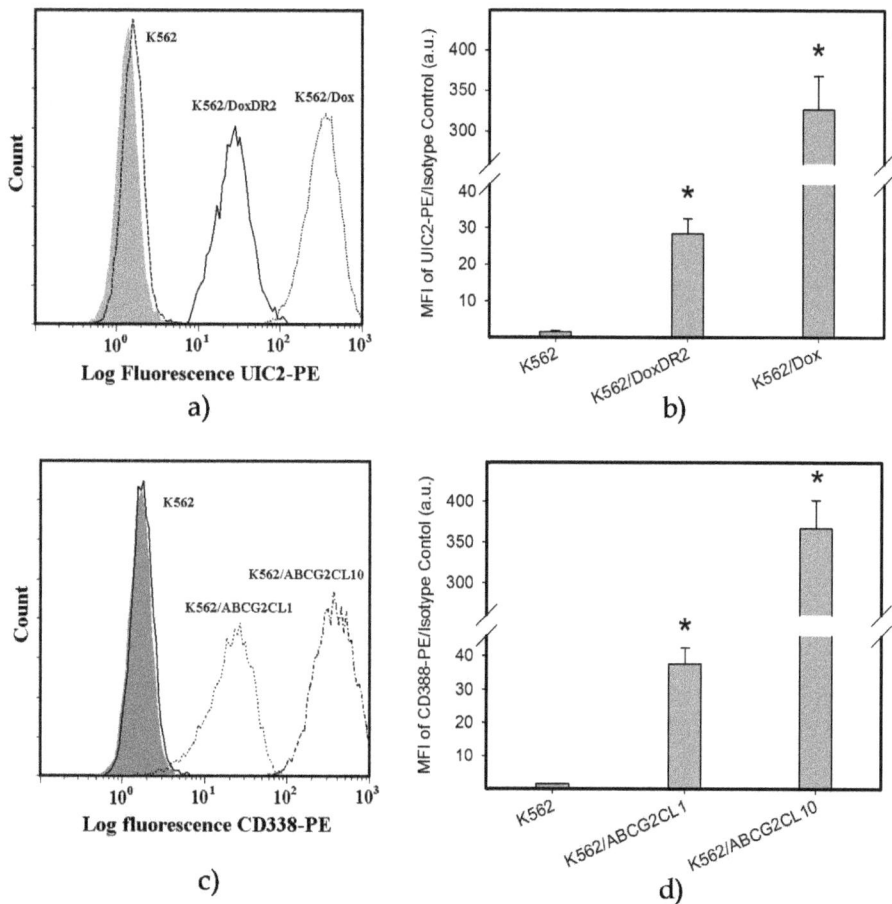

Figure 8. Analysis of ABCB1 and ABCG2 expression: (**a**) Flow cytometric analysis of ABCB1 expression. Isotype control (grey histogram); K562 parental cell line (dash line); K562/Dox cells (dot line); K562/DoxDR2 cells (solid line). (**b**) Quantitative analysis of ABCB1 expression. ABCB1 expression was quantified as the mean fluorescence intensity (MFI) shift (ratio of MFI of UIC2-PE antibody and isotype control). The experimental points represent mean values from three replicate experiments, with standard deviations; * denotes significant change in ABCB1 expression ($p < 0.05$) between K562 cells and cells expressing various levels of ABCB1 (K562/Dox, K562/DoxDR2). (**c**) Flow cytometric analysis of ABCG2 expression. Isotype control (grey histogram); K562 parental cell line (solid line); K562/ABCG2CL10 cells (dash-dot line); K562/ABCGCL1 cells (dot line). (**d**) Quantitative analysis of ABCG2 expression. ABCG2 expression was quantified as the mean fluorescence intensity (MFI) shift (ratio of MFI of CD338-PE antibody and isotype control). The experimental points represent mean values from three replicate experiments, with standard deviations; * denotes significant change in ABCG2 expression ($p < 0.05$) between K562 cells and cells expressing various levels of ABCG2 (K562/ABCGCL10, K562/ABCGCL1).

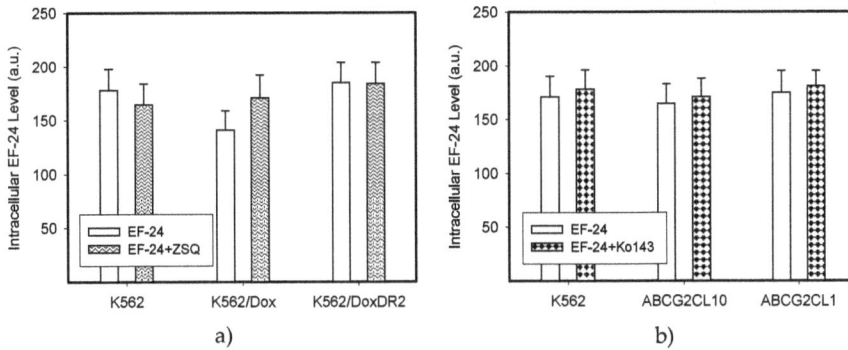

Figure 9. Intracellular levels of EF-24 in cells expressing ABCB1 or ABCG2. The parental cell line K562, which does not express any of the transporters was used as a control. (**a**) Relative intracellular levels of EF-24 in cells overexpressing ABCB1 transporter. Cells were treated with EF-24 or with EF-24 + inhibitor zosuquidar trihydrochloride (ZSQ), as indicated. The experimental points represent mean values from three replicate experiments, with standard deviations. (**b**) Relative intracellular levels of EF-24 in cells overexpressing ABCG2 transporter. Cells were treated with EF-24 or with EF-24 + inhibitor Ko143, as indicated. The experimental points represent mean values from three replicate experiments, with standard deviations.

Table 2. Effect of ABCB1 or ABCG2 expression on cell proliferation and viability in EF-24 treated K562 cells. ZSQ, zosuquidar trihydrochloride.

	EF-24	EF-24 + ZSQ	EF-24 + Ko143
K562	$0.74 \pm 0.10\ \mu M$	$0.71 \pm 0.10\ \mu M$	$0.77 \pm 0.12\ \mu M$
K562/Dox	$0.87 \pm 0.15\ \mu M$	$0.72 \pm 0.11\ \mu M$	-
K562/DoxDR2	$0.75 \pm 0.11\ \mu M$	$0.69 \pm 0.10\ \mu M$	-
K562/ABCG2CL10	$0.71 \pm 0.11\ \mu M$	-	$0.75 \pm 0.11\ \mu M$
K562/ABCG2CL1	$0.72 \pm 0.10\ \mu M$	-	$0.76 \pm 0.11\ \mu M$

3. Discussion

A large number of studies have demonstrated aberrant NF-κB signaling in solid cancers, as well as in various types of hematologic malignancies. NF-κB is involved in regulating the expression of a number of genes that affect proliferation, cell survival, tumor metastasis, angiogenesis and inflammation. Therefore, targeting aberrant NF-κB activation together with its upstream and downstream interacting regulatory molecules using low molecular weight inhibitors, may be useful in clinical settings for the treatment of solid cancers and hematological malignancies [18].

Our results indicate the strong proapoptotic potential of EF-24 in K562 cells (Figure 1, Table 1). Its proapoptotic effect is much higher than that demonstrated in colorectal or gastric cancer cells [16,17]. This is not surprising, since chronic myelogenous leukemia cells rely on aberrant activation of NF-κB [18], and EF-24 efficiently suppresses NF-κB signaling through direct inhibition of IKK [5,12]. However, several laboratories have shown that the proapoptotic effects of EF-24 are also mediated by a redox-dependent mechanism [14–17]. A recent publication concluded that EF-24 induces apoptosis via ROS-dependent mitochondrial dysfunction in human colorectal cancer cells [17]. This is supported by the findings that the EF-24-induced cell death was preceded by production of ROS, GSH depletion and mainly by the observation that NAC, a well-known radical scavenger and a precursor of GSH synthesis, prevented cell death [23]. Our results suggest that the EF-24-induced cell death with apoptotic features (Figure 1) was preceded by transient production of ROS (not shown) with a peak at 3 h after the EF-24 addition, and by a decrease in GSH levels in human leukemia K562 cells (Figure 2).

However, in contrast to the conclusions published by He and co-workers [17], we do not think that ROS production itself is the direct cause of cell death. This conclusion stems from following findings. First, although CAT prevented ROS production (Figure 3a), it failed to reduce the antiproliferative effects of EF-24 (Table 1, Figure 3c). Second, even though ROS production and reduced intracellular level of GSH are typical signs of oxidative stress, these were not accompanied by elevated levels of GSSG (Figure 2d). In contrast to expectations, we observed that, both GSH and GSSG decreased, however, the decrease in GSSG was somewhat lower than that of GSH for 2 µM EF-24 (Figure 2d). There was no evidence for any other mechanisms of GSH depletion, such as leakage of GSH and/or GSSG into the growth medium (not shown). Third, we also failed to find any signs of Nrf2 activation. Nrf2 is a master regulator of cell response to oxidative stress [19,20]. Under "normal" conditions it is maintained at low levels and resides predominantly in the cytoplasm. Upon oxidative stress, Nrf2 levels increase due to diminished proteasomal degradation, and it translocates to the nucleus to trigger transcription of a plethora of genes to cope with the oxidative stress [19,26]. However, no elevated Nrf2 level (Figure 4a,b) or increased expression of antioxidant genes, including HO-1 (Figure 4c,d) or NQO1 (Figure 4e,f) were found in the present study.

Similarly, the decrease in GSH level itself is probably not a direct cause of cell death. Indeed, EF-24, up to 1 µM concentration, had no effect on inducing a significant decrease in GSH, while it induced significant cell death (Figures 1 and 2). In addition, NAC at sub-millimolar concentrations, affected cell survival only partially, despite significantly reduced the drop in GSH level (not shown).

Instead, EF-24 may serve as a Michael acceptor and form adducts with thiols [15]. Similar to others [15], we found that EF-24 forms mono-adducts (Figure 5) and di-adducts (not shown) with GSH. Since the di-adducts represented only a minor part of the adducts, we analyzed mono-adducts. We hypothesize that the formation of EF-24-GSH adducts may contribute to the decrease in GSH levels (Figures 2c and 3b), since we found no evidence for any other mechanism of GSH depletion, such as leakage of GSH and/or GSSG into the growth medium (not shown).

NAC prevented the adverse effects of EF-24, including cell death (Figures 1 and 2). However, we do not agree with the interpretation that NAC prevented EF-24-induced cell death due to the abrogation of ROS production and restoration of GSH synthesis [14,15,17]. Our results strongly suggest that addition of NAC at millimolar concentrations into the culture medium induces a rapid conversion of EF-24 into EF-24-NAC adduct (Figure 6) which is non-cytotoxic (Figure 7). Based on our results, we believe that this is the main mechanism of NAC protection against EF-24 cytotoxicity. A summary of our proposed mechanism is shown in Figure 10. It is necessary to note that EF-24 forms mono-adducts (Figure 6c,d) and di-adducts (not shown) with NAC. The di-adducts represented only a minor percentage of adducts. The results presented refer to the mono-adducts for this reason.

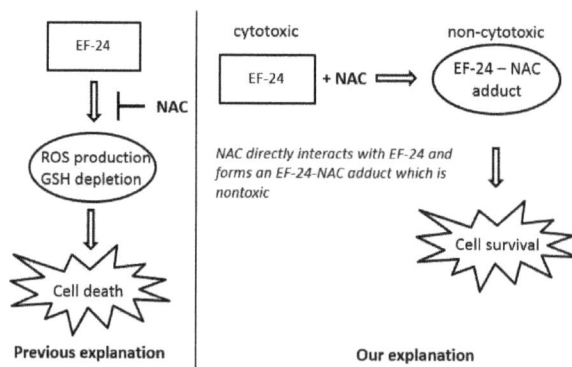

Figure 10. Proposed mechanism of NAC protection against EF-24 cytotoxicity in K562 cells.

We believe that this mechanism of NAC protection plays a crucial role in experimental systems with other cytotoxic agents. For example, cell death induced by geldanamycin (GDN) or by carbonyl cyanide-4-(trifluoromethoxy)phenylhydrazone (FCCP) at high micromolar concentrations is preceded by ROS production and GSH depletion, and can be prevented by NAC addition to the culture medium. In these experimental systems, NAC prevents GDN and FCCP cytotoxicity by conversion of these cytotoxic drugs into corresponding non-cytotoxic adducts with NAC [27,28].

EF-24 may have promise as an anticancer compound applicable in cases of drug resistance, specifically for those which express main drug transporters, ABCB1 and ABCG2 [29]. Indeed, neither ABCB1 nor ABCG2 overexpression significantly reduced the intracellular level of EF-24 in K562 cells (Figure 9). Correspondingly, cells overexpressing ABCB1 or ABCG2 were sensitive to EF-24 in a way that is similar to the parent K562 cells (Table 2). Since drug resistance strongly depends on transporter expression levels [30–33], we used cells with various expression levels of ABCB1 or ABCG2 in this study. Importantly, cells with high expression levels of drug transporters, which are often used in laboratory experiments, might exhibit a distinct resistance to the studied drug. In contrast, cells expressing lower levels of drug transporters, which may occur in clinical samples [34–36], might exhibit a much lower degree of resistance to the particular drug, or the resistance can be completely lost. However, neither moderate nor high expression levels of either transporter failed to mediate resistance to EF-24 in our study (Figures 8 and 9, Table 2). These results indicate that the antiproliferative potential of EF-24 cannot easily be decreased by overexpression of the main drug transporters, ABCB1 and ABCG2, neither in clinical settings nor in the laboratory. In this context, the properties of EF-24 are very similar to those of curcumin, which is not transported by ABCB1 or ABCG2 [37,38].

In conclusion, the results strongly suggest only marginal contribution of a redox-dependent mechanism to EF-24 induced apoptosis in human leukemia K562 cells. The main mechanism of NAC protection against EF-24-induced apoptosis is the conversion of cytotoxic EF-24 into the non-cytotoxic EF-24-NAC adduct. Drug transporters, ABCB1 and ABCG2, do not reduce the antiproliferative effects of EF-24 in K562 cells.

4. Material and Methods

4.1. Chemicals and Cell Treatment

EF-24 (3,5-bis[(2-fluorophenyl)methylene]-4-piperidinone) and DL-sulforaphane were obtained from Sigma-Aldrich (Saint Louis, MO, USA) and dissolved in DMSO. The residual concentration of DMSO in growth medium was approximately 0.1–0.2%. Catalase (CAT) from bovine liver was dissolved in 25 mM Tris/HCl buffer, pH = 7.0. Cell treatment was done using 50 units of CAT per mL. N-acetylcystein (NAC) was dissolved in distilled water. All chemicals were purchased from Sigma-Aldrich (St. Louis, MO, USA). Zosuquidar trihydrochloride (ZSQ; LY335979) and Ko143 (3S,6S,12aS)-1,2,3,4,6,7,12,12a-Octahydro-9-methoxy-6-(2-methylpropyl)-1,4-dioxopyrazino-[1',2':1,6] pyrido [3,4-b]indole-3-propanoic acid 1,1-dimethylethyl ester were obtained from Enzo Life Sciences AG (Lausen, Switzerland).

4.2. Cell Culture

Human chronic myelogenous leukaemia K562 cells were grown in RPMI-1640 medium supplemented with 10% calf fetal serum and antibiotics in 5% CO_2 atmosphere at 37 °C. K562 cells were obtained from European Collection of Authenticated Cell Culture (ECACC).

K562/Dox cells overexpressing ABCB1 were kindly provided by Prof J.P. Marie (University of Paris 6, Paris, France). Characterization of the K562/Dox cell line is given in details elsewhere [39]. K562/DoxDR2 cells with decreased expression of ABCB1 were established by stable transfection of K562/Dox cells with a plasmid vector expressing shRNA targeting the *ABCB1* gene [39,40].

K562/ABCG2 cells overexpressing wild type ABCG2 [41] were kindly provided by Prof B. Sarkadi (National Blood Center and Semmelweis University, Budapest, Hungary). Sub-clones

K562/ABCG2CL10 and K562/ABCG2CL1 with high and low expression level of ABCG2, respectively, were used in this study. They were established by a single cell cloning by limiting dilution of K562/ABCG2 cells [31].

Resistant cells were cultured under the same conditions as maternal K562 cells.

4.3. Determination of Cell Survival and Proliferation

Cell viability and proliferation was determined using the MTT assay as described previously [42].

4.4. Analysis of Cell Cycle and Apoptotic Cells

Flow cytometric measurements of DNA content were used to analyze the cell cycle as well as for identification of apoptotic cells (fraction of cells in the sub G1 phase), as described previously [27,43]. Apoptotic cells are expressed as a percentage of cells in sub G1 phase.

4.5. Measurement of ROS Production

Intracellular ROS were detected in $2',7'$-dichlorodihydrofluorescein diacetate (H_2DCF-DA)-loaded cells (Molecular Probe, Leiden, The Netherlands) as described previously [28].

4.6. Morphological Analysis of Apoptosis

Fixed cells were stained with Hoechst 33342 Sigma-Aldrich (St. Louis, MO, USA) and morphology of cell nuclei was examined using an Olympus BX60 (Olympus, Hamburg, Germany) fluorescence microscope as described previously [44].

4.7. Measurement of Caspase-3 Enzymatic Activity

Caspase-3 enzymatic (DEVDase) activity was measured in cytoplasmic extracts using the fluorescent substrate Ac-DEVD-AMC [45]. It is necessary to note that Ac-DEVD-AMC substrate is efficiently cleaved also by caspase-7. Therefore, we used the term "DEVDase activity" when refering to the cleavage of Ac-DEVD-AMC substrate in the text.

4.8. Measurement of Caspase-3 Processing

Western blot analysis was used to directly demonstrate capase-3 processing. Briefly, protein extracts and sample preparation were done as described previously [46]. Caspase-3 processing was determined using Western blot analysis with a polyclonal anti-caspase-3 antibody (1:1000; Cell Signaling Technology, Danvers, MA, USA) recognizing both pro- and active protease forms and polyclonal anti-HSP90 antibody (1:2000; Cell Signaling Technology, Danvers, MA, USA) for detection of reference protein. A horseradish peroxidase-conjugated secondary anti-rabbit antibody (1:2000; Dako, Glostrup, Denmark) in combination with an enhanced chemiluminiscence (ECL; Amersham, Little Chalfont, UK) was used for signal detection.

4.9. Western Blot Analysis of Activation of Nrf2 and Stress-Response Pathway

Western blot analysis was used to demonstrate the activation of Nrf2 and its downstream regulated genes, including HO-1 and NQO1 [20]. Protein extracts were done as described previously [46]. Mouse monoclonal anti-Nrf2 antibody (A-10; 1:1000; Santa Cruz Biotechnology, Dallas, TX, USA) was used to determine human ~60 kDa forms of Nrf2. Rabbit polyclonal anti-HO-1 antibody (1:1000; Thermo Fisher Scientific, Waltham, MA, USA) for detection of HO-1 and rabbit polyclonal anti-NQO1 antibody (1:1000; Thermo Fisher Scientific, Waltham, MA, USA) for the detection of HO-1 and NQO1, respectively. Polyclonal anti-HSP90 antibody and polyclonal anti-HSP70 antibody (both 1:2000; Cell Signaling Technology, Danvers, MA, USA) was used for detection of reference protein. Goat anti-rabbit IgG-HRP or goat anti-mouse IgG-HRP secondary antibody (1:2000 or 1:10,000; Dako, Glostrup, Denmark) in combination with an enhanced chemiluminiscence (ECL; Amersham, Little Chalfont, UK) was used for signal detection.

4.10. Preparation of Cell Extracts

An optimized acidic extraction of cells after their separation from the culture medium by centrifugation through a layer of silicone oil with a slight modification was used [27,28,47]. Briefly, cells at a density of 5×10^5/mL were incubated in the culture medium (with or without EF-24) for appropriate time periods at 37 °C. Afterwards, cells were centrifuged through silicone oil and cell pellets were extracted as follows: ice cold 5% (v/v) formic acid was used for GSH and EF-24 analysis; or ice cold 4% (v/v) formic acid in 40% (v/v) methanol in water was used for EF-24-GSH and EF-24-NAC adduct analysis. Clarified cell extracts (centrifugation: $40,000 \times g$ 10 min at 4 °C) were diluted with distilled water and analyzed by liquid chromatography coupled with a low-energy collision tandem mass spectrometer (LC/MS/MS). Alternatively, clarified cell extracts were stored at −80 °C.

4.11. High-Performance Liquid Chromatography (HPLC) Analysis of Glutathione (GSH) and Oxidized Glutathione (GSSG)

Quantitative analysis of GSH and GSSG was done using LC/MS/MS during one run with the specific parameters. The chromatographic separations were performed using the high-performance liquid chromatography (HPLC) tower system UltiMate 3000 (Dionex, Germering, Germany), a Polaris C18-A, 5 μm, 250 × 2.0 mm HPLC column (Varian Inc., Lake Forest, CA, USA), and a guard C18, 4.0 × 2.0 mm precolumn (Phenomenex, Torrance, CA, USA). The chromatographic parameters were as follows: the binary gradient of mobile phase A (95% methanol in 0.25% formic acid, v/v) and B (0.25% formic acid in water, v/v) from 0–3 min (5 → 23% of solvent A), from 3–4 min (23 → 95% of solvent A), from 4–6 min (95 → 5% of solvent A) and from 6–10 min (5% of solvent A); the flow rate at 0.3 mL/min; the sample injection volume at 5 μL. The API 3200 triple quadrupole mass spectrometer (MDS SCIEX, Concord, ON, Canada) with the TurboIonSpray interface in the positive ion mode was applied for quantification of analytes. The Product Ion Scan mode (GSH: Q1 quadrupole 308.1 amu, Q3 quadrupole at scale 178.95–179.05 amu and GSSG: Q1 quadrupole 613.1 amu, Q3 quadrupole at scale 230.6–231.4 amu) was used. The mass-dependent parametres were optimized: the collision energy and the declustering potential for GSH standard were 17 V and 26 V, and for GSSG standard were 45 V and 51 V, respectively. Ion spray probe parameters were set for GSH and GSSG standards: needle voltage 5500 V and temperature 450 °C. Data were acquired using Analyst® software, ver. 1.5.1 (MDS SCIEX, Concord, ON, Canada).

4.12. Determination of EF-24

The HPLC/MS/MS analytical system was described in Section 4.11. A Kinetex HILIC, 2.6 μm, 150 × 2.1 mm column (Phenomenex, Torrance, CA, USA) and a guard Kinetex HILIC, 4.0 × 2.0 mm precolumn (Phenomenex, Torrance, CA, USA) were installed. The binary linear gradient of mobile phase C (95% acetonitrile in 0.25% formic acid, v/v) and B (0.25% formic acid in water, v/v) from 0–6 min (95 → 25% of solvent C), from 6–7 min (25 → 95% of solvent C), and from 7–11 min (95% of solvent C) was programmed. The flow rate at 0.15 mL/min and the sample injection volume at 10 μL were set. The mass spectrometer was operated in the multiple-reaction monitoring (MRM) mode. MRM transition mode of 312 > 149 amu (dwell-time = 100 ms) was optimized for quantifying of EF-24. Mass spectrometric parameters were set to the following values: needle voltage, temperature, collision energy, declustering potential and entrance potential at 5500 V, 400 °C, 33.0 V, 46.0 V and 5.0 V, respectively.

4.13. Determination of EF-24-GSH and EF-24-NAC Adducts

The HPLC/MS/MS analytical system was described in Section 4.11. For chromatography columns see Section 4.12. The binary gradient of mobile phase C (95% acetonitrile in 0.25% formic acid, v/v) and B (0.25% formic acid in water, v/v) from 0–6 min (95 → 75% of solvent C), from 6–7 min (75 → 50%

of solvent C), from 7–10 min (50% of solvent C), from 10–11 min (50 → 95% of solvent C), and from 11 → 15 min (95% of solvent C) was used. The flow rate at 0.2 mL/min and the sample injection volume at 10 μL were adjusted. The Product Ion Scan mode was optimized (EF-24–GSH adduct: Q1 quadrupole 619.0 amu, Q3 quadrupole at range 148.8–149.6 amu and EF-24–NAC adduct: Q1 quadrupole 475.4 amu, Q3 quadrupole at range 148.8–149.6 amu). Mass spectrometric parameters were adjusted to the values: needle voltage, temperature, collision energy, declustering potential and entrance potential at 5500 V, 400 °C, 70.0 V, 30.0 V and 5.0 V for EF-24-GSH adduct and 5500 V, 400 °C, 65 V, 35 V and 5.0 V for EF-24-NAC adduct, respectively.

4.14. Statistical Analysis

Data are reported as means ± S.D. All statistical analyses were performed using SigmaPlot 11.0 software package (Systat Software Inc., San Jose, CA, USA). Statistical significance of differences was determined by Student's *t*-tests and one-way ANOVA. *p* values equal to or less than 0.05 were considered significant.

Acknowledgments: This work was supported by an Internal grant of Palacky University Olomouc (IGA_LF_2017_038).

Author Contributions: Research design: Petr Mlejnek; experiments: Nikola Skoupa and Eliska Ruzickova; analytical tools and data analysis: Petr Dolezel; manuscript preparation: Mlejnek Petr.

Conflicts of Interest: The authors declare there are no conflicts of interest.

References

1. Cheng, A.L.; Hsu, C.H.; Lin, J.K.; Hsu, M.M.; Ho, Y.F.; Shen, T.S.; Ko, J.Y.; Lin, J.T.; Lin, B.R.; Ming-Shiang, W.; et al. Phase I clinical trial of curcumin, a chemopreventive agent, in patients with high-risk or pre-malignant lesions. *Anticancer Res.* **2001**, *21*, 2895–2900. [PubMed]
2. Sharma, R.A.; Euden, S.A.; Platton, S.L.; Cooke, D.N.; Shafayat, A.; Hewitt, H.R.; Marczylo, T.H.; Morgan, B.; Hemingway, D.; Plummer, S.M.; et al. Phase I clinical trial of oral curcumin: Biomarkers of systemic activity and compliance. *Clin. Cancer Res.* **2004**, *10*, 6847–6854. [CrossRef] [PubMed]
3. Adams, B.K.; Ferstl, E.M.; Davis, M.C.; Herold, M.; Kurtkaya, S.; Camalier, R.F.; Hollingshead, M.G.; Kaur, G.; Sausville, E.A.; Rickles, F.R.; et al. Synthesis and biological evaluation of novel curcumin analogs as anti-cancer and anti-angiogenesis agents. *Bioorg. Med. Chem.* **2004**, *12*, 3871–3883. [CrossRef] [PubMed]
4. Tan, X.; Sidell, N.; Mancini, A.; Huang, R.P.; Shenming, W.; Horowitz, I.R.; Liotta, D.C.; Taylor, R.N.; Wieser, F. Multiple anticancer activities of EF24, a novel curcumin analog, on human ovarian carcinoma cells. *Reprod. Sci.* **2010**, *17*, 931–940. [CrossRef] [PubMed]
5. Kasinski, A.L.; Du, Y.; Thomas, S.L.; Zhao, J.; Sun, S.Y.; Khuri, F.R.; Wang, C.Y.; Shoji, M.; Sun, A.; Snyder, J.P.; et al. Inhibition of IkappaB kinase-nuclear factor-kappaB signaling pathway by 3,5-bis(2-flurobenzylidene)piperidin-4-one (EF24), a novel monoketone analog of curcumin. *Mol. Pharmacol.* **2008**, *74*, 654–661. [CrossRef] [PubMed]
6. Selvendiran, K.; Tong, L.; Vishwanath, S.; Bratasz, A.; Trigg, N.J.; Kutala, V.K.; Hideg, K.; Kuppusamy, P. EF24 induces G2/M arrest and apoptosis in cisplatin-resistant human ovarian cancer cells by increasing PTEN expression. *J. Biol. Chem.* **2007**, *282*, 28609–28618. [CrossRef] [PubMed]
7. Thomas, S.L.; Zhong, D.; Zhou, W.; Malik, S.; Liotta, D.; Snyder, J.P.; Hamel, E.; Giannakakou, P. EF24, a novel curcumin analog, disrupts the microtubule cytoskeleton and inhibits HIF-1. *Cell Cycle* **2008**, *7*, 2409–2417. [CrossRef] [PubMed]
8. Liang, Y.; Zheng, T.; Song, R.; Wang, J.; Yin, D.; Wang, L.; Liu, H.; Tian, L.; Fang, X.; Meng, X.; et al. Hypoxia-mediated sorafenib resistance can be overcome by EF24 through Von Hippel-Lindau tumor suppressor-dependent HIF-1α inhibition in hepatocellular carcinoma. *Hepatology* **2013**, *57*, 1847–1857. [CrossRef] [PubMed]
9. Chen, W.; Zou, P.; Zhao, Z.; Chen, X.; Fan, X.; Vinothkumar, R.; Cui, R.; Wu, F.; Zhang, Q.; Liang, G.; et al. Synergistic antitumor activity of rapamycin and EF24 via increasing ROS for the treatment of gastric cancer. *Redox Biol.* **2016**, *10*, 78–89. [CrossRef] [PubMed]

10. Subramaniam, D.; May, R.; Sureban, S.M.; Lee, K.B.; George, R.; Kuppusamy, P.; Ramanujam, R.P.; Hideg, K.; Dieckgraefe, B.K.; Houchen, C.W.; et al. Diphenyl difluoroketone: A curcumin derivative with potent in vivo anticancer activity. *Cancer Res.* **2008**, *68*, 1962–1969. [CrossRef] [PubMed]

11. Liu, H.; Liang, Y.; Wang, L.; Tian, L.; Song, R.; Han, T.; Pan, S.; Liu, L. In vivo and in vitro suppression of hepatocellular carcinoma by EF24, a curcumin analog. *PLoS ONE* **2012**, *7*, e48075. [CrossRef] [PubMed]

12. Yang, C.H.; Yue, J.; Sims, M.; Pfeffer, L.M. The curcumin analog EF24 targets NF-κB and miRNA-21, and has potent anticancer activity in vitro and in vivo. *PLoS ONE* **2013**, *8*, e71130. [CrossRef] [PubMed]

13. Mosley, C.A.; Liotta, D.C.; Snyder, J.P. Highly active anticancer curcumin analogues. *Adv. Exp. Med. Biol.* **2007**, *595*, 77–103. [CrossRef] [PubMed]

14. Adams, B.K.; Cai, J.; Armstrong, J.; Herold, M.; Lu, Y.J.; Sun, A.; Snyder, J.P.; Liotta, D.C.; Jones, D.P.; Shoji, M. EF-24, a novel synthetic curcumin analog, induces apoptosis in cancer cells via a redox-dependent mechanism. *Anticancer Drugs* **2005**, *16*, 263–275. [CrossRef] [PubMed]

15. Sun, A.; Lu, Y.J.; Hu, H.; Shoji, M.; Liotta, D.C.; Snyder, J.P. Curcumin analog cytotoxicity against breast cancer cells: Exploitation of a redox-dependent mechanism. *Bioorg. Med. Chem. Lett.* **2009**, *19*, 6627–6631. [CrossRef] [PubMed]

16. Zou, P.; Xia, Y.; Chen, W.; Chen, X.; Ying, S.; Feng, Z.; Chen, T.; Ye, Q.; Wang, Z.; Qiu, C.; et al. EF24 induces ROS-mediated apoptosis via targeting thioredoxin reductase 1 in gastric cancer cells. *Oncotarget* **2016**, *7*, 18050–18064. [CrossRef] [PubMed]

17. He, G.; Feng, C.; Vinothkumar, R.; Chen, W.; Dai, X.; Chen, X.; Ye, Q.; Qiu, C.; Zhou, H.; Wang, Y.; et al. Curcumin analog EF24 induces apoptosis via ROS-dependent mitochondrial dysfunction in human colorectal cancer cells. *Cancer Chemother. Pharmacol.* **2016**, *78*, 1151–1161. [CrossRef] [PubMed]

18. Braun, T.; Carvalho, G.; Fabre, C.; Grosjean, J.; Fenaux, P.; Kroemer, G. Targeting NF-kappaB in hematologic malignancies. *Cell Death Differ.* **2006**, *13*, 748–758. [CrossRef] [PubMed]

19. Suzuki, T.; Yamamoto, M. Molecular basis of the Keap1-Nrf2 system. *Free Radic. Biol. Med.* **2015**, *88*, 93–100. [CrossRef] [PubMed]

20. Gorrini, C.; Harris, I.S.; Mak, T.W. Modulation of oxidative stress as an anticancer strategy. *Nat. Rev. Drug Discov.* **2013**, *12*, 931–947. [CrossRef] [PubMed]

21. Myzak, M.C.; Dashwood, R.H. Chemoprotection by sulforaphane: Keep one eye beyond Keap1. *Cancer Lett.* **2006**, *233*, 208–218. [CrossRef] [PubMed]

22. Ketterer, B.; Coles, B.; Meyer, D.J. The role of glutathione in detoxication. *Environ. Health Perspect.* **1983**, *49*, 59–69. [CrossRef] [PubMed]

23. Franco, R.; Cidlowski, J.A. Apoptosis and glutathione: Beyond an antioxidant. *Cell Death Differ.* **2009**, *16*, 1303–1314. [CrossRef] [PubMed]

24. Gottesman, M.M. How cancer cells evade chemotherapy: Sixteenth Richard and Hinda Rosenthal Foundation Award Lecture. *Cancer Res.* **1993**, *53*, 747–754. [PubMed]

25. Ambudkar, S.V.; Kimchi-Sarfaty, C.; Sauna, Z.E.; Gottesman, M.M. P-glycoprotein: From genomics to mechanism. *Oncogene* **2003**, *22*, 7468–7485. [CrossRef] [PubMed]

26. Nguyen, T.; Sherratt, P.J.; Huang, H.C.; Yang, C.S.; Pickett, C.B. Increased protein stability as a mechanism that enhances Nrf2-mediated transcriptional activation of the antioxidant response element. Degradation of Nrf2 by the 26 S proteasome. *J. Biol. Chem.* **2003**, *278*, 4536–4541. [CrossRef] [PubMed]

27. Mlejnek, P.; Dolezel, P. N-acetylcysteine prevents the geldanamycin cytotoxicity by forming geldanamycin-N-acetylcysteine adduct. *Chem. Biol. Interact.* **2014**, *220*, 248–254. [CrossRef] [PubMed]

28. Mlejnek, P.; Dolezel, P. Loss of mitochondrial transmembrane potential and glutathione depletion are not sufficient to account for induction of apoptosis by carbonyl cyanide 4-(trifluoromethoxy)phenylhydrazone in human leukemia K562 cells. *Chem. Biol. Interact.* **2015**, *239*, 100–110. [CrossRef] [PubMed]

29. Türk, D.; Szakács, G. Relevance of multidrug resistance in the age of targeted therapy. *Curr. Opin. Drug Discov. Dev.* **2009**, *12*, 246–252.

30. Kosztyu, P.; Dolezel, P.; Mlejnek, P. Can P-glycoprotein mediate resistance to nilotinib in human leukaemia cells? *Pharmacol. Res.* **2013**, *67*, 79–83. [CrossRef] [PubMed]

31. Kosztyu, P.; Bukvova, R.; Dolezel, P.; Mlejnek, P. Resistance to daunorubicin, imatinib, or nilotinib depends on expression levels of ABCB1 and ABCG2 in human leukemia cells. *Chem. Biol. Interact.* **2014**, *219*, 203–210. [CrossRef] [PubMed]

32. Mlejnek, P.; Dolezel, P.; Ruzickova, E. Drug resistance of cancer cells is crucially affected by expression levels of ABC-transporters. *BioDiscovery* **2017**, *20*, e11211. [CrossRef]

33. Ruzickova, E.; Janska, R.; Dolezel, P.; Mlejnek, P. A clinically relevant study of interactions of inhibitors of anti-apoptotic Bcl-2 proteins with the ABC transporters. *Pharmazie* **2017**, in press.

34. Fojo, A.T.; Ueda, K.; Slamon, D.J.; Poplack, D.G.; Gottesman, M.M.; Pastan, I. Expression of a multidrug-resistance gene in human tumors and tissues. *Proc. Natl. Acad. Sci. USA* **1987**, *84*, 265–269. [CrossRef] [PubMed]

35. Goldstein, L.J.; Galski, H.; Fojo, A.; Willingham, M.; Lai, S.L.; Gazdar, A.; Pirker, R.; Green, A.; Crist, W.; Brodeur, G.M.; et al. Expression of a multidrug resistance gene in human cancers. *J. Natl. Cancer Inst.* **1989**, *81*, 116–124. [CrossRef] [PubMed]

36. Mlejnek, P.; Kosztyu, P.; Dolezel, P.; Bates, S.E.; Ruzickova, E. Reversal of ABCB1 mediated efflux by imatinib and nilotinib in cells expressing various transporter levels. *Chem. Biol. Interact.* **2017**, *273*, 171–179. [CrossRef] [PubMed]

37. Chearwae, W.; Anuchapreeda, S.; Nandigama, K.; Ambudkar, S.V.; Limtrakul, P. Biochemical mechanism of modulation of human P-glycoprotein (ABCB1) by curcumin I, II, and III purified from Turmeric powder. *Biochem. Pharmacol.* **2004**, *68*, 2043–2052. [CrossRef] [PubMed]

38. Chearwae, W.; Shukla, S.; Limtrakul, P.; Ambudkar, S.V. Modulation of the function of the multidrug resistance-linked ATP-binding cassette transporter ABCG2 by the cancer chemopreventive agent curcumin. *Mol. Cancer Ther.* **2006**, *5*, 1995–2006. [CrossRef] [PubMed]

39. Tang, R.; Faussat, A.M.; Perrot, J.Y.; Marjanovic, Z.; Cohen, S.; Storme, T.; Morjani, H.; Legrand, O.; Marie, J.P. Zosuquidar restores drug sensitivity in P-glycoprotein expressing acute myeloid leukemia (AML). *BMC Cancer* **2008**, *8*, 51–59. [CrossRef] [PubMed]

40. Mlejnek, P.; Dolezel, P.; Kosztyu, P. P-glycoprotein mediates resistance to A3 adenosine receptor agonist 2-chloro-N6-(3-iodobenzyl)-adenosine-5′-N-methyluronamide in human leukemia cells. *J. Cell. Physiol.* **2012**, *227*, 676–685. [CrossRef] [PubMed]

41. Elkind, N.B.; Szentpétery, Z.; Apáti, Á.; Özvegy-Laczka, C.; Várady, G.; Ujhelly, O.; Szabó, K.; Homolya, L.; Váradi, A.; Buday, L.; et al. Multidrug transporter ABCG2 prevents tumor cell death induced by the epidermal growth factor receptor inhibitor Iressa (ZD1839, Gefitinib). *Cancer Res.* **2005**, *65*, 1770–1777. [CrossRef] [PubMed]

42. Mosmann, T. Rapid colorimetric assay for cellular growth and survival: Application to proliferation and cytotoxicity assays. *J. Immunol. Methods* **1983**, *65*, 55–63. [CrossRef]

43. Nicoletti, I.; Migliorati, G.; Pagliacci, M.C.; Grignani, F.; Riccardi, C. A rapid and simple method for measuring thymocyte apoptosis by propidium iodide staining and flow cytometry. *J. Immunol. Methods* **1991**, *139*, 271–279. [CrossRef]

44. Mlejnek, P.; Kuglik, P. Induction of apoptosis in HL-60 cells by N6-Benzyladenosine. *J. Cell. Biochem.* **2000**, *77*, 6–17. [CrossRef]

45. Frydrych, I.; Mlejnek, P. Serine protease inhibitors N-alpha-tosyl-L-lysinyl-chloromethylketone (TLCK) and N-tosyl-L-phenylalaninyl-chloromethylketone (TPCK) are potent inhibitors of activated caspase proteases. *J. Cell. Biochem.* **2008**, *103*, 1646–1656. [CrossRef] [PubMed]

46. Frydrych, I.; Mlejnek, P.; Dolezel, P.; Zoumpourlis, V.; Krumpochova, P. The broad-spectrum caspase inhibitor Boc-Asp-CMK induces cell death in human leukaemia cells. *Toxicol. In Vitro* **2008**, *22*, 1356–1360. [CrossRef] [PubMed]

47. Mlejnek, P.; Novak, O.; Dolezel, P. A non-radioactive assay for precise determination of intracellular levels of imatinib and its main metabolite in Bcr-Abl positive cells. *Talanta* **2011**, *283*, 1466–1471. [CrossRef] [PubMed]

© 2017 by the authors. Licensee MDPI, Basel, Switzerland. This article is an open access article distributed under the terms and conditions of the Creative Commons Attribution (CC BY) license (http://creativecommons.org/licenses/by/4.0/).

International Journal of
Molecular Sciences

MDPI

Article

T315 Decreases Acute Myeloid Leukemia Cell Viability through a Combination of Apoptosis Induction and Autophagic Cell Death

Chang-Fang Chiu [1,2,3], Jing-Ru Weng [4,5], Appaso Jadhav [6], Chia-Yung Wu [5], Aaron M. Sargeant [7] and Li-Yuan Bai [1,3,*]

[1] Division of Hematology and Oncology, Department of Internal Medicine,
     China Medical University Hospital, Taichung 40447, Taiwan; d5686@mail.cmuh.org.tw
[2] Cancer Center, China Medical University Hospital, Taichung 40447, Taiwan
[3] College of Medicine, School of Medicine, China Medical University, Taichung 40402, Taiwan
[4] Department of Marine Biotechnology and Resources, National Sun Yat-sen University,
     Kaohsiung 80424, Taiwan; columnster@gmail.com
[5] Department of Biological Science and Technology, China Medical University, Taichung 40402, Taiwan;
     kelen0523@yahoo.com.tw
[6] Division of Medicinal Chemistry, College of Pharmacy, The Ohio State University,
     Columbus, OH 43210, USA; appaso06@gmail.com
[7] Charles River Laboratories, Preclinical Services, Spencerville, OH 45887, USA; aaron.Sargeant@crl.com
*   Correspondence: lybai6@gmail.com; Tel.: +886-4-2205-2121 (ext. 5051); Fax: +886-4-2233-7675

Academic Editor: Geoffrey Brown
Received: 23 June 2016; Accepted: 10 August 2016; Published: 15 August 2016

Abstract: T315, an integrin-linked kinase (ILK) inhibitor, has been shown to suppress the proliferation of breast cancer, stomach cancer and chronic lymphocytic leukemia cells. Here we demonstrate that T315 decreases cell viability of acute myeloid leukemia (AML) cell lines (HL-60 and THP-1) and primary leukemia cells from AML patients in a dose-responsive manner. Normal human bone marrow cells are less sensitive than leukemia cells to T315. T315 down regulates protein kinase B (Akt) and p-Akt and induces caspase activation, poly-ADP-ribose polymerase (PARP) cleavage, apoptosis and autophagy through an ILK-independent manner. Interestingly, pretreatment with autophagy inhibitors rescues cells from apoptosis and concomitant PARP cleavage, which implicates a key role of autophagic cell death in T315-mediated cytotoxicity. T315 also demonstrates efficacy in vivo, suppressing the growth of THP-1 xenograft tumors in athymic nude mice when administered intraperitoneally. This study shows that autophagic cell death and apoptosis cooperatively contribute to the anticancer activity of T315 in AML cells. In conclusion, the complementary roles of apoptotic and autophagic cell death should be considered in the future assessment of the translational value of T315 in AML therapy.

Keywords: T315; acute myeloid leukemia; apoptosis; autophagy; autophagic cell death

1. Introduction

Acute myeloid leukemia (AML) is a hematological malignancy characterized by the proliferation of clonal neoplastic hematopoietic cells and diverse clinical presentations. Chemotherapy with or without hematopoietic stem cell transplantation remains the mainstay of AML treatment. While advances in medicine and supportive care have led to complete remission for 70%–80% of adult AML patients, only 20%–30% of these patients have long term disease-free survival [1]. The major cause of this discrepancy is the acquisition of chemoresistance in refractory or relapsed AML. Alternative compounds or strategies are therefore needed to more effectively manage patients with AML.

Int. J. Mol. Sci. **2016**, *17*, 1337

T315, N-methyl-3-(1-(4-(piperazin-1-yl)phenyl)-5-(4′-(trifluoromethyl)-[1,1′-biphenyl]-4-yl)-1H-pyrazol-3-yl)propanamide, was originally identified as an integrin-linked kinase (ILK) inhibitor and characterized by Lee et al. [2]. Subsequent studies demonstrate the efficacy of T315 against several types of cancers. In breast cancer, T315 suppressed γ-secretase-mediated Notch1 activation in caveolae of IL-6-abundant cells through inhibition of ILK [3]. Amelioration of NF-κB (nuclear factor κ-light-chain-enhancer of activated B cells) was thought to be responsible for the anticancer activity mediated by T315 in human gastric cancer cells [4]. T315 also has antitumor activity independent of the canonical ILK inhibition. In chronic lymphocytic leukemia, Liu et al. demonstrated that T315 directly abrogated protein kinase B (Akt) activation by preventing translocation of Akt into lipid rafts, and induced caspase-dependent apoptosis by suppressing B-cell receptor, CD49d, CD40, and Toll-like receptor 9-mediated Akt activation in an ILK-independent manner [5].

In the present study, we examine the anticancer activity and possible underlying mechanisms of T315 against two AML cell lines and primary leukemia cells from patients with AML. In addition, the ability of T315 to inhibit leukemia growth is demonstrated in athymic nude mice bearing THP-1 xenografts.

2. Results

2.1. T315 Increases Apoptotic Cells and Reduces Viability of Acute Myeloid Leukemia (AML) Cell Lines and Primary Leukemia Cells from AML Patients

The annexin-V/PI staining and the MTS assay were used to determine the effect of T315 on the viability of HL-60 and THP-1 cells which were treated with 0, 1, 2, 3 or 4 μmol/L T315 for 24 or 48 h. There was a dose-dependent increase of apoptotic cells in both HL-60 and THP-1 cells treated with T315 (Figure 1A). Figure 1B demonstrates the dose and time-dependent decrease in cell viability induced by T315. The IC_{50} values were 2.53 and 2.72 μmol/L at 24 h, and 2.01 and 2.90 μmol/L at 48 h for HL-60 and THP-1, respectively.

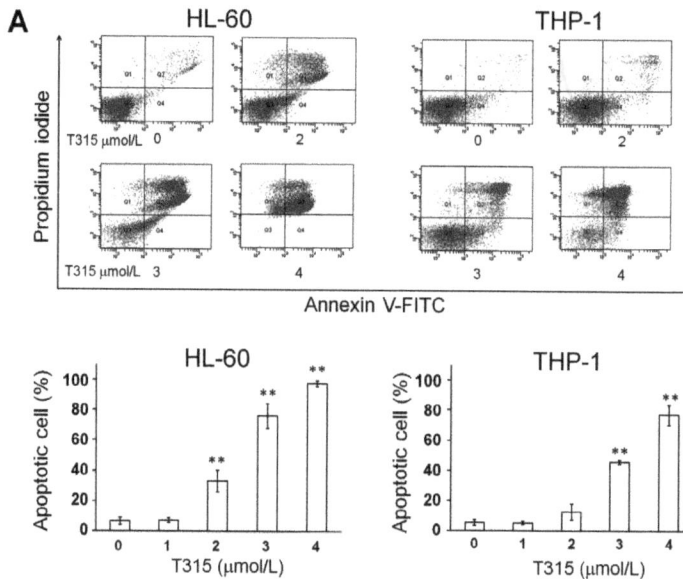

Figure 1. *Cont.*

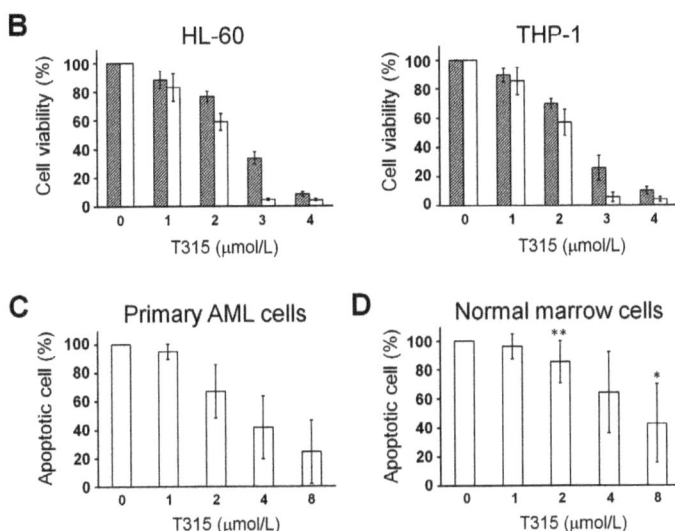

Figure 1. Cell viability inhibition study of T315 in acute myeloid leukemia (AML) cell lines, primary AML cells and normal marrow cells. (**A**) HL-60 and THP-1 cells (0.25×10^6 cells/mL) were incubated with T315 or dimethyl sulfoxide (DMSO) vehicle for 24 h. The apoptotic cells were analyzed by annexin V-FITC and propidium iodide (PI) staining, as described in Materials and Methods. **Upper** panel: one example; **Lower** panel: apoptotic cell percentage ($n = 3$); (**B**) HL-60 and THP-1 cells (0.25×10^6 cells/mL) were incubated with T315 or DMSO vehicle for 24 h ( ) or 48 h (□). The cells were analyzed by MTS assay, as described in Materials and Methods; (**C**) Primary AML cells (0.25×10^6 cells/mL) were incubated with T315 or DMSO for 24 h. The cells were stained with annexin V-FITC and PI to assess apoptotic cells percentage ($n = 26$); (**D**) Normal bone marrow nucleated cells (0.25×10^6 cells/mL) were incubated with T315 or DMSO for 24 h. The cells were stained with annexin V-FITC and PI to assess apoptotic cells percentage ($n = 16$). * denotes $p < 0.05$; ** denotes $p < 0.01$ compared to the control group (in panel **A**) or compared to the primary AML cells at the same concentration of T315 (in panel **D**).

In order to determine the efficacy of T315 on primary AML cell viability, freshly isolated AML cells were treated with T315 (ranging from 0, 1, 2, 4 and 8 μmol/L) and the cell viability was evaluated by annexin-V/PI staining analysis. The mean IC_{50} at 24 h for 26 patients was 4.2 ± 1.6 μmol/L (Figure 1C). Importantly, the normal bone marrow nucleated cells were less sensitive to T315 with an IC_{50} of 6 ± 1.9 μmol/L at 24 h ($n = 16$, Figure 1D). The IC_{50} of T315 for AML cells was significantly lower than the IC_{50} for normal marrow cells ($p = 0.003$).

2.2. T315 Induces Down-Regulation of Protein Kinase B (Akt) and Phosphorylated Akt in AML Cell Lines

T315 has been reported as an ILK inhibitor [2]. We evaluated the influence of T315 on the expression of p^{Thr173}-ILK and total ILK, as well as proteins regulating cell proliferation and survival in AML cells (Figure 2). T315 treatment did not change the protein expression of p^{Thr173}-ILK and total ILK in either HL-60 or THP-1 cells (Figure 2A). This suggested that T315 induced cytotoxicity of AML cells through an ILK-independent manner. Although the expression of Akt did not change (Figure 2B), cells treated with T315 exhibited down regulation of both p^{Thr308}-Akt and p^{Ser473}-Akt which was in contrast with the effect of T315 on prostate and breast cancer cells [2]. There was no change in protein expression of extracellular signal–regulated kinase 1 and 2 (ERK1/2) and phosphorylated ERK1/2 after T315 treatment.

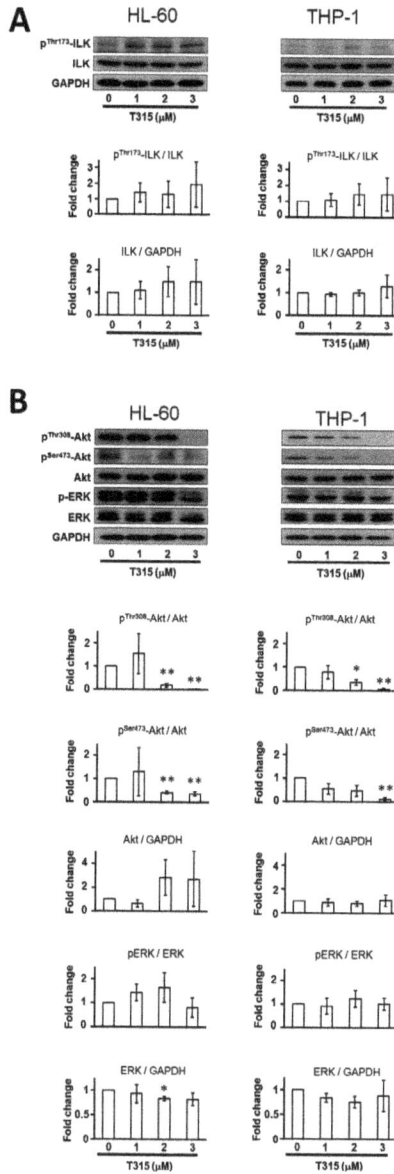

Figure 2. T315 induces dephosphorylation of protein kinase B (Akt) without change of integrin-linked kinase (ILK) in AML cell lines. Cells (0.25×10^6 cells/mL) were treated with T315 at the indicated concentration or DMSO for 24 h, and 20 μg protein extract from cell lysates in each condition were used for Western blot analysis. (**A**) T315 did not change the p^{Thr173}-ILK and total ILK expression. Histogram of fold change of p^{Thr173}-ILK/ILK and ILK/glyceraldehyde 3-phosphate dehydrogenase (GAPDH) were shown in lower panels (*n* = 3); (**B**) T315 down regulated both p^{Thr308}-Akt and p^{Ser473}-Akt, but not Akt, p-ERK and ERK expression. Histogram of fold change of p^{Thr308}-Akt/Akt, p^{Ser473}-Akt/Akt, Akt/GAPDH, p-ERK/ERK, and ERK/GAPDH were shown in lower panels (*n* = 3). * denotes $p < 0.05$; ** denotes $p < 0.01$ compared to the control group.

2.3. T315 Induces Apoptosis, Caspase Activation and Poly-ADP-Ribose Polymerase (PARP) Cleavage in AML Cell Lines

In order to determine if PARP cleavage and caspase activation occur in T315-mediated cytotoxicity, HL-60 and THP-1 cells were incubated with T315 at 0, 1, 2 or 3 μmol/L for 24 h. Western blotting showed that T315 induced PARP cleavage and caspase-3 and caspase-7 activation in HL-60 and THP-1 cell lines in a dose-dependent manner (Figure 3A). The histogram of cleaved PARP versus β-actin, cleaved caspase-3 versus β-actin, and cleaved caspase-7 versus β-actin change folds are shown in Figure 3B ($n = 3$). The time course of PARP cleavage and caspase-3 activation induced by T315 is shown in Figure 3C.

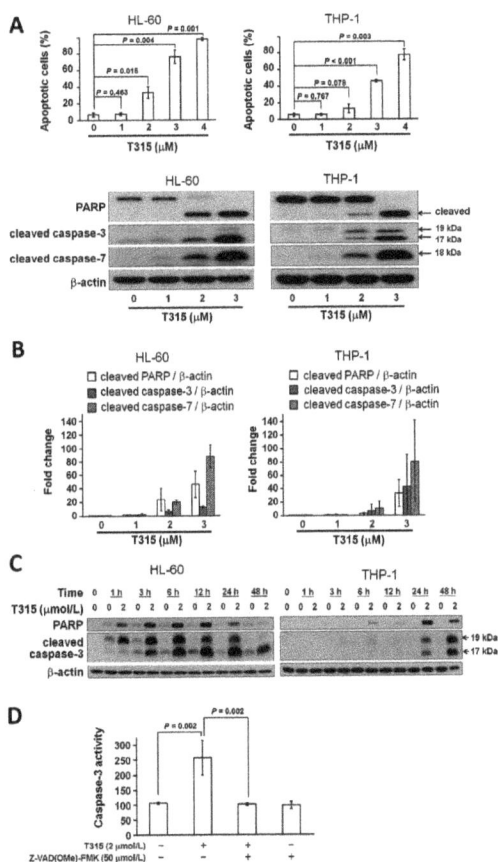

Figure 3. T315-mediated cytotoxicity is dependent on caspase activation and apoptosis. (**A**) T315 induced poly-ADP-ribose polymerase (PARP) cleavage and activation of caspase-3 and caspase-7 in HL-60 and THP-1 cells at 24 h. Protein extract of 20 μg from cell lysates were used for Western blot analysis; (**B**) Fold change of cleaved PARP/β-actin, cleaved caspase-3/β-actin, and cleaved caspase-7/β-actin in treatment with T315 of 1, 2 or 3 μM compared with DMSO control ($n = 3$); (**C**) Time course change of PARP cleavage and caspase-3 activation induced by T315 of 2 μM or DMSO control; (**D**) The increased caspase-3 activity in HL-60 cells treated with T315 for 24 h was rescued by pretreatment of 50 μmol/L Z-Val-Ala-Asp(OMe)-fluoromethyl ketone (Z-VAD(OMe)-FMK).

In order to further validate the caspase-3 activation induced by T315, HL-60 cells were incubated with T315 for 24 h with or without pretreatment of 50 µmol/L Z-Val-Ala-Asp(OMe)-fluoromethyl ketone (Z-VAD(OMe)-FMK), a pan-caspase inhibitor (Figure 3D). The increased caspase-3 activity was completely prevented by Z-VAD(OMe)-FMK treatment.

2.4. T315 Induces Autophagic Cell Death in AML Cell Lines

Autophagy is a physiological process in which cellular components are degraded by lysosomal activity. Either autophagic cytoprotection or autophagic cell death has been shown to be important for the antileukemic effect of different chemotherapeutic agents [6]. Therefore, in addition to apoptosis, we investigated if autophagy was involved in T315-mediated cytotoxicity. Treatment with T315 for 24 h induced dose-dependent increases in microtubule-associated protein 1A/1B light chains 3B (LC3B)-II expression in HL-60 and THP-1 cells (Figure 4A). For comparison, histograms of fold changes of LC3B-II/glyceraldehyde 3-phosphate dehydrogenase (GAPDH) protein expression are shown in Figure 4B.

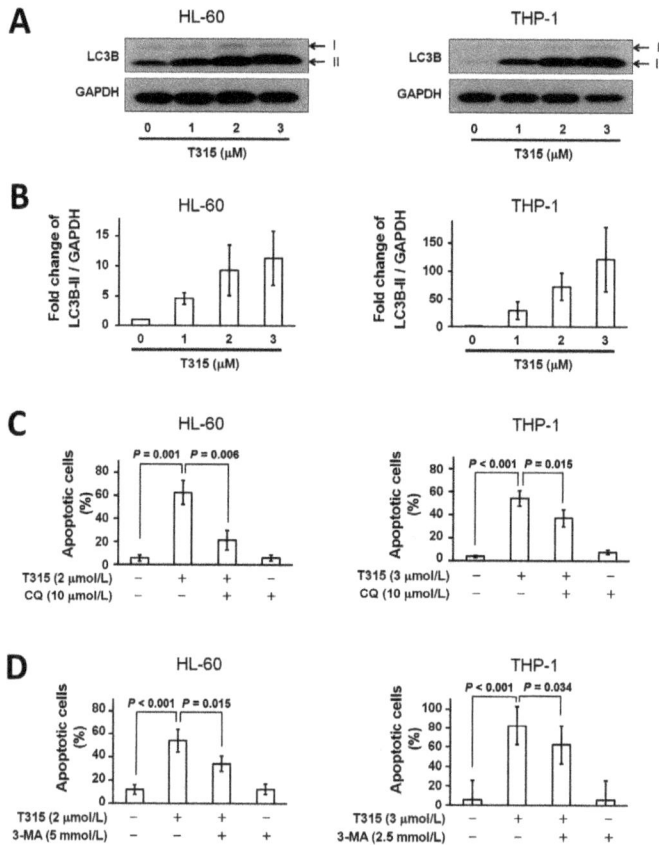

Figure 4. *Cont.*

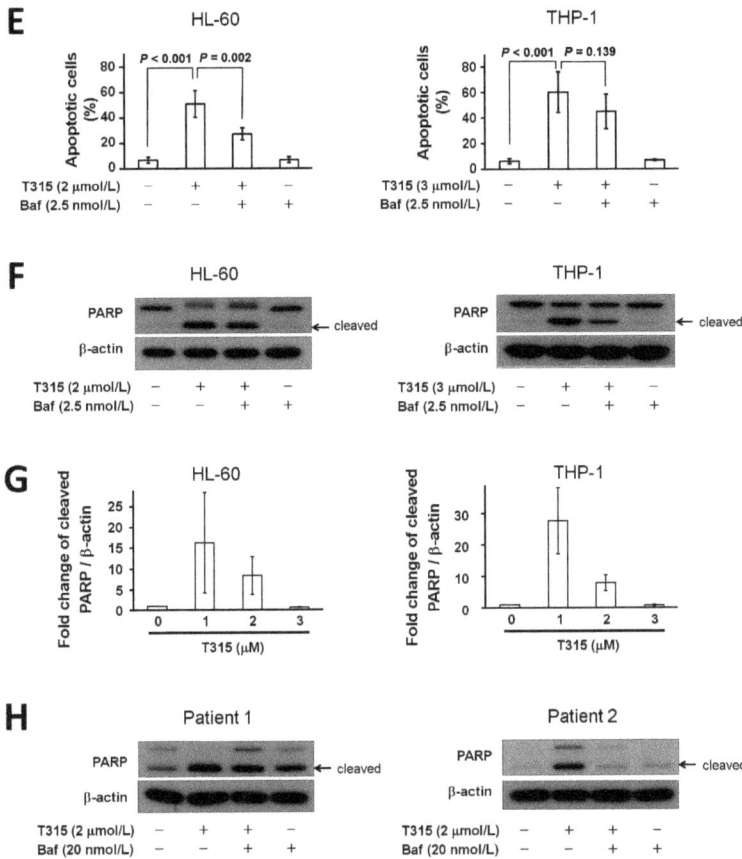

Figure 4. T315 induces autophagic cell death but not protective autophagy in AML cells. (**A**) T315 induced upregulation of LC3B-II in HL-60 and THP-1 cells. Cells (0.25 × 10^6 cells/mL) were treated with indicated concentrations of T315 for 24 h. 20 μg protein from cell lysates were used for Western blot analysis; (**B**) Histogram of fold change of LC3B-II/GAPDH protein expression in cells treated with T315 for 24 h (*n* = 3); (**C**) T315-induced apoptosis was partially rescued by chloroquine (CQ), an autophagy inhibitor. Cells were treated with DMSO vehicle or T315 for 24 h with or without pretreatment of CQ for 1 h, and then analyzed by a flow cytometer. (*n* = 3 for HL-60 and *n* = 4 for THP-1 cells); (**D**) T315-induced apoptosis was partially rescued by 3-methyladenosine (3-MA), an autophagy inhibitor. Cells were treated with DMSO vehicle or T315 for 24 h with or without pretreatment of 3-MA for 1 h, and then analyzed by a flow cytometer. (*n* = 4 for HL-60 and n = 5 for THP-1 cells); (**E**) T315-induced apoptosis was partially rescued by bafilomycin-A1 (Baf), an autophagy inhibitor. Cells were treated with DMSO vehicle or T315 for 24 h with or without pretreatment of Baf for 1 h, and then analyzed by a flow cytometer. (*n* = 5 for HL-60 and *n* = 5 for THP-1 cells); (**F**) T315-induced PARP cleavage was partially rescued by Baf. Cells were treated with DMSO vehicle or T315 for 24 h with or without pretreatment of Baf for 1 h, and then analyzed by Western blotting; (**G**) Histogram of fold change of cleaved PARP/β-actin protein expression in cells treated with T315 with or without pretreatment of Baf for 1 h (*n* = 3); (**H**) T315-induced PARP cleavage in primary AML cells was partially rescued by Baf. Primary AML cells were treated with DMSO vehicle or T315 for 24 h with or without pretreatment of Baf for 1 h, and then analyzed by Western blotting (two patients' data shown here).

Next, to see if autophagic cell death contributed to T315-mediated cytotoxicity, HL-60 and THP-1 cells were treated with dimethyl sulfoxide (DMSO) vehicle control or T315 for 24 h with or without pretreatment of 3 kinds of autophagy inhibitors, chloroquine (CQ), 3-methyladenosine (3-MA), and bafilomycin-A1, and then analyzed for apoptosis (Figure 4C–E). Although the degree of apoptosis rescue varied, all 3 autophagy inhibitors lessened the cell apoptosis induced by T315. These findings implied that autophagic cell death contributed to T315-mediated cell apoptosis.

Compatible with the autophagy inhibitor-mediated rescue of cell apoptosis in flow cytometric analysis, pretreatment with bafilomycin-A1 for 1 h also lessened the PARP cleavage in AML cell lines (Figure 4F,G) and, more important, in primary AML cells (Figure 4H). In summary, T315 induced autophagic cell death, not protective autophagy, in AML cells.

2.5. T315-Mediated Cytotoxicity Is Rescued by Combination of an Apoptosis Inhibitor and an Autophagy Inhibitor

In light of the generation of both apoptosis and autophagic cell death, we further examined the combinatorial effect of an apoptosis inhibitor and an autophagy inhibitor on cell death induced by T315 (Figure 5A,B). For HL-60 cells, the combination of Z-VAD(OMe)-FMK and bafilomycin-A1 rescued more cells than Z-VAD(OMe)-FMK alone ($p = 0.001$). However, the difference of apoptosis rescued between Z-VAD(OMe)-FMK plus bafilomycin-A1 treatment and bafilomycin-A1 alone were less significant ($p = 0.414$). This suggests that autophagic cell death plays a more important role than apoptosis in T315-mediated death of HL-60 cells.

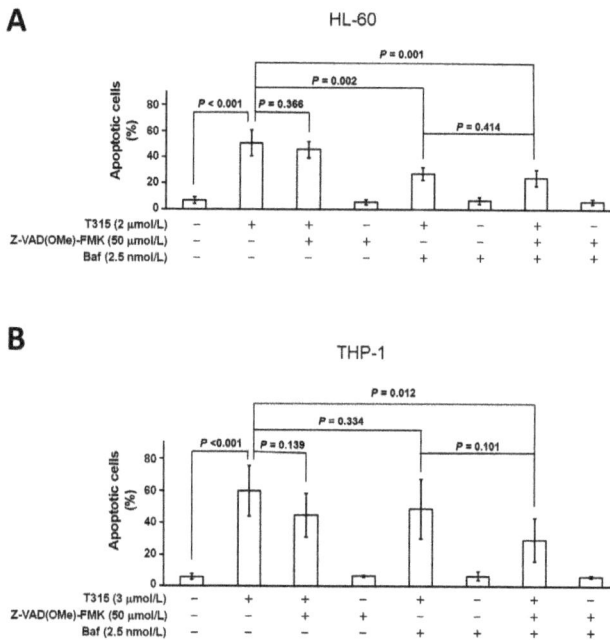

Figure 5. T315-mediated cytotoxicity is rescued by combination of an apoptosis inhibitor and an autophagy inhibitor. Cells were treated with DMSO vehicle or T315 for 24 h with or without pretreatment of Z-VAD(OMe)-FMK and/or bafilomycin-A1 (Baf) for 1 h, and then analyzed by a flow cytometer. (**A**) For HL-60 cells ($n = 5$); (**B**) For THP-1 cells ($n = 5$).

2.6. T315 Slows the Growth of THP-1 Xenografts and Prolongs the Survival of Tumor-Bearing Athymic Nude Mice

To investigate the anti-leukemia effect of T315 in vivo, thirteen male athymic nude mice were xenografted with THP-1 cells. Six mice in the treatment group received T315 intraperitoneally at a dose of 37.5 mg/kg per day, and seven mice in the placebo-control group received the DMSO vehicle daily. T315 had a trend to delay the growth of xenograft tumors (Figure 6A). Although mice in the T315 group had less body weight compared with those in the placebo-controlled group in the first days after initiation of treatment, the loss of body weight did not exceed the 20% endpoint criterion (Figure 6B). In terms of survival time, five mice in the placebo-treated and 4 mice in the T315-treated group reached the humane sacrifice criterion of tumor size ($\geq$2000 mm^3). Although most mice were sacrificed early due to tumor size, a T315-mediated delay in tumor growth was still evident. T315 prolonged the tumor-defined survival time by approximately eight days compared with controls, with median survival times of 20.0 $\pm$ 5.2 days and 28.0 $\pm$ 6.5 days in placebo control mice and T315 treated mice, respectively (Figure 6C, $p = 0.373$).

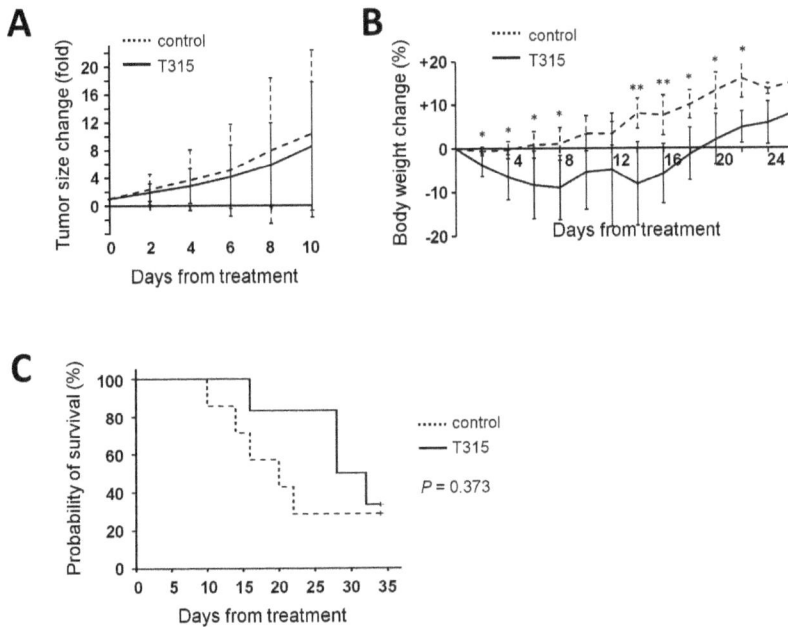

Figure 6. T315 mitigates the growth of THP-1 xenografts and prolongs the survival of tumor-bearing athymic nude mice. (**A**) Mice bearing THP-1 xenografts were treated with DMSO vehicle ($\cdots\cdots$, $n = 7$) or T315 ($\longrightarrow$, $n = 6$) at 37.5 mg/kg/day intraperitoneally. The data represent group means and were plotted until day 10 when one mouse in the control group reached the endpoint tumor size ($\geq$2000 mm^3) and was sacrificed; (**B**) Body weight change of mice. The data were plotted until day 26 when the control group and treated group had 2 and 3 mice remaining, respectively; (**C**) Overall survival curve plotted by Kaplan-Meier method. * denotes $p < 0.05$; ** denotes $p < 0.01$ compared to the control group.

3. Discussion

We have described here the anticancer activity of an ILK inhibitor, T315, in both AML cell lines and primary AML cells. The T315-mediated decrease in cell viability is through both apoptosis and autophagic cell death. Akt and p308-Akt are also down-regulated. In addition, the tumor inhibitory effect of T315 is demonstrated in a THP-1 xenograft mouse model.

Autophagy is a cellular process in which intracellular components are engulfed, digested and recycled via the formation of autophagosomes and autolysosomes, important for cell survival under stress and harmful conditions [6]. This anti-apoptosis function of autophagy has important biological and pathological implications including ischemic injury, cancer therapy and chemoresistance [7]. In the context of cancer, this protective role of autophagy may actually promote tumor survival in a cellular environment of inadequate nutrition or during therapy. Over periods of prolonged stress or poor nutrition, however, autophagy may signal cell death by apoptosis when a cell can no longer survive by recycling organelles. Therefore, the process of autophagy is a double-edged sword in cancer and can either facilitate cancer cell survival or promote cell death depending on other internal and exernal stimuli [8,9].

Cell death mediated by autophagy, referred to as autophagic cell death or type II cell death, has been induced by cancer therapies and was thought to contribute to the death of leukemia [10,11], malignant glioma [12] and lung cancer cells [13]. Indeed, methods used to interfere with the autophagic cell death in these studies rescued the treated cells. Our study shows that T315 induces autophagic cell death but not protective autophagy in AML cells.

Regarding the induction of both apoptosis and autophagic cell death in the present study, it is interesting to note that pretreatment with different autophagy inhibitor rescues both HL-60 and THP-1 cells from apoptosis. This observation implies crosstalk between these modes of cell death rather than 2 independent pathways in AML cells (Figure 4). However, the interplay of apoptosis and autophagic cell death remains undefined.

Various studies have demonstrated an overlap in the regulatory machinery for apoptosis and autophagic cell death. Beclin 1, for example is a protein required for autophagy and also belongs to an apoptosis-requlating domain of proteins. Stressed cells can undergo autophagy induced by Beclin 1 or can undergo apoptosis [8]. Caspase-mediated Beclin 1 cleavage and Beclin 1-Bcl-2 interaction are 2 examples of nodes of crosstalk between autophagy and apoptosis reviewed by Su et al. [9]. One of the most studied and characterized molecular regulator of autophagy and apoptosis is p53 localization [7]. It has been reported that cytoplasmic p53 inhibits autophagy and induces apoptosis while nuclear localization of p53 stimulates both apoptosis and autophagy via the transactivation of target genes [14,15]. In our experiment, the rescue of cell apoptosis by autophagy inhibitors also suggests a crosstalk between autophagy and apoptosis (Figure 4). Our data provides evidence that autophagic cell death and apoptosis can act cooperatively to achieve a cell killing effect. Further mechanistic studies are needed to better characterize the crosstalk between autophagy and apoptosis in AML.

Although our results show a convincing antileukemia effect of T315 in vitro and in vivo, some limitations of our study are noteworthy. First, while the combination of Z-VAD(OMe)-FMK and an autophagy inhibitor resulted in greater rescue from apoptosis induced by T315, the rescue was not complete. This partial rescue of apoptosis in AML cells by both Z-VAD(OMe)-FMK and autophagy inhibitors suggests the existence of other mechanisms of T315-mediated cell death in AML cells. Second, even though T315 delayed the growth of THP-1 xenografts compared to the vehicle control group, most animals in both groups were sacrificed early due to the tumor size reaching the pre-set size criterion. Further modification of T315 to improve the efficacy and to reduce the toxicity is necessary for clinical application.

In conclusion, T315 exhibits a potent antileukemia effect in both AML cell lines and primary AML cells with cell death mediated, at least in part, through generation of apoptosis and autophagic cell death. In vivo, T315 inhibits AML xenograft tumor growth. Collectively, this study provides additional clarity to the anticancer activity of T315 that will be useful in furthering its development for the treatment of AML and possibly other hematological malignancies.

4. Materials and Methods

4.1. Cells and Culture Conditions

Primary AML cells were isolated from freshly collected bone marrow using Ficoll-Paque™ PLUS (GE Healthcare Bio-Sciences AB, Uppsala, Sweden) according to the manufacturer's instructions if the leukemia cells accounted for more than 90% of non-erythroid mononucleated cells of bone marrow. Normal bone marrow nucleated cells were harvested using Ficoll-Paque™ PLUS from patients with treatment-naive non-Hodgkin's lymphoma for whom bone marrow examination for lymphoma staging was performed but determined to be normal. All bone marrow samples were obtained under a protocol approved by the China Medical University Hospital internal review board (CMUH102-REC1-124 issued on 26 May 2014). Written informed consent was obtained from all patients in accordance with the Declaration of Helsinki. Human AML cell lines HL-60 (ATCC CCL-240) and THP-1 (ATCC TIB-202) were from American Type Culture Collection (ATCC, Manassas, VA, USA). All cells were incubated in RPMI-1640 media (Invitrogen, Carlsbad, CA, USA) supplemented with 10% heat-inactivated fetal bovine serum (FBS; Invitrogen) and penicillin (100 U/mL)/streptomycin (100 µg/mL) (Invitrogen) at 37 °C in the presence of 5% CO_2.

4.2. Reagents

T315 {*N*-methyl-3-(1-(4-(piperazin-1-yl)phenyl)-5-(4'-(trifluoromethyl)-[1,1'-biphenyl]-4-yl)-1*H*-pyrazol-3-yl)propanamide} was synthesized as previously described [2], with identity and purity ($\geq$99%) verified by proton nuclear magnetic resonance, high-resolution mass spectrometry, and elemental analysis. For in vitro experiments, T315 was dissolved in dimethyl sulfoxide (DMSO), and added to the culture medium with a final DMSO concentration less than 0.1%. The pharmacological agents were purchased from the respective vendors: bafilomycin-A1 (Cayman Chemical, Ann Arbor, MI, USA); chloroquine (Sigma-Aldrich, St. Louis, MO, USA); 3-methyladenine (3-MA; Sigma-Aldrich); Z-VAD(OMe)-FMK (Santa Cruz Biotechnology, Santa Cruz, CA, USA).

4.3. MTS Assay

Measurement of cell growth was performed using CellTiter 96 Aqueous Non-radioactive Cell Proliferation Assay kit purchased from Promega (Madison, WI, USA). Cells (0.25×10^6/mL) were placed in 200 µL volume in 96-well microtiter plates with the indicated reagent and incubated at 37 °C [16]. MTS solution [3-(4,5-dimethylthiazol-2-yl)-5-(3-carboxymethoxyphenyl)-2-(4-sulfophenyl)-2*H*-tetrazolium] and PMS (phenazine methosulfate) solution were mixed 20:1 by volume. The colorimetric measurements were performed 4 h later at 490-nm wavelength by a VersaMax tunable microplate reader (Molecular Devices, Sunnyvale, CA, USA). The cell viability was expressed as a percentage of absorbance value in treated samples compared to that observed in control vehicle-treated samples (subtract the blank in both conditions).

4.4. Cell Viability and Apoptosis Assay by Flow Cytometry

Cell viability was assessed by dual staining with annexin V conjugated to fluorescein isothiocyanate (FITC) and propidium iodide (PI) [17]. Cells (0.5×10^6) were stained by annexin V-FITC (BD Pharmingen, San Diego, CA, USA) and PI (BD Pharmingen) according to the manufacturer's instructions. Cells were analyzed by a flow cytometer BD FACSCanto II (BD, Franklin Lakes, NJ, USA). Viable cells were those with both annexin V-FITC negative and PI negative staining. The viable cells in each sample were expressed as % by normalizing annexin V-/PI- cells to control. Annexin V-FITC positive cells were identified as apoptotic cells [18].

4.5. Western Blotting

Cell lysates were prepared using RIPA buffer (150 mmol/L NaCl, 50 mmol/L Tris pH 8.0, 1% NP40, 0.5% sodium deoxycholate and 0.1% sodium dodecyl sulfate) supplemented with protease inhibitor (Sigma-Aldrich) and phosphatase inhibitor cocktail (Calbiochem, Darmstadt, Germany) [19]. Antibodies against various proteins were obtained from the following sources: poly-ADP-ribose polymerase (PARP), p^{Thr308}-Akt, p^{Ser473}-Akt, cleaved caspase-3, LC3B, cleaved caspase-7 (Cell Signaling, Danvers, MA, USA); Akt, ERK1/2, $p^{Thr202Tyr204}$-ERK1/2, GAPDH, ILK, p^{Thr173}-ILK (Santa Cruz Biotechnology); β-actin (Sigma-Aldrich). The goat anti-rabbit IgG-horseradish peroxidase (HRP) conjugates and goat anti-mouse IgG-HRP conjugates were purchased from Jackson ImmunoResearch Laboratories, Inc. (West Grove, PA, USA).

4.6. Analysis of Caspase-3 Activity

Caspase-3 activity was assessed using a FITC rabbit anti-active caspase-3 kit (BD Pharmingen) according the manufacturer's protocol.

4.7. In Vivo Therapeutic Efficacy Evaluation of T315 in the THP-1 Xenograft Model

The in vivo efficacy evaluation of T315 was carried out using a xenograft model in athymic nude mice [16]. Thirteen male nude mice of 5 to 7 weeks of age were obtained from the National Laboratory Animal Center (Taipei, Taiwan). The mice were housed under conditions of constant photoperiod (12 h light and 12 h dark) with ad libitum access to sterilized food and water. THP-1 cells were cultured in RPMI-1640 supplemented with 10% heat-inactivated FBS. Before inoculation, THP-1 cells were washed with PBS twice and resuspended in a mixture of RPMI-1640 and Matrigel (BD Matrigel™ Basement Membrane Matrix; BD) with a 1:1 volume ratio. Each mouse was inoculated over the flank subcutaneously with 1×10^7 THP-1 cells in a total volume of 0.2 mL. Tumor diameter was measured every three days using calipers and the tumor volume was calculated using a standard formula: $width^2 \times length \times 0.52$. Body weights of the mice were measured every three days. When the mean tumor volume had reached 50 mm^3, mice were randomized to two groups (seven mice and six mice in placebo-control group and treatment group, respectively). The mice in the treatment group received T315 (concentration 18.75 mg/mL = 35.14 mmol/L) intraperitoneally once daily at a dose of 37.5 mg/kg per day (for example, volume injected is 50 μL for a mouse weighting 25 g), and the mice in the placebo-control group received the DMSO vehicle. All mice received treatments daily until reaching the endpoint. Humane endpoint criteria included body weight loss more than 20% or tumor size more than 2000 mm^3. Scheduled terminal sacrifice for surviving mice occurred on day 35 after initiation of T315 or placebo. The in vivo experiment protocol was approved by the Institutional Animal Care and Use Committee of China Medical University (Taichung, Taiwan, IACUC Approval no.: 104-87-N, period of protocol valid from 1 August 2015 to 31 July 2017).

4.8. Statistical Analysis

Nonlinear mixed models were used to obtain IC_{50}. Two-tailed unpaired *t*-test was used for comparisons of two sets of data. Kaplan-Meier overall survival curve of mice was analyzed using log rank test. All statistical analysis was performed with SPSS for Windows (SPSS, Inc., Chicago, IL, USA).

Acknowledgments: We thank Tse-Yen Yang of China Medical University Hospital for the statistical assistance. This work was supported in part by grants from the Ministry of Health and Welfare, China Medical University Hospital Cancer Research Center of Excellence (MOHW105-TDU-B-212-134003), Ministry of Science and Technology, R.O.C. (MOST 103-2314-B-039-022, MOST 104-2314-B-039-049), China Medical University (CMU104-S-33) and China Medical University Hospital (DMR-105-018, DMR-105-141).

Author Contributions: Chang-Fang Chiu, Jing-Ru Weng and Li-Yuan Bai designed the research study; Chang-Fang Chiu, Appaso Jadhav and Chia-Yung Wu performed the study; Aaron M. Sargeant and Li-Yuan Bai performed data analysis; Jing-Ru Weng, Aaron M. Sargeant and Li-Yuan Bai prepared the manuscript; all authors approved the final version of the manuscript.

Int. J. Mol. Sci. **2016**, *17*, 1337

Conflicts of Interest: The authors declare no conflict of interest.

Abbreviations

AML	acute myeloid leukemia
DMSO	dimethyl sulfoxide
FITC	fluorescein isothiocyanate
HRP	horseradish peroxidase
ILK	integrin-linked kinase
PARP	poly-ADP-ribose polymerase
PI	propidium iodide

References

1. Tallman, M.S.; Gilliland, D.G.; Rowe, J.M. Drug therapy for acute myeloid leukemia. *Blood* **2005**, *106*, 1154–1163. [CrossRef] [PubMed]
2. Lee, S.L.; Hsu, E.C.; Chou, C.C.; Chuang, H.C.; Bai, L.Y.; Kulp, S.K.; Chen, C.S. Identification and characterization of a novel integrin-linked kinase inhibitor. *J. Med. Chem.* **2011**, *54*, 6364–6374. [CrossRef] [PubMed]
3. Hsu, E.C.; Kulp, S.K.; Huang, H.L.; Tu, H.J.; Salunke, S.B.; Sullivan, N.J.; Sun, D.; Wicha, M.S.; Shapiro, C.L.; Chen, C.S. Function of integrin-linked kinase in modulating the stemness of IL-6-abundant breast cancer cells by regulating γ-secretase-mediated NOTCH1 activation in caveolae. *Neoplasia* **2015**, *17*, 497–508. [CrossRef] [PubMed]
4. Tseng, P.C.; Chen, C.L.; Shan, Y.S.; Chang, W.T.; Liu, H.S.; Hong, T.M.; Hsieh, C.Y.; Lin, S.H.; Lin, C.F. An increase in integrin-linked kinase non-canonically confers NF-κB-mediated growth advantages to gastric cancer cells by activating ERK1/2. *Cell Commun. Signal.* **2014**, *12*, 69. [CrossRef] [PubMed]
5. Liu, T.M.; Ling, Y.; Woyach, J.A.; Beckwith, K.; Yeh, Y.Y.; Hertlein, E.; Zhang, X.; Lehman, A.; Awan, F.; Jones, J.A.; et al. OSU-T315: A novel targeted therapeutic that antagonizes AKT membrane localization and activation of chronic lymphocytic leukemia cells. *Blood* **2015**, *125*, 284–295. [CrossRef] [PubMed]
6. Ekiz, H.A.; Can, G.; Baran, Y. Role of autophagy in the progression and suppression of leukemias. *Crit. Rev. Oncol. Hematol.* **2012**, *81*, 275–285. [CrossRef] [PubMed]
7. Chaabane, W.; User, S.D.; El-Gazzah, M.; Jaksik, R.; Sajjadi, E.; Rzeszowska-Wolny, J.; Los, M.J. Autophagy, apoptosis, mitoptosis and necrosis: Interdependence between those pathways and effects on cancer. *Arch. Immunol. Ther. Exp. (Warsz)* **2013**, *61*, 43–58. [CrossRef] [PubMed]
8. Mathew, R.; Karantza-Wadsworth, V.; White, E. Role of autophagy in cancer. *Nat. Rev. Cancer* **2007**, *7*, 961–967. [CrossRef] [PubMed]
9. Su, M.; Mei, Y.; Sinha, S. Role of the crosstalk between autophagy and apoptosis in cancer. *J. Oncol.* **2013**, *2013*, 102735. [CrossRef] [PubMed]
10. Bredholt, T.; Dimba, E.A.; Hagland, H.R.; Wergeland, L.; Skavland, J.; Fossan, K.O.; Tronstad, K.J.; Johannessen, A.C.; Vintermyr, O.K.; Gjertsen, B.T. Camptothecin and khat (*Catha edulis* Forsk.) induced distinct cell death phenotypes involving modulation of c-FLIP$_L$, Mcl-1, procaspase-8 and mitochondrial function in acute myeloid leukemia cell lines. *Mol. Cancer* **2009**, *8*, 101. [CrossRef] [PubMed]
11. Crazzolara, R.; Bradstock, K.F.; Bendall, L.J. RAD001 (Everolimus) induces autophagy in acute lymphoblastic leukemia. *Autophagy* **2009**, *5*, 727–728. [CrossRef] [PubMed]
12. Ito, H.; Aoki, H.; Kuhnel, F.; Kondo, Y.; Kubicka, S.; Wirth, T.; Iwado, E.; Iwamaru, A.; Fujiwara, K.; Hess, K.R.; et al. Autophagic cell death of malignant glioma cells induced by a conditionally replicating adenovirus. *J. Natl. Cancer Inst.* **2006**, *98*, 625–636. [CrossRef] [PubMed]
13. Peng, P.L.; Kuo, W.H.; Tseng, H.C.; Chou, F.P. Synergistic tumor-killing effect of radiation and berberine combined treatment in lung cancer: The contribution of autophagic cell death. *Int. J. Radiat. Oncol. Biol. Phys.* **2008**, *70*, 529–542. [CrossRef] [PubMed]
14. Tasdemir, E.; Maiuri, M.C.; Galluzzi, L.; Vitale, I.; Djavaheri-Mergny, M.; D'Amelio, M.; Criollo, A.; Morselli, E.; Zhu, C.; Harper, F.; et al. Regulation of autophagy by cytoplasmic p53. *Nat. Cell Biol.* **2008**, *10*, 676–687. [CrossRef] [PubMed]

15. Tasdemir, E.; Maiuri, M.C.; Orhon, I.; Kepp, O.; Morselli, E.; Criollo, A.; Kroemer, G. p53 represses autophagy in a cell cycle-dependent fashion. *Cell Cycle* **2008**, *7*, 3006–3011. [CrossRef] [PubMed]

16. Bai, L.Y.; Weng, J.R.; Chiu, C.F.; Wu, C.Y.; Yeh, S.P.; Sargeant, A.M.; Lin, P.H.; Liao, Y.M. OSU-A9, an indole-3-carbinol derivative, induces cytotoxicity in acute myeloid leukemia through reactive oxygen species-mediated apoptosis. *Biochem. Pharmacol.* **2013**, *86*, 1430–1440. [CrossRef] [PubMed]

17. Bai, L.Y.; Weng, J.R.; Lo, W.J.; Yeh, S.P.; Wu, C.Y.; Wang, C.Y.; Chiu, C.F.; Lin, C.W. Inhibition of hedgehog signaling induces monocytic differentiation of HL-60 cells. *Leuk. Lymphoma* **2012**, *53*, 1196–1202. [CrossRef] [PubMed]

18. Bai, L.Y.; Chiu, C.F.; Chu, P.C.; Lin, W.Y.; Chiu, S.J.; Weng, J.R. A triterpenoid from wild bitter gourd inhibits breast cancer cells. *Sci. Rep.* **2016**, *6*, 22419. [CrossRef] [PubMed]

19. Bai, L.Y.; Chiu, C.F.; Kapuriya, N.P.; Shieh, T.M.; Tsai, Y.C.; Wu, C.Y.; Sargeant, A.M.; Weng, J.R. BX795, a TBK1 inhibitor, exhibits antitumor activity in human oral squamous cell carcinoma through apoptosis induction and mitotic phase arrest. *Eur. J. Pharmacol.* **2015**, *769*, 287–296. [CrossRef] [PubMed]

© 2016 by the authors. Licensee MDPI, Basel, Switzerland. This article is an open access article distributed under the terms and conditions of the Creative Commons Attribution (CC BY) license (http://creativecommons.org/licenses/by/4.0/).

International Journal of
Molecular Sciences

MDPI

Review

Immunotherapeutic Concepts to Target Acute Myeloid Leukemia: Focusing on the Role of Monoclonal Antibodies, Hypomethylating Agents and the Leukemic Microenvironment

Olumide Babajide Gbolahan [1], Amer M. Zeidan [2], Maximilian Stahl [2], Mohammad Abu Zaid [3], Sherif Farag [3], Sophie Paczesny [4] and Heiko Konig [1,*]

1   Department of Medicine, Division of Hematology/Oncology, Indiana University School of Medicine, Indianapolis, IN 46202, USA; obgbolah@iu.edu
2   Department of Medicine, Section of Hematology, Yale University School of Medicine, New Haven, CT 06510, USA; amer.zeidan@yale.edu (A.M.Z.); maximilian.stahl@yale.edu (M.S.)
3   Department of Medicine, Bone Marrow and Stem Cell Transplantation, Indiana University School of Medicine, Indianapolis, IN 46202, USA; mabuzaid@iu.edu (M.A.Z.); ssfarag@iu.edu (S.F.)
4   Wells Center for Pediatric Research, Riley Hospital for Children, Indiana University School of Medicine, Indianapolis, IN 46202, USA; sophpacz@iu.edu
*   Correspondence: hkonig@iupui.edu; Tel.: +1-317-274-3590

Received: 27 June 2017; Accepted: 24 July 2017; Published: 31 July 2017

Abstract: Intensive chemotherapeutic protocols and allogeneic stem cell transplantation continue to represent the mainstay of acute myeloid leukemia (AML) treatment. Although this approach leads to remissions in the majority of patients, long-term disease control remains unsatisfactory as mirrored by overall survival rates of approximately 30%. The reason for this poor outcome is, in part, due to various toxicities associated with traditional AML therapy and the limited ability of most patients to tolerate such treatment. More effective and less toxic therapies therefore represent an unmet need in the management of AML, a disease for which therapeutic progress has been traditionally slow when compared to other cancers. Several studies have shown that leukemic blasts elicit immune responses that could be exploited for the development of novel treatment concepts. To this end, early phase studies of immune-based therapies in AML have delivered encouraging results and demonstrated safety and feasibility. In this review, we discuss opportunities for immunotherapeutic interventions to enhance the potential to achieve a cure in AML, thereby focusing on the role of monoclonal antibodies, hypomethylating agents and the leukemic microenvironment.

Keywords: acute myeloid leukemia; immunotherapy, monoclonal antibodies; hypomethylating agents; microenvironment

1. Introduction

Acute myeloid leukemia (AML) remains one of the greatest therapeutic challenges in the field of hematologic malignancies. Despite significant progress in understanding AML at the molecular level, current AML treatments almost generally fail following an initial remission and have remained largely unchanged for almost 40 years [1]. Only about 35–40% of adult patients aged 60 years or younger and approximately 5–15% of elderly patients are currently cured by the means of conventional anti-leukemic treatments, including intensive chemotherapy and allogeneic stem cell transplantation (allo-SCT) [2]. Systemic AML treatment has long been shaped by the prevailing belief that leukemic cells can only be eliminated by a "direct hit" against the malignant cell itself. In consequence of this dogma, cell-cycle active compounds such as cytosine arabinosides have been established as the backbone of most treatment protocols. Depending on the ability to tolerate such treatment, up to 80% of patients achieve a complete remission (CR) in response to these regimens [3] . However, without additional therapy virtually all patients relapse within a matter of months. Post-remission therapy in the form of additional chemotherapy or allo-SCT is therefore mandatory and frequently employed with the goal to eliminate residual leukemia cells that survive induction chemotherapy. Yet, many patients still relapse after post-remission therapy which highlights the need for novel strategies to more effectively combat AML.

Against the background of the "direct hit" dogma, harnessing the immune system to systemically attack AML cells has initially been considered to be of little benefit. This reckoning was fueled by the results of several AML vaccination studies which showed only a few significant clinical responses [4,5]. However, the success of allo-SCT foregrounded the importance of immunotherapeutic concepts in the management of this fatal disease. In recent years, an increasing number of immune system targeted agents have gained access to the clinical arena. With the advent of rituximab in the treatment of Non-Hodgkin lymphomas [6], passive immunotherapies targeting defined targets on tumor cells have become an essential component in the treatment of various hematologic malignancies. In addition, the dramatic impact of checkpoint inhibitors such as ipilimumab [7] and nivolumab [8] on the outcome of advanced melanomas have clearly shown that immunotherapy can result in durable cancer remissions, and that immunogenic cells represent promising, "tumor cell independent" therapeutic targets. Most recently, the bispecific T-cell engager blinatumomab was granted full approval by the Food and Drug Administration (FDA) to treat relapsed/refractory B-cell precursor acute lymphoblastic leukemia in adults and children after a phase 3 study showed a significant survival benefit for patients treated with blinatumomab compared to traditional chemotherapy [9]. This approval marks the first time the FDA has approved an immunotherapeutic agent for the treatment of acute leukemia since the approval of gemtuzumab ozogamicin, and rings in the beginning of a paradigm change in the management of this disease.

The goal of this review is to provide insight into novel immunotherapeutic principles that holds the promise of a paradigm shift in the management of AML.

2. Monoclonal Antibodies (mAbs)

2.1. CD33

CD33, a glycosylated transmembranous protein and member of the "sialic acid-binding Ig-related lectins" (siglecs, siglec-3), functions as an important mediator of cellular adhesion and interaction. High levels of CD33 expression have been reported on myeloid precursor cells in the bone marrow (BM) and on AML blasts, where expression of the CD33 antigen is found in up to 90% of cases [10]. CD33 therefore represents a promising target for AML therapy. Gemtuzumab ozogamicin (GO), a conjugate of a recombinant humanized CD33 antibody and the antitumor antibiotic calicheamicin, is one of various antibody-cytotoxic agent complexes that was initially designed to selectively target CD33 expressing leukemic cells. Due to its encouraging activity in single agent and combination clinical trials, GO was granted accelerated approval in 2001 but was then voluntarily withdrawn from the US market in 2010 after considerable toxicities, mainly consisting of substantial liver toxicity, were reported [11]. In 2011, the United Kingdom Medical Research Council published the results of a clinical trial (MRC AML 15) in which 1,113 de novo AML patients aged less than 60 years were randomized to receive induction chemotherapy with or without GO (3 mg/m^2). Upon remission, 948 patients were randomized to receive consolidation chemotherapy alone or combined with GO. The investigators reported that the addition of GO to chemotherapy was safe but did not lead to any improvement in response or survival rates. However, predefined analysis by cytogenetics showed a significant survival benefit for AML patients with favorable, and a trend for benefit in patients with intermediate risk disease [12]. Another randomized trial conducted by the same group (MCR AML 16) randomized

1115 elderly patients with previously untreated AML or high risk myelodysplastic syndrome (MDS) to induction chemotherapy with or without GO (3 mg/m^2). In this study, the addition of GO was not associated with any improvements in remission rate, 30- or 60-day mortality, or toxicity. However, the 3-year cumulative incidence of relapse and 3-year survival rate were both significantly improved in the GO arm [13]. Another trial investigated the benefits of low dose Ara-C (LDAC) compared to LDAC combined with GO (5 mg) in 495 elderly AML patients. The authors found that the addition of GO was associated with a significant improvement in CR rates (30% [LDAC+GO] vs. 17% [LDAC]) but that the 12-month overall survival (OS) was not improved [14]. Recently, a meta-analysis of individual patient data derived from five randomized trials, including 3325 patients with a median age of 58 years, demonstrated that the addition of GO to conventional induction chemotherapy is safe and associated with a significant survival benefit for patients without adverse cytogenetic risk factors [15]. Targeting CD33 on leukemic blasts with bispecific T-cell engagers (BiTEs) has shown promising activity in preclinical studies. BiTEs were specifically designed to bind to a surface target antigen on cancer cells and to CD3 on T-cells, thus bringing both cells to close proximity with the goal to induce T-cell activation and lysis of the attached cancer cell by cytotoxic granule fusion, cytokines release, and membrane perforation. AMG330, a CD3/CD33- bispecific T-cell engaging antibody showed promising pre-clinical activity against AML cells in vitro as well as in mouse models [16]. A phase 1 study of AMG330 in subjects with relapsed/refractory AML is currently ongoing (ClinicalTrials.gov Identifier: NCT02520427). Several other trials targeting CD33 are actively recruiting patients (Table 1).

2.2. CD123

Another promising target is represented by the Interleukin-3 receptor α chain (IL3Rα/CD123) which is overexpressed on leukemic stem cells and AML blasts compared to normal hematopoietic cells [17]. CSL362, a monoclonal antibody to CD123 demonstrated significant, antibody-dependent cell mediated cytotoxicity against AML blasts and was highly effective in reducing the leukemic burden in AML xenograft mouse models [18]. Results from a first in man, phase 1 study of CSL362 in patients with CD123 positive AML in complete remission at high risk for early relapse showed that CSL362 is safe, well tolerated, and durably depletes CD123 positive cells [19]. Phase 2 studies of CSL362 are currently under way (Table 1). SL-401, a novel CD123 targeted antibody, is currently being evaluated as a consolidation strategy in AML patients in CR1 or CR2 with high risk of relapse. Data from a multicenter, single-arm phase 2 trial showed that SL-401 is well tolerated and confers potent anti-leukemic activity with the potential to eliminate drug-resistant AML cells in patients with minimal residual disease (MRD) [20] (NCT02270463) (Table 1). XmAb14045 is a bispecific antibody that contains both a CD123 and a CD3 binding domain to activate T-cells for effective killing of CD123 expressing AML cells. A Phase 1 clinical trial of XmAb14045 for the treatment of AML and other hematologic malignancies is currently ongoing (NCT02730312) (Table 1). Similarly, flotetuzumab (also known as MGD006 or S80880) both recognizes CD123 and CD3 to redirect T-lymphocytes to eliminate CD123 expressing cells. Flotetuzumab effectively reduced leukemic burden in mouse AML xenograft models and was well tolerated in monkeys who were treated with continuous infusion of up to 1 μg/kg per day during a 4-week period [21]. In January 2017, flotetuzumab was granted orphan drug designation by the FDA. A phase 1 study of flotetuzumab in relapsed/refractory AML or intermediate-2/ high risk Myelodysplastic Syndrome (MDS) patients is under way (NCT02152956) (Table 1). JNJ-63709178, another CD3/CD123 targeted bispecific antibody was in early phase studies when a complete hold was instituted after several patients suffered serious adverse events.

Table 1. Immune-based therapeutic concepts under active development for the treatment of AML.

Target	Drug	Trial Phase	Patient Population	Single Agent/Combination	Ref./Identifier	Status
CD33	IMGN779	I	Adult patients with relapsed/refractory CD33+ AML.	Single agent	NCT02674763	Recruiting
CD33	Gemtuzumab ozogamicin	II	Patients up to 70 years with AML induction/re-induction failure, AML in CR1 with poor cytogenetics, AML in 2nd CR with MRD, AML in 3rd CR, AML in refractory relapse but ≤25% BM blasts, MDS with >6% BM blasts at diagnosis, secondary MDS with ≤5% BM blasts at diagnosis Note: disease must express >/=10% CD33+ for patients with AML	Combination with busulfan and cyclophosphamide	NCT02221310	Recruiting
CD33	AMV564	I	Adult patients with relapsed/refractory AML	Single agent	NCT03144245	Recruiting
CD33	SGN-CD33A	III	Adult patients with newly diagnosed, previously untreated intermediate or adverse risk de novo or secondary AML.	Combination with azacitidine or decitabine	NCT02785900	Recruiting
CD123	SGN-CD123A	I	Adult patients up to 74 years with relapsed/refractory CD123-detectable AML following at least 2 but no more than 3 prior regimens; patients may be eligible after only 1 previous regimen if in a high risk category	Single agent	NCT02848248	Recruiting
CD123	XmAb14045	I	Adult patients with primary or secondary AML, B-cell Acute lymphocytic leukemia (ALL), blastic plasmacytoid dendritic cell neoplasm (BPDCN), Chronic myeloid leukemia (CML) in blast phase, resistant or intolerant to tyrosine kinase inhibitors; patients with relapsed or refractory disease with no available standard therapy	Single agent	NCT02730312	Recruiting
CD123	JNJ-56022473 (CSL362)	II/III	Elderly patients, 65 years or older with de novo or secondary AML.	Combination with decitabine	NCT02472145	Recruiting
CD123	MGD006	I	Adult patients with primary or secondary AML or MDS with an International prognostic scoring system (IPSS) category of intermediate 2 or high risk	Single agent	NCT02152956	Recruiting
CD123	SL-401	I/II	Adult patients with AML in first or second CR or CRi	Single agent	NCT02270463	Recruiting
PD-L1	Durvalumab (MEDI4736)	II	Adult patients with MDS or elderly patients (≥65 years) with newly diagnosed de novo AML or secondary AML.	Combination with azacitidine	NCT02775903	Recruiting
PD-L1	Atezolizumab	I	Adult patients with relapsed refractory AML; elderly patients with treatment naïve AML who are unfit for induction chemotherapy	Combination with guadecitabine	NCT02892318	Recruiting
PD-1	Nivolumab	II	Adult patients with relapsed/refractory AML	Combination with azacitidine; Combination with ipilimumab and azacitidine	NCT02397720	Recruiting
PD-1	Pembrolizumab	II	Adult patients with relapsed/refractory AML	Combination with azacitidine	NCT02845297	Recruiting
CTLA-4	Ipilimumab	I	Adult patients with relapsed/refractory AML or MDS; Elderly patients (≥75 years) with treatment naïve de novo or secondary AML.	Combination with decitabine	NCT02890329	Recruiting
IDO	Indoximod	I/II	Adult patients with newly diagnosed AML.	Combination with "7+3"	NCT02835729	Recruiting

2.3. CD133

CD133, also known as prominin-1, is a transmembrane glycoprotein that is primarily localized to the plasma membrane [22]. Expression of CD133 has been reported for a wide variety of tumor cells, including AML progenitors [23], and has been associated with resistance to chemotherapy and radiation [24]. Strategies to effectively target CD133 positive tumor cells are currently being explored in an effort to eliminate residual tumor cells that remain after conventional treatment. Recently, Rothfelder et al. reported that 293C3-SDIE, an FC-engineered CD133 monoclonal antibody, induced degranulation and lysis of primary CD133 positive AML cells by natural killer cells in allogeneic and autologous ex vivo settings. Interestingly, 293C3-SDIE exhibited no relevant toxicity against healthy BM cells. In xenotransplantation models, treatment with 293C3-SDIE led to the elimination of patient AML cells by NK cells from a matched human donor [25].

2.4. CD64

Another potential target for the delivery of cytotoxic agents is represented by CD64 (FcγRI), the high affinity receptor for IgG, which is expressed in several types of AML [26,27]. In vitro studies using the AML-related lymphoma cell line U937 showed that low nanomolar doses of Gb-H22/scFv (granzyme B fused to H22, a humanized single chain antibody fragment [scFv] specific to CD64) bound to CD64 positive U937 cells and induced apoptosis. Similar results were obtained in primary CD64 positive AML cells whereas CD64 negative AML cells were unaffected [28].

2.5. C-Type Lectin-Like Molecule 1

C-type lectin-like molecule 1 (CLL-1) is a transmembrane receptor that is expressed on the majority of myeloid blasts and stem cells derived from AML patients but not on normal tissues [29]. In a series of in vitro and in vivo studies, Zhao et al. showed that CLL-1 directed antibodies induced complement-dependent cytotoxicity against AML cell lines and primary cells, and furthermore reduced the tumor burden in AML xenograft mouse models [30].

2.6. Other Targets for Antibody-Directed Therapy

Recent evidence suggests that the CD98 glycoprotein plays a key role in mediating cellular interaction of AML cells with their microenvironment thus promoting leukemogenesis and leukemic cell maintenance. Using patient derived AML cells and mouse models, Bajaj et al. demonstrated that the CD98 directed antibody IGN523 blocks AML cell growth and reduces leukemic burden. Their findings strongly indicate that targeting CD98 could serve as a powerful tool to improve therapeutic targeting of AML cells [31]. Another promising target is represented by the fatty acid transporter CD36. Ye at al. showed that CD36 expressing leukemic progenitors are enriched in adipose tissue, such as in gonadal adipose tissue, where they are protected from the cytotoxic effects of chemotherapy suggesting a role for targeting CD36 as a novel strategy to hinder the development of treatment resistance [32,33]. Lines of evidence suggest that CD25 (Interleukin-2 receptor α) expression is increased in AML blasts and that its expression may be associated with an adverse outcome [34,35]. Further, methotrexate resistant subpopulations of the leukemic cell lines HL60 and MOLT4 exhibited elevated CD25 and TRAIL receptor 2 (TRAILR2)/Death receptor 5 (DR5) levels. Targeting CD25 as well as TRAILR2 or DR5 resulted in substantial cytotoxicity against leukemic cell lines and leukemic cells derived from AML, ALL, CML and CLL patient samples. CD25-based targeting may therefore represent and valuable strategy for targeting chemo-resistant leukemic cells [36]. CD38 (also known as cyclic ADP ribose hydroxylase) is a transmembrane glycoprotein that is expressed in most hematopoietic cells where it is involved in cell adhesion and signal transduction. When investigating CD38 expression in 304 AML patients, Keyhani et al. found that increased CD38 expression is associated with a favorable prognosis [37]. When investigating CD38 expression in AML cell lines, AML patient and healthy donor bone marrow samples, Dos Santos et al. reported a considerable variation among cell lines and AML patients whereas CD38 expression was more consistent in healthy bone marrow samples. The same group also investigated the effects of the CD38-targeted monoclonal antibody daratumumab against a range of AML cell lines and found that daratumumab-induced apoptosis was not correlated with CD38 expression levels. In AML patient xenografts, daratumumab significantly reduced leukemic burden in the peripheral blood and spleen, but not in the bone marrow. Intriguingly, the authors observed that daratumumab treated AML blasts displayed diminished CD38 surface expression indicating that the bone marrow microenvironment stalls the anti-leukemic effects of daratumumab [38]. Further studies to enhance the anti-leukemic activity of daratumumab might thus need to focus on disrupting the protective effects of the bone marrow microenvironment.

2.7. Targeting AML Stem Cells

Leukemia stem cells (LSCs), capable of giving rise to identical daughter cells and differentiated cells, perpetuate and maintain AML. It has become increasingly clear that LSCs are difficult to eliminate by the means of standard chemotherapy for which reason they represent the source of treatment resistance and relapse. The advent of immunotherapy in hematological malignancies has sparked hope for improved targeting of AML stem cells but many obstacles remain. For example, a clear characterization of the human AML stem cell phenotype is lacking. In recent years, several putative LSC targets such as the hedgehog and NFkB signaling pathways, MLL, CD44, CD47, CD33, CD96, CD123 and c-KIT have been proposed and tested in clinical trials [39]. To this end, encouraging experimental studies have recently been published for CD123. Han et al., for example, reported that SL-101, a CD123 targeted antibody conjugate, conferred remarkable anti-leukemic activity by sustained inhibition of protein synthesis, induction of apoptosis, blockade of IL3-induced STAT5 and AKT signaling, and inhibition of colony formation. In addition, the authors found that SL-101 significantly hindered repopulation of LSCs in patient derived xenograft models [40]. Similarly, dual targeting of CD19 and CD123 on leukemic blasts with chimeric antigen receptor T cells (CART) exhibited striking anti-leukemic activity in animal studies [41].

3. Increase Antigenicity of AML Cells by Hypomethylating Agents (HMAs)

The hypomethylating agents azacitidine (AZA) and decitabine (DEC) are the standard of care for management of patients with higher risk MDS [42] and are also frequently used off-label in AML patients deemed ineligible for intensive chemotherapy [43]. Despite clinical activity, only half of MDS patients respond to HMAs and the response is usually limited as most patient progress within 2 years of therapy with dismal survival [44]. In addition to their off-label upfront use in elderly and unfit patients with AML, HMAs are used in the setting of refractory or relapsed AML with a complete response (CR)/CR with incomplete count recovery (CRi) rate of approximately 16% [45]. Furthermore, HMA-based combinations with other forms of epigenetic therapy, particularly histone deacetylase inhibitors (HDACi), have been extensively studied in both AML and MDS but without clinical evidence of increased potency or synergism [46,47]. One important reason for suboptimal responses is that the underlying mechanism of action and resistance to HMAs are poorly understood [48].

HMAs do not only lead to promoter hypomethylation of silenced tumor suppressor genes but were also found to have pleiotropic effects on cell differentiation, senescence, apoptosis, angiogenesis and the immune system [48]. Further, HMAs stimulate multiple aspects of the immune response against malignant cells by enhancing antigenicity (e.g., tumor antigen expression, processing and presentation) as well as T-cell priming and effector function [49]. HMAs were shown to limit hyper-activation of the immune system against malignant cells by up-regulating the PD-1/PD-L1 and CTLA-4/CD80/86 axis, as well as by expansion of T regulatory cell subsets (Figure 1) [49,50]. Their important role in modulating the immune response to AML holds promise for potential synergism with different forms of immune therapy and several HMA based combination are currently being evaluated in clinical trials (Table 1).

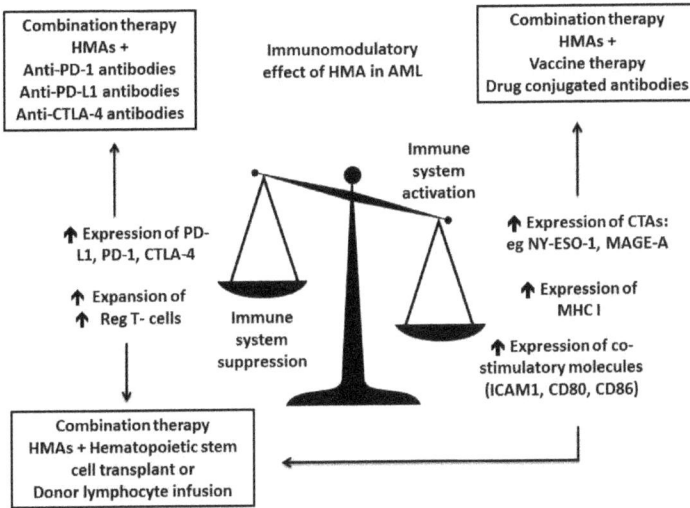

Figure 1. Hypomethylating agents (HMAs) and their role as immunomodulatory drugs in AML. HMAs possess both immuno-stimulatory, as well as immunosuppressive properties. They stimulate an immune response against AML blasts by increasing the expression of cancer testis antigens (e.g., NY-ESO-1 and MAGE-A), as well as important elements of the antigen-presenting machinery like the MHC I molecule and costimulatory molecules like ICAM1, as well as CD80 and CD86. On the other hand, HMAs can lead to immune escape of AML blasts through upregulation of immune checkpoints and their ligands, as well as regulatory T-cells. Combining the immunomodulatory effects of HMAs with other forms of immunotherapy holds the promise of a synergistic effect on the immune system. HMAs are currently combined with vaccines and drug-conjugated antibodies with the goal of increasing antigenicity and therefore AML blasts recognition and elimination by the immune system. Furthermore, combining HMA with checkpoint inhibitors might enhance the effect of checkpoint inhibitors in restoring immune surveillance. Lastly, combining HMAs with allo-HSCT or donor lymphocyte infusion is based on the hope that HMAs will enhance the graft versus leukemia effect (GVL) via enhanced antigenicity while limiting graft versus host disease (GVHD) by expansion of regulatory T-cells.

3.1. HMA Enhance Antigen Presentation

Cancer testis antigens (CTA) are expressed in adult testes and fetal ovaries but also in a variety of solid tumors where they trigger endogenous anti-tumor immune responses [51]. In contrast, AML blasts rarely express CTA as these genes are silenced by dense promoter hypermethylation. One exception, however, is the CTA preferentially expressed antigen in melanoma (PRAME), which is detected on AML blasts [52–54]. AZA, DEC and SGI-110 have been shown to induce the expression of multiple, previously suppressed CTAs (such as MAGE-A, NY-ESO-1 and SSX-2) per hypo-methylation of their promoter regions in AML cell lines and in AML xenograft models, as well as in primary AML blasts (Figure 1) [52,55–58]. HMA-induced expression of CTA in AML cells is sufficient for recognition of AML cells by CTA specific CD8+ cytotoxic T cells [55,57,58]. Furthermore, HMAs up-regulated the expression of important co-stimulatory proteins participating in antigen presentation, including the MHC class I molecule and the co-stimulatory molecules CD80 and CD86, as well as the intercellular adhesion molecule 1 (ICAM1) (Figure 1) [57–59]. However, HMAs alone are unlikely to trigger an adequate, endogenous immune response to eliminate leukemic cells. Therefore, several clinical trials are focusing on the combination between HMAs and vaccine therapy against CTA, drug conjugated antibodies directed towards tumor antigens, and adoptive cell therapy (Figure 1). Importantly, prior studies have not found an association of HMA-induced CTA expression and clinical responses to HMAs, suggesting that combining HMA with immunotherapy might be beneficial, including in patients who have previously failed HMA therapy [57]. A phase 1 study of an NY-ESO-1 peptide vaccine utilized protein CDX-1401, a fusion of a full length NY-ESO-1 protein sequence with a monoclonal antibody against DEC-205 (a surface marker present on many antigen presenting cells) combined with DEC in patients with AML and MDS (NCT01834248) [60]. Fusing NY-ESO-1 with an antibody against DEC-205 is supposed to enhance targeted delivery of NY-ESO-1 to the antigen processing machinery and subsequently enhance the efficacy of dendritic cell mediated T-cell priming. These NY-ESO-1 primed T cells will then more effectively target AML blasts expressing increased levels of NY-ESO-1 after being treated with DEC. For each cycle, patients were vaccinated 14 days prior and 14 days after receiving a 5-day course of DEC with the plan to administer 4 cycles. No significant side effects were observed and of a total of 9 patients included in this study, 5 patients had evidence of NY-ESO-1 specific CD4+ T-cell responses and 4 patients were found to have NY-ESO-1 specific CD8+ T-cell responses following vaccination. Another phase 1 trial examined the combination of the anti-CD33 targeted antibody drug conjugate SGN-CD33A with AZA or DEC as frontline therapy in 53 AML patients deemed unfit for intensive chemotherapy [61]. The response rate for the combination of SGN-CD33A and HMAs was 73% and of all responding patients, 47% achieved a minimal residual disease negative state. Based on these results, a phase 3 trial to test the combination between SGN-CD33A and HMAs (NCT02785900) is in progress (Table 1). Unfortunately, the FDA imposed a hold on clinical trials evaluating SGN-CD33A due to concerns of hepatotoxicity in patients who were treated with SGN-CD33A and received allo-SCT.

3.2. HMA Enhance Checkpoint Inhibition

Overexpression of the immune checkpoints PD-L1 on AML blast as well as PD-1 on BM stromal cells was found to lead to diminished antitumor T-cell responses and subsequent immune escape of AML blasts [50,62]. Furthermore, interferon-γ induced PD-L1 expression on AML cells is most prominent after initial treatment with chemotherapy possibly explaining high rates of relapse in AML despite initial remissions [63]. Importantly, HMAs also increased expression levels of PD-1, PD-L1 and CTLA-4 on peripheral blood (PB) mononuclear cells from MDS and AML patients as well as on leukemia cell lines (Figure 1) [50]. Patients resistant to HMA had relative higher increments in expression of the immune checkpoint genes compared to patients who responded to HMAs. This could not only explain the development of HMA resistance but also presented an opportunity for drug synergism when combining HMA with checkpoint inhibitors. Recently, early results of a phase 1 clinical trial combining AZA with the anti-PD1 antibody nivolumab in 51 AML patients, who had failed prior therapy, were presented at the American Society of Hematology 2016 Annual Meeting [64]. In this study, patients received AZA on days 1–7 and nivolumab on Day 1 and 14 with courses repeated indefinitely as tolerated. Median OS was 9.3 months, which compared favorably to historical survival with AZA-based salvage protocols in a similar patient population. Patients who responded had higher ratio of PD1 positive effector T-cells to PD1 positive regulatory T-cells in the BM microenvironment at baseline compared to non-responders. Several other clinical trials combining HMA with immune checkpoint inhibition are currently recruiting refractory and relapsed AML patients as well as elderly AML patients, who are unfit for intense chemotherapy (Figure 1). Some of these trials examine the combination of the anti-PD-1 antibodies nivolumab+AZA (NCT02397720), pembrolizumab+AZA (NCT02845297), as well as the combination of the anti-PD-L1 antibodies durvalumab+AZA (NCT02775903) and atezolizumab+ guadecitabine (NCT02892318) (Table 1).

3.3. HMA Might Enhance GVL Effect While Reducing GVHD

The goal of many allo-SCT trials is to enhance the graft versus leukemia effect (GVL) while limiting toxicities associated with graft versus host disease (GVHD). Sequential combination of allo-SCT followed by HMA (preemptively or at time of minimal/clinical relapse) could serve both goals by increasing the GVL through enhanced antigenicity of AML blasts as well as limiting GVHD by expanding regulatory T-cells, in addition to directly targeting leukemic cells by HMAs [65–69] (Figure 1). In a study of 27 patients with AML who had underwent reduced intensity allo-SCT, monthly courses of AZA were given and cytotoxic T-cell responses to candidate CTAs and circulating regulatory T-cells were measured [70]. HMA administration was found to not only increase the cytotoxic T-cell response to several CTA (including MAGE-A1 and WT-1) and expand the number for regulatory T-cells, but was also associated with low incidence of GVHD. A Phase 1/2 clinical trial of DEC followed by donor lymphocyte infusion in patients with AML, who relapsed after allo-SCT is currently recruiting participants (NCT01758367).

4. Modulation of the Leukemic Immune Microenvironment

Tumors generate a disabling immunosuppressive tumor microenvironment that limits the ability of the immune system to act against malignant cells. The immunosuppressive microenvironment begins forming as early as the first pre-malignant change. Thus, developing strategies to disrupt the immunosuppressive properties of the tumor microenvironment are necessary.

4.1. Small Molecule Immunomodulatory Drugs (IMiDs)

Small molecule immunomudulary drugs (IMiDs) are thalidomide analogues with immune-modulary, anti-angiogenic, anti-inflammatory and anti-proliferative properties. Lenalidomide, one of the best known thalidomide derivatives, is an orally bioavailable compound that has been FDA approved for the treatment of patients with multiple myeloma, MDS and mantle cell lymphoma. In vitro co-culture systems of endothelial cells and chronic lymphocytic leukemia cells showed that lenalidomide effectively altered the leukemic microenvironment and inhibited CLL cell survival by disruption of the cross-talk between leukemic and endothelial cells [71]. Recent studies suggest that lenalidomide has clinical activity in AML although the patient population that would benefit the most from lenalidomide needs to be better defined [72–74]. Several studies of lenalidomide in AML are currently ongoing, including the evaluation of lenalidomide as a consolidation and maintenance strategy in elderly AML patients following standard induction (NCT01578954).

4.2. Immunosuppressive Factors Expressed and Secreted by the Tumor or Tumor Microenvironment

Most of soluble immunosuppressive factors are secreted by the cells in the tumor microenvironment (stromal cells or infiltrating myeloid cells as described below) in response to the tumor or the tumor itself. In addition to cell–cell interactions that can inhibit effector lymphocytes, soluble factors can suppress the local immune response, which creates a hostile environment for infiltrating effector cells within the tumor. Indoleamine 2,3-dioxygenase (IDO), transforming growth factor-β (TGFβ), interleukin-10 (IL-10), vascular endothelial growth factor (VEGF), galectins, and IL-33 have been the most studied so far. IDO is the rate-limiting enzyme for the catabolism of the essential amino acid tryptophan. High levels of IDO reduce tryptophan levels and generate tryptophan metabolites which suppress T cell activity [75]. Phase 1 and 2 trials with IDO inhibitors are currently being carried out in melanoma, MDS as well as in AML patients (Table 1). Increased local and systemic TGFβ levels are associated with progression and poor clinical outcome in several tumors which make it an attractive therapeutic target. A small molecule kinase inhibitor blocking TGFβ receptor (LY2157299, Lilly®) is in clinical development in melanoma. The hypoxic environment of tumor also induces other modulators such as IL-10, IL-27, IL-33 and VEGF [76]. IL-33 can be detected not only in the tumor environment, but also in the serum of cancer patients. IL-33 levels are increased in the serum of lung and gastric cancer patients and correlate with disease stage, and expression levels of IL-33 and ST2 correlate with tumor grade and inferior survival of glioblastoma patients, suggesting that IL-33 may be a negative prognostic marker for these types of cancer [77–79]. As for hematological diseases, the IL-33/ST2 signaling pathway has been shown to increase the development of both BCR-ABL1-positive and -negative MPNs. Increased levels of nuclear IL-33 protein are present in biopsies of BCR-ABL1-negative MPN patients, and high amounts of circulating soluble ST2 levels were detected in the plasma from CML patients, compared to controls [80–82]. While cytokine blockade is currently applied for the treatment of inflammatory disorders, it remains to be investigated whether a blockade of the IL-33/ST2 pathway may represent a valid approach for the therapy of established IL-33-dependent tumors.

4.3. Myeloid-Derived Suppressor Cells

Myeloid-derived Suppressor Cells (MDSC) are a heterogeneous group of immature myeloid cells with potent immune suppressing activity [83]. MDSCs are characterized by the expression of the myeloid markers CD11b and CD33, and absence of HLA-DR8. There are two subsets of MDSCs: the monocytic MDSCs that are CD15 negative and the granulocytic MDSCs that are CD15 positive [83]. Presence of high numbers and frequencies of MDSCs in the tumor microenvironment has been associated with tumor progression [84], worse outcomes [85] and poor response to new immunotherapeutic approaches [86]. MDSCs directly suppress effector CD8+ T cells via T cell receptor (TCR) downregulation, expression of immunomodulatory enzymes such as Arginase-1, iNOS and the production of ROS [83,87]. Although MDSCs have mostly been studied in solid tumors growing evidence suggests that MDSCs are increased in patients with MDS [88]. Further, Pyzer et al. demonstrated that MDSCs are elevated in the PB of AML patients. Using tracking studies, the authors also showed that tumor-derived extracellular vesicles (EVs) are taken up by myeloid progenitors which leads to the selective proliferation of MDSCs. This process is largely driven by the Mucin 1 (MUC1) onco-protein which induces the expression of c-myc in AML cells and EVs [89]. Strategies to suppress MDSCs in vivo represent a promising approach to improve the virtue of immune-based therapies in AML [90].

4.4. Tumor Associated Macrophages

Tumor-associated macrophages (TAMs) are abundant in most human and murine cancers and are pro-tumorigenic [91]. TAMs express high levels of IL-10 and low levels of IL-12 with expression of the mannose receptor and scavenger receptor class A (SR-A) [92]. There is clinical evidence that an abundance of TAMs in the tumor microenvironment is correlated with poor prognosis [93]. It has recently been shown that AML cells polarize macrophages towards a leukemia supporting state in a Growth factor independence-1 (GFI1) dependent manner [94]. Targeting TAM with Anti-CSF-1R Antibody has been proposed in experimental models of colon carcinoma with promising results [95].

4.5. Tumor Associated Neutrophils

There is evidence that tumor associated neutrophils (TANs) promote primary tumor growth in mouse cancer models by enhancing angiogenesis and increasing immune suppression [96–98]. The role of TANs in the malignant BM niche has yet to be defined.

4.6. Regulatory T Cells

Regulatory T cells (Tregs) play a crucial role in maintaining peripheral tolerance and preventing autoimmunity. However, they also represent a major barrier to effective antitumor immunity and immunotherapy. Consequently, there has been considerable interest in developing approaches that can selectively or preferentially target Tregs in tumors, while not impacting their capacity to maintain peripheral immune homeostasis. Suppressive cytokines, such as IL-10, IL-35, and TGFb, are secreted by Tregs and required for their maximal suppressive function [99]. IL-35 contributes to the optimal suppressive activity of Tregs. In an experimental melanoma model, tumor growth was reduced in mice treated with an IL-35-neutralizing mAb or in mice harboring a Treg-restricted deletion of Ebi3 that prevents IL-35 production [100]. Treg-mediated cytolysis of NKs and CD8+ T cells via granzyme B and perforin may also be a relevant mechanism within tumors [101]. Lines of evidence suggest that the frequency of BM and PB Tregs are greater in AML patients compared to healthy controls. In mice, Tregs accumulate in leukemic sites and impede the proliferative and cytolytic capacity of adoptively transferred anti-AML reactive CTLs [102]. Depletion of CD25 (IL-2 receptor α-chain)–expressing

Tregs by the administration of IL-2 diphtheria toxin results in temporary tumor regression associated with increased CTLs at tumor sites. Combination therapy with IL-2 diphtheria toxin and anti-AML adoptive CTL transfer not only reduces tumor mass but also improves survival in mice. Moreover, mice display resistance to AML cells on re-challenge, implying the development of effective adaptive immunity [102,103]. Clinical studies with anti-CD25 monoclonal antibodies in patients with several cancers have led to no or only transient reductions in circulating Tregs following its administration, which has tempered enthusiasm for this strategy [104]. In AML, clinical studies are currently underway to investigate whether T cell responses to tumor vaccines are augmented in AML patients following depletion of Tregs [104].

4.7. Tumor Expressing Inhibitory Molecules, Cytotoxic CD8+ T Cells Exhaustion and Checkpoint Inhibitors

Therapeutic blockade of immune checkpoint pathways, in particular cytotoxic T-lymphocyte associated protein 4 (CTLA-4) and programmed-death 1 (PD-1), has become a paradigm-shifting treatment in solid tumor oncology. This therapeutic approach evolved from the recognition that tumors can evade the host immune system by high-jacking immune checkpoint pathways, such as CTLA-4 and PD-1 pathways [105]. CTLA-4, PD-L1 and PD-L2 are expressed by either stromal or immune cells in the microenvironment of tumor cells [106]. The engagement of checkpoint receptors on the surface of CD8+ T cells by their cognate ligands [B7-1 and B7-2 for CTLA-4, PD ligand 1 (PD-L1) and PD-L2 for PD-1] leads to the temporary downregulation of CD8+ T cell function, and to their exhaustion [106,107]. Of note, inhibition of T-cell activity by PD-1/PD-L1 engagement appears to be stronger than by CTLA-4 engagement [108]. Hematologic malignancies, many of which are known to have clinically exploitable immune sensitivity, are a natural target for this type of treatment [109]. Several clinical trials of checkpoint blockade have been carried out in hematologic malignancies, with preliminary results strongly suggesting therapeutic usefulness of this approach across several tumor types. In particular, the results of PD-1 blockade in Hodgkin lymphoma have been remarkable [110,111]. In non-Hodgkin lymphomas, response rates of 36% in diffuse large B-cell lymphoma (DLBCL) and 40% in follicular lymphoma (FL) were achieved with nivolumab, however, it is to be noted that these results were achieved in combination with Rituximab [112]. In multiple myeloma (MM), there has been no objective response in an initial phase 1 trial for checkpoint blockade as single agent. A phase 2 study of the humanized anti–PD-1 mAb pembrolizumab (MK-3475) with lenalidomide post-autologous HCT (NCT02331368) and a phase 1/2 study of pembrolizumab plus pomalidomide/dexamethasone in refractory MM (NCT02289222) are ongoing. In addition, the anti–PD-1 mAb pidilizumab (CT-011) is currently being evaluated in combination with vaccination post- autologous HCT (NCT01067287), as well as with lenalidomide in refractory MM patients (NCT02077959). Activity in myeloid malignancies has only recently been evaluated knowing that PD-L1 is expressed on MDS blasts, possibly at a higher level in high-risk and in more refractory disease. Furthermore, it has been shown that PD-L1 expression is enhanced by treatment with hypomethylating agents [50,113]. In consequence, a phase 1 trial to evaluate PD-L1 blockade in subjects with MDS is currently accruing. Although there are no dedicated AML trials yet, the anti-CTL-4 antibody ipilimumab has been evaluated in patients with persistent or progressive hematologic malignancies, including AML, after allo-SCT [114]. In this study, among 22 patients who received a dose of 10 mg ipilimumab per kilogram, 5 (23%) had a complete response, 2 (9%) had a partial response, and 6 (27%) had decreased tumor burden. The best responses were seen in patients with leukemia cutis and Hodgkin lymphoma. Most patients showed immunological responses as well. However, numerous questions remain and toxicities were not negligible. For example, the authors observed immune-related adverse events in 6 patients (21%), including one death. GVHD that precluded further administration of ipilimumab was observed in 4 patients (14%). Comprehensive reviews of all immune-related adverse events observed with immune checkpoint blockade has recently been published [8,115]. Toxicities encompass severe auto-immune reactions that can affect all organs/tissues. The most frequent organs affected are skin, gastrointestinal tract, pulmonary, and joints, while the most severe immune-related adverse events involve the gastrointestinal tract and endocrine glands. Toxicities are less severe with PD-1 blockade as compared to CTLA-4 blockade and less with PD-L1 blockade as compared to PD-1 blockade. Of note, hematological syndromes are possible in patients with solid tumors but appears to occur more frequently in patients with lymphoma and include red cell aplasia, autoimmune neutropenia or pancytopenia.

5. Conclusions

Recent advances in our understanding of the complex interactions between the immune system and cancer cells has opened the door to novel treatment strategies for AML. These concepts have mostly emerged from the knowledge that the immune system can inhibit, shape and/or support cancer in a multistep fashion (commonly referred to as "cancer immuno-editing") [116], and the realization that standard chemotherapeutic agents can exhaust immunosuppressive cells, including regulatory Tregs and myeloid-derived suppressive cells, thereby augmenting antitumor immune responses. To this end, multi-parameter flow cytometry on AML bone marrow specimens demonstrated that total T-cell and T-cell subpopulations are largely preserved in newly diagnosed and relapsed AML patients and therefore subject to therapeutic manipulation via checkpoint receptors such as PD-1 and OX40 [117]. Against this background, combining immunotherapy with conventional AML treatments holds the promise to more effectively eradicate leukemic cells compared to standard therapy alone. Several studies of immune-based therapies in AML and related diseases showed encouraging clinical activity in the setting of acceptable side effects and illustrated that these approaches have the potential to significantly enhance the current armamentarium for AML treatment. Data that was recently presented at ASH showed that inhibition of PD-1 with nivolumab in combination with azacitidine is safe and clinically beneficial in previously untreated patients with high risk MDS [118], a disease closely related to AML. Despite encouraging progress, major challenges on the way to establish immunotherapy as an integral element of AML therapy remain and are, at least in part, rooted in patient specific disparities in age, comorbidity and functional status, paired with the enormous molecular heterogeneity and complex clonal architecture of AML. For example, several immunotherapy trials have reported heterogenous and unconventional response patterns with regard to tumor regression and response times. Specifically, when compared to standard cytotoxic therapies which directly target cancer cells and therefore frequently lead to rapid reductions in tumor burden, responses to immunotherapy may take several months and frequently occur after an initial episode of "stable disease" or a temporary increase in total tumor burden ("pseudo-progression") [119]. Therefore, accounting for the unique modes of action of immunotherapeutic agents, novel response evaluation criteria are needed to (i) accurately capture and report immunotherapeutic responses in a standardized fashion and (ii) avoid premature/inappropriate discontinuation of therapy. Additional challenges are represented by the identification and management of immune-related side effects. Although these events can occur in any organ, gastrointestinal, hepatic and dermatologic side effects, as well as endocrinopathies, are most commonly reported. Standard approaches to identify and treat immune-related adverse events, however, are lacking and any inflammatory process is therefore almost uniformly managed by the early initiation of corticosteroids. Additional hurdles are represented by staggering treatment costs. Immunotherapies belong to the most expensive group of cancer therapeutics and it remains unclear how patient- and disease-specific characteristics should guide the choice of immunotherapy. In addition, questions remain regarding their implementation in the curative, adjuvant and maintenance setting. In order to establish a role for immune-based therapies in the management of a highly heterogenous disease as AML, costly and large, randomized trials are needed which requires the identification of adequate biomarkers to help predict treatment response and toxicities, and to accurately select patients for accrual.

Conflicts of Interest: The authors declare no conflict of interest.

References

1. Estey, E. Why is progress in acute myeloid leukemia so slow? *Semin. Hematol.* **2015**, *52*, 243–248. [CrossRef] [PubMed]
2. Dohner, H.; Weisdorf, D.J.; Bloomfield, C.D. Acute myeloid leukemia. *N. Engl. J. Med.* **2015**, *373*, 1136–1152. [CrossRef] [PubMed]
3. Dombret, H.; Gardin, C. An update of current treatments for adult acute myeloid leukemia. *Blood* **2016**, *127*, 53–61. [CrossRef] [PubMed]
4. Schmitt, M.; Schmitt, A.; Rojewski, M.T.; Chen, J.; Giannopoulos, K.; Fei, F.; Yu, Y.; Götz, M.; Heyduk, M.; Ritter, G.; et al. RHAMM-R3 peptide vaccination in patients with acute myeloid leukemia, myelodysplastic syndrome, and multiple myeloma elicits immunologic and clinical responses. *Blood* **2008**, *111*, 1357–1365. [CrossRef] [PubMed]
5. Rezvani, K.; Yong, A.S.; Mielke, S.; Savani, B.N.; Jafarpour, B.; Eniafe, R.; Quan Le, R.; Musse, L.; Boss, C.; Childs, R.; et al. Lymphodepletion is permissive to the development of spontaneous T-cell responses to the self-antigen PR1 early after allogeneic stem cell transplantation and in patients with acute myeloid leukemia undergoing WT1 peptide vaccination following chemotherapy. *Cancer Immunol. Immunother.* **2012**, *61*, 1125–1136. [CrossRef] [PubMed]
6. Saini, K.S.; Azim, H.A., Jr.; Cocorocchio, E.; Vanazzi, A.; Saini, M.L.; Raviele, P.R.; Pruneri, G.; Peccatori, F.A. Rituximab in Hodgkin lymphoma: Is the target always a hit? *Cancer Treat. Rev.* **2011**, *37*, 385–390. [CrossRef] [PubMed]
7. Hodi, F.S.; O'Day, S.J.; McDermott, D.F.; Weber, R.W.; Sosman, J.A.; Haanen, J.B.; Gonzalez, R.; Robert, C.; Schadendorf, D.; Hassel, J.C.; et al. Improved survival with ipilimumab in patients with metastatic melanoma. *N. Engl. J. Med.* **2010**, *363*, 711–723. [CrossRef] [PubMed]
8. Topalian, S.L.; Sznol, M.; McDermott, D.F.; Kluger, H.M.; Carvajal, R.D.; Sharfman, W.H.; Brahmer, J.R.; Lawrence, D.P.; Atkins, M.B.; Powderly, J.D.; et al. Survival, durable tumor remission, and long-term safety in patients with advanced melanoma receiving nivolumab. *J. Clin. Oncol.* **2014**, *32*, 1020–1030. [CrossRef] [PubMed]
9. Kantarjian, H.; Stein, A.; Gokbuget, N.; Fielding, A.K.; Schuh, A.C.; Ribera, J.M.; Wei, A.; Dombret, H.; Foà, R.; Bassan, R.; et al. Blinatumomab versus chemotherapy for advanced acute lymphoblastic leukemia. *N. Engl. J. Med.* **2017**, *376*, 836–847. [CrossRef] [PubMed]
10. Jilani, I.; Estey, E.; Huh, Y.; Joe, Y.; Manshouri, T.; Yared, M.; Giles, F.; Kantarjian, H.; Cortes, J.; Thomas, D.; et al. Differences in CD33 intensity between various myeloid neoplasms. *Am. J. Clin. Pathol.* **2002**, *118*, 560–566. [CrossRef] [PubMed]
11. O'Hear, C.; Rubnitz, J.E. Recent research and future prospects for gemtuzumab ozogamicin: Could it make a comeback? *Expert Rev. Hematol.* **2014**, *7*, 427–429. [CrossRef] [PubMed]
12. Burnett, A.K.; Hills, R.K.; Milligan, D.; Kjeldsen, L.; Kell, J.; Russell, N.H.; Yin, J.A.L.; Hunter, A.; Goldstone, A.H.; Wheatley, K. Identification of patients with acute myeloblastic leukemia who benefit from the addition of gemtuzumab ozogamicin: Results of the MRC AML15 trial. *J. Clin. Oncol.* **2011**, *29*, 369–377. [CrossRef] [PubMed]
13. Burnett, A.K.; Russell, N.H.; Hills, R.K.; Kell, J.; Freeman, S.; Kjeldsen, L.; Hunter, A.E.; Yin, J.; Craddock, C.F.; Dufva, I.H.; et al. Addition of gemtuzumab ozogamicin to induction chemotherapy improves survival in older patients with acute myeloid leukemia. *J. Clin. Oncol.* **2012**, *30*, 3924–3931. [CrossRef] [PubMed]
14. Burnett, A.K.; Hills, R.K.; Hunter, A.E.; Milligan, D.; Kell, W.J.; Wheatley, K.; Yin, J.; McMullin, M.F.; Dignum, H.; Bowen, D.; et al. The addition of gemtuzumab ozogamicin to low-dose Ara-C improves remission rate but does not significantly prolong survival in older patients with acute myeloid leukaemia: Results from the LRF AML14 and NCRI AML16 pick-a-winner comparison. *Leukemia* **2013**, *27*, 75–81. [CrossRef] [PubMed]
15. Hills, R.K.; Castaigne, S.; Appelbaum, F.R.; Delaunay, J.; Petersdorf, S.; Othus, M.; Estey, E.H.; Dombret, H.; Chevret, S.; Ifrah, N.; et al. Addition of gemtuzumab ozogamicin to induction chemotherapy in adult patients with acute myeloid leukaemia: A meta-analysis of individual patient data from randomised controlled trials. *Lancet Oncol.* **2014**, *15*, 986–996. [CrossRef]

16. Friedrich, M.; Henn, A.; Raum, T.; Bajtus, M.; Matthes, K.; Hendrich, L.; Wahl, J.; Hoffmann, P.; Kischel, R.; Kvesic, M.; et al. Preclinical characterization of AMG 330, a CD3/CD33-bispecific T-cell-engaging antibody with potential for treatment of acute myelogenous leukemia. *Mol. Cancer Ther.* **2014**, *13*, 1549–1557. [CrossRef] [PubMed]

17. Testa, U.; Pelosi, E.; Frankel, A. CD 123 is a membrane biomarker and a therapeutic target in hematologic malignancies. *Biomark. Res.* **2014**, *2*, 4. [CrossRef] [PubMed]

18. Busfield, S.J.; Biondo, M.; Wong, M.; Ramshaw, H.S.; Lee, E.M.; Ghosh, S.; Braley, H.; Panousis, C.; Roberts, A.W.; He, S.Z.; et al. Targeting of acute myeloid leukemia in vitro and in vivo with an anti-CD123 mAb engineered for optimal ADCC. *Leukemia* **2014**, *28*, 2213–2221. [CrossRef] [PubMed]

19. Smith, B.D.; Roboz, G.J.; Walter, R.B.; Altman, J.K.; Ferguson, A.; Curcio, T.J.; Orlowski, K.F.; Garrett, L.; Busfield, S.J.; Barnden, M.; et al. First-in man, phase 1 study of CSL362 (anti-IL3Rα / anti-CD123 monoclonal antibody) in patients with CD123+ acute myeloid leukemia (AML) in CR at high risk for early relapse. *Blood* **2014**, *124*, 120.

20. Lane, A.A.; Sweet, K.L.; Wang, E.S.; Donnellan, W.B.; Walter, R.B.; Stein, A.S.; Rizzieri, D.A.; Carraway, H.E.; Mantzaris, L.; Prebet, T.; et al. Results from ongoing phase 2 trial of SL-401 as consolidation therapy in patients with acute myeloid leukemia (AML) in remission with high relapse risk including minimal residual disease (MRD). *Blood* **2016**, *128*, 215.

21. Chichili, G.R.; Huang, L.; Li, H.; Burke, S.; He, L.; Tang, Q.; Jin, L.; Gorlatov, S.; Ciccarone, V.; Chen, F.; et al. A CD3 × CD123 bispecific DART for redirecting host T cells to myelogenous leukemia: Preclinical activity and safety in nonhuman primates. *Sci. Transl. Med.* **2015**, *7*, 289ra82. [PubMed]

22. Li, J.; Zhong, X.Y.; Li, Z.Y.; Cai, J.F.; Zou, L.; Li, J.M.; Yang, T.; Liu, W. CD133 expression in osteosarcoma and derivation of CD133+ cells. *Mol. Med. Rep.* **2013**, *7*, 577–584. [PubMed]

23. Vercauteren, S.M.; Sutherland, H.J. CD133 (AC133) expression on AML cells and progenitors. *Cytotherapy* **2001**, *3*, 449–459. [PubMed]

24. Ferrandina, G.; Petrillo, M.; Bonanno, G.; Scambia, G. Targeting CD133 antigen in cancer. *Expert Opin. Ther. Targets* **2009**, *13*, 823–837. [PubMed]

25. Rothfelder, K.; Koerner, S.; Andre, M.; Leibold, J.; Kousis, P.; Buehring, H.J.; Haen, S.P.; Kuebler, A.; Kanz, L.; Grosse-Hovest, L.; et al. Induction of NK cell reactivity against myeloid leukemia by a novel Fc-optimized CD133 antibody. *Blood* **2015**, *126*, 3793.

26. Ball, E.D.; McDermott, J.; Griffin, J.D.; Davey, F.R.; Davis, R.; Bloomfield, C.D. Expression of the three myeloid cell-associated immunoglobulin G Fc receptors defined by murine monoclonal antibodies on normal bone marrow and acute leukemia cells. *Blood* **1989**, *73*, 1951–1956. [PubMed]

27. Krasinskas, A.M.; Wasik, M.A.; Kamoun, M.; Schretzenmair, R.; Moore, J.; Salhany, K.E. The usefulness of CD64, other monocyte-associated antigens, and CD45 gating in the subclassification of acute myeloid leukemias with monocytic differentiation. *Am. J. Clin. Pathol.* **1998**, *110*, 797–805. [PubMed]

28. Stahnke, B.; Thepen, T.; Stocker, M.; Rosinke, R.; Jost, E.; Fischer, R.; Tur, M.K.; Barth, S. Granzyme B-H22(scFv), a human immunotoxin targeting CD64 in acute myeloid leukemia of monocytic subtypes. *Mol. Cancer Ther.* **2008**, *7*, 2924–2932. [PubMed]

29. Bakker, A.B.; van den Oudenrijn, S.; Bakker, A.Q.; Feller, N.; van Meijer, M.; Bia, J.A.; Jongeneelen, M.A.C.; Visser, T.J.; Bijl, N.; Geuijen, C.A.W.; et al. C-type lectin-like molecule-1: A novel myeloid cell surface marker associated with acute myeloid leukemia. *Cancer Res.* **2004**, *64*, 8443–8450. [PubMed]

30. Zhao, X.; Singh, S.; Pardoux, C.; Zhao, J.; His, E.D.; Abo, A.; Korver, W. Targeting C-type lectin-like molecule-1 for antibody-mediated immunotherapy in acute myeloid leukemia. *Haematologica* **2010**, *95*, 71–78. [PubMed]

31. Bajaj, J.; Konuma, T.; Lytle, N.K.; Kwon, H.Y.; Ablack, J.N.; Cantor, J.M.; Rizzieri, D.; Chuah, C.; Oehler, V.G.; Broome, E.H.; et al. CD98-mediated adhesive signaling enables the establishment and propagation of acute myelogenous leukemia. *Cancer Cell* **2016**, *30*, 792–805. [CrossRef] [PubMed]

32. Ye, H.; Adane, B.; Khan, N.; Sullivan, T.; Minhajuddin, M.; Gasparetto, M.; Stevens, B.; Pei, S.; Balys, M.; Ashton, J.M.; et al. leukemic stem cells evade chemotherapy by metabolic adaptation to an adipose tissue niche. *Cell Stem Cell* **2016**, *19*, 23–37. [CrossRef] [PubMed]

33. Farge, T.; Saland, E.; de Toni, F.; Aroua, N.; Hosseini, M.; Perry, R.; Bosc, C.; Sugita, M.; Stuani, L.; Fraisse, M.; et al. Chemotherapy-resistant human acute myeloid leukemia cells are not enriched for leukemic stem cells but require oxidative metabolism. *Cancer Discov.* **2017**, *7*, 716–735. [CrossRef] [PubMed]

34. Ikegawa, S.; Doki, N.; Kurosawa, S.; Yamaguchi, T.; Sakaguchi, M.; Harada, K.; Yamamoto, K.; Hino, Y.; Shingai, N.; Senoo, Y.; et al. CD25 expression on residual leukemic blasts at the time of allogeneic hematopoietic stem cell transplant predicts relapse in patients with acute myeloid leukemia without complete remission. *Leuk Lymphoma* **2016**, *57*, 1375–1381. [CrossRef] [PubMed]

35. Gonen, M.; Sun, Z.; Figueroa, M.E.; Patel, J.P.; Abdel-Wahab, O.; Racevskis, J.; Ketterling, R.P.; Fernandez, H.; Rowe, J.M.; Tallman, M.S.; et al. CD25 expression status improves prognostic risk classification in AML independent of established biomarkers: ECOG phase 3 trial, E1900. *Blood* **2012**, *120*, 2297–2306. [CrossRef] [PubMed]

36. Madhumathi, J.; Sridevi, S.; Verma, R.S. CD25 targeted therapy of chemotherapy resistant leukemic stem cells using DR5 specific TRAIL peptide. *Stem Cell Res.* **2017**, *19*, 65–75. [CrossRef] [PubMed]

37. Keyhani, A.; Huh, Y.O.; Jendiroba, D.; Pagliaro, L.; Cortez, J.; Pierce, S.; Pearlman, M.; Estey, E.; Kantarjian, H.; Freireich, E.J. Increased CD38 expression is associated with favorable prognosis in adult acute leukemia. *Leuk Res.* **2000**, *24*, 153–159. [CrossRef]

38. Dos Santos, C.; Xiaochuan, S.; Chenghui, Z.; Ndikuyeze, G.H.; Glover, J.; Secreto, T.; Doshi, P.; Sasser, K.; Danet-Desnoyers, G. Anti-leukemic activity of daratumumab in acute myeloid leukemia cells and patient-derived xenografts. *Blood* **2014**, *124*, 2312.

39. Pollyea, D.A.; Gutman, J.A.; Gore, L.; Smith, C.A.; Jordan, C.T. Targeting acute myeloid leukemia stem cells: A review and principles for the development of clinical trials. *Haematologica* **2014**, *99*, 1277–1284. [CrossRef] [PubMed]

40. Han, L.; Jorgensen, J.L.; Brooks, C.; Shi, C.; Zhang, Q.; Nogueras Gonzalez, G.M.; Cavazos, A.; Pan, R.; Mu, H.; Wang, S.A.; et al. Anti-leukemia efficacy and mechanisms of action of SL-101, a novel anti-CD123 antibody-conjugate, in acute myeloid leukemia. *Clin. Cancer Res.* **2015**, *15*, S14. [CrossRef]

41. Ruella, M.; Barrett, D.M.; Kenderian, S.S.; Shestova, O.; Hofmann, T.J.; Perazzelli, J.; Klichinsky, M.; Aikawa, V.; Nazimuddin, F.; Kozlowski, M.; et al. Dual CD19 and CD123 targeting prevents antigen-loss relapses after CD19-directed immunotherapies. *J. Clin. Investig.* **2016**, *126*, 3814–3826. [CrossRef] [PubMed]

42. Zeidan, A.M.; Stahl, M.; Komrokji, R. Emerging biological therapies for the treatment of myelodysplastic syndromes. *Expert Opin. Emerg. Drugs* **2016**, *21*, 283–300. [CrossRef] [PubMed]

43. Podoltsev, N.A.; Stahl, M.; Zeidan, A.M.; Gore, S.D. Selecting initial treatment of acute myeloid leukaemia in older adults. *Blood Rev.* **2016**, *31*, 43–62. [CrossRef] [PubMed]

44. Prebet, T.; Gore, S.D.; Esterni, B.; Gardin, C.; Itzykson, R.; Thepot, S.; Dreyfus, F.; Rauzy, O.B.; Recher, C.; Adès, L.; et al. Outcome of high-risk myelodysplastic syndrome after azacitidine treatment failure. *J. Clin. Oncol.* **2011**, *29*, 3322–3327. [CrossRef] [PubMed]

45. Stahl, M.P.N.; de Veaux, M.; Perreault, S.; Itzykson, R.; Ritchie, E.K.; Sekeres, M.A.; Fathi, A.T.; Komrokji, R.S.; Bhatt, V.R.; Al-Kali, A.; et al. The use of hypomethylating agents (HMAs) in patients with relapsed and refractory acute myeloid leukemia (RR-AML): Clinical outcomes and their predictors in a large international patient cohort. *Blood* **2016**, *128*, 1063.

46. Stahl, M.; Gore, S.D.; Vey, N.; Prebet, T. Lost in translation? Ten years of development of histone deacetylase inhibitors in acute myeloid leukemia and myelodysplastic syndromes. *Expert Opin. Investig. Drugs* **2016**, *25*, 307–317. [CrossRef] [PubMed]

47. Stahl, M.; Zeidan, A.M. Hypomethylating agents in combination with histone deacetylase inhibitors in higher risk myelodysplastic syndromes: Is there a light at the end of the tunnel? *Cancer* **2017**, *123*, 911–914. [CrossRef] [PubMed]

48. Stahl, M.; Kohrman, N.; Gore, S.D.; Kim, T.K.; Zeidan, A.M.; Prebet, T. Epigenetics in cancer: A hematological perspective. *PLoS Genet.* **2016**, *12*, e1006193. [CrossRef] [PubMed]

49. Heninger, E.; Krueger, T.E.; Lang, J.M. Augmenting antitumor immune responses with epigenetic modifying agents. *Front. Immunol.* **2015**, *6*, 29. [PubMed]

50. Yang, H.; Bueso-Ramos, C.; DiNardo, C.; Estecio, M.R.; Davanlou, M.; Geng, Q.R.; Fang, Z.; Nguyen, M.; Pierce, S.; Wei, Y.; et al. Expression of PD-L1, PD-L2, PD-1 and CTLA4 in myelodysplastic syndromes is enhanced by treatment with hypomethylating agents. *Leukemia* **2014**, *28*, 1280–1288. [CrossRef] [PubMed]

51. Akers, S.N.; Odunsi, K.; Karpf, A.R. Regulation of cancer germline antigen gene expression: Implications for cancer immunotherapy. *Future Oncol.* **2010**, *6*, 717–732. [CrossRef] [PubMed]

52. Atanackovic, D.; Luetkens, T.; Kloth, B.; Fuchs, G.; Cao, Y.; Hildebrandt, Y.; Meyer, S.; Bartels, K.; Reinhard, H.; Lajmi, N.; et al. Cancer-testis antigen expression and its epigenetic modulation in acute myeloid leukemia. *Am. J. Hematol.* **2011**, *86*, 918–922. [CrossRef] [PubMed]
53. Chambost, H.; van Baren, N.; Brasseur, F.; Olive, D. MAGE-A genes are not expressed in human leukemias. *Leukemia* **2001**, *15*, 1769–1771. [CrossRef] [PubMed]
54. Ortmann, C.A.; Eisele, L.; Nuckel, H.; Klein-Hitpass, L.; Fuhrer, A.; Duhrsen, U.; Zeschnigk, M. Aberrant hypomethylation of the cancer-testis antigen PRAME correlates with PRAME expression in acute myeloid leukemia. *Ann. Hematol.* **2008**, *87*, 809–818. [CrossRef] [PubMed]
55. Goodyear, O.; Agathanggelou, A.; Novitzky-Basso, I.; Siddique, S.; McSkeane, T.; Ryan, G.; Vyas, P.; Cavenagh, J.; Stankovic, T.; Moss, P.; et al. Induction of a CD8+ T-cell response to the MAGE cancer testis antigen by combined treatment with azacitidine and sodium valproate in patients with acute myeloid leukemia and myelodysplasia. *Blood* **2010**, *116*, 1908–1918. [CrossRef] [PubMed]
56. Almstedt, M.; Blagitko-Dorfs, N.; Duque-Afonso, J.; Karbach, J.; Pfeifer, D.; Jager, E.; Lübbert, M. The DNA demethylating agent 5-aza-2′-deoxycytidine induces expression of NY-ESO-1 and other cancer/testis antigens in myeloid leukemia cells. *Leuk. Res.* **2010**, *34*, 899–905. [CrossRef] [PubMed]
57. Srivastava, P.; Paluch, B.E.; Matsuzaki, J.; James, S.R.; Collamat-Lai, G.; Blagitko-Dorfs, N.; Ford, L.A.; Naqash, R.; Lübbert, M.; Karpf, A.R.; et al. Induction of cancer testis antigen expression in circulating acute myeloid leukemia blasts following hypomethylating agent monotherapy. *Oncotarget* **2016**, *7*, 12840–12856. [CrossRef] [PubMed]
58. Srivastava, P.; Paluch, B.E.; Matsuzaki, J.; James, S.R.; Collamat-Lai, G.; Karbach, J.; Nemeth, M.J.; Taverna, P.; Karpf, A.R.; Griffiths, E.A. Immunomodulatory action of SGI-110, a hypomethylating agent, in acute myeloid leukemia cells and xenografts. *Leuk. Res.* **2014**, *38*, 1332–1341. [CrossRef] [PubMed]
59. Wang, L.X.; Mei, Z.Y.; Zhou, J.H.; Yao, Y.S.; Li, Y.H.; Xu, Y.H.; Li, J.X.; Gao, X.N.; Zhou, M.H.; Jiang, M.M.; et al. Low dose decitabine treatment induces CD80 expression in cancer cells and stimulates tumor specific cytotoxic T lymphocyte responses. *PLoS ONE* **2013**, *8*, e62924. [CrossRef] [PubMed]
60. Srivastava, P.M.J.; Paluch, B.E.; Brumberger, Z.; Kaufman, S.; Karpf, A.R.; Odunsi, K.; Miller, A.; Kocent, J.; Wang, E.S.; Nemeth, M.J.; et al. NY-ESO-1 vaccination in combination with decitabine for patients with MDS induces CD4+ and CD8+ T-cell responses. *Blood* **2015**, *126*, 2873.
61. Fathi, A.T.; Erba, H.P.; Lancet, J.E.; Stein, E.M.; Ravandi, F.; Faderl, S.; Walter, R.B.; Advani, A.; DeAngelo, D.J.; Kovacsovics, T.J.; et al. Vadastuximab talirine plus hypomethylating agents: A well-tolerated regimen with high remission rate in frontline older patients with acute myeloid leukemia (AML). *Blood* **2016**, *128*, 591.
62. Zhang, L.; Gajewski, T.F.; Kline, J. PD-1/PD-L1 interactions inhibit antitumor immune responses in a murine acute myeloid leukemia model. *Blood* **2009**, *114*, 1545–1552. [CrossRef] [PubMed]
63. Kronig, H.; Kremmler, L.; Haller, B.; Englert, C.; Peschel, C.; Andreesen, R.; Blank, C.U. Interferon-induced programmed death-ligand 1 (PD-L1/B7-H1) expression increases on human acute myeloid leukemia blast cells during treatment. *Eur. J. Haematol.* **2014**, *92*, 195–203. [CrossRef] [PubMed]
64. Daver, N.; Basu, S.; Garcia-Manero, G.; Cortes, J.E.; Ravandi, F.; Jabbour, E.J.; Hendrickson, S.; Pierce, S.; Ning, J.; Konopleva, M.; et al. Phase IB/II study of nivolumab in combination with azacytidine (AZA) in patients (pts) with relapsed acute myeloid leukemia (AML). *Blood* **2016**, *128*, 763.
65. Czibere, A.; Bruns, I.; Kroger, N.; Platzbecker, U.; Lind, J.; Zohren, F.; Fenk, R.; Germing, U.; Schröder, T.; Gräf, T.; et al. 5-Azacytidine for the treatment of patients with acute myeloid leukemia or myelodysplastic syndrome who relapse after allo-SCT: A retrospective analysis. *Bone Marrow Transplant.* **2010**, *45*, 872. [CrossRef] [PubMed]
66. Lal, G.; Zhang, N.; van der Touw, W.; Ding, Y.; Ju, W.; Bottinger, E.P.; Reid, S.P.; Levy, D.E.; Bromberg, J.S. Epigenetic regulation of Foxp3 expression in regulatory T cells by DNA methylation. *J. Immunol.* **2009**, *182*, 259–273. [CrossRef] [PubMed]
67. Polansky, J.K.; Kretschmer, K.; Freyer, J.; Floess, S.; Garbe, A.; Baron, U.; Olek, S.; Hamann, A.; von Boehmer, H.; Huehn, J. DNA methylation controls Foxp3 gene expression. *Eur. J. Immunol.* **2008**, *38*, 1654–1663. [CrossRef] [PubMed]
68. Choi, J.; Ritchey, J.; Prior, J.L.; Holt, M.; Shannon, W.D.; Deych, E.; Piwnica-Worms, D.R.; DiPersio, J.F. In vivo administration of hypomethylating agents mitigate graft-versus-host disease without sacrificing graft-versus-leukemia. *Blood* **2010**, *116*, 129–139. [CrossRef] [PubMed]

69. Sanchez-Abarca, L.I.; Gutierrez-Cosio, S.; Santamaria, C.; Caballero-Velazquez, T.; Blanco, B.; Herrero-Sanchez, C.; García, J.L.; Carrancio, S.; Hernández-Campo, P.; González, F.J.; et al. Immunomodulatory effect of 5-azacytidine (5-azaC): Potential role in the transplantation setting. *Blood* **2010**, *115*, 107–121. [CrossRef] [PubMed]

70. Goodyear, O.C.; Dennis, M.; Jilani, N.Y.; Loke, J.; Siddique, S.; Ryan, G.; Nunnick, J.; Khanum, R.; Raghavan, M.; Cook, M.; et al. Azacitidine augments expansion of regulatory T cells after allogeneic stem cell transplantation in patients with acute myeloid leukemia (AML). *Blood* **2012**, *119*, 3361–3369. [CrossRef] [PubMed]

71. Maffei, R.; Fiorcari, S.; Bulgarelli, J.; Rizzotto, L.; Martinelli, S.; Rigolin, G.M.; Debbia, G.; Castelli, I.; Bonacorsi, G.; Santachiara, R.; et al. Endothelium-mediated survival of leukemic cells and angiogenesis-related factors are affected by lenalidomide treatment in chronic lymphocytic leukemia. *Exp. Hematol.* **2014**, *42*, 126–136. [CrossRef] [PubMed]

72. Chen, Y.; Kantarjian, H.; Estrov, Z.; Faderl, S.; Ravandi, F.; Rey, K.; Cortes, J.; Borthakur, G. A phase II study of lenalidomide alone in relapsed/refractory acute myeloid leukemia or high-risk myelodysplastic syndromes with chromosome 5 abnormalities. *Clin. Lymphoma Myeloma Leuk.* **2012**, *12*, 341–344. [CrossRef] [PubMed]

73. Sekeres, M.A.; Gundacker, H.; Lancet, J.; Advani, A.; Petersdorf, S.; Liesveld, J.; Mulford, D.; Norwood, T.; Willman, C.L.; Appelbaum, F.R.; et al. A phase 2 study of lenalidomide monotherapy in patients with deletion 5q acute myeloid leukemia: Southwest oncology group study S0605. *Blood* **2011**, *118*, 523–528. [CrossRef] [PubMed]

74. Fehniger, T.A.; Uy, G.L.; Trinkaus, K.; Nelson, A.D.; Demland, J.; Abboud, C.N.; Cashen, A.F.; Stockerl-Goldstein, K.E.; Westervelt, P.; DiPersio, J.F.; et al. A phase 2 study of high-dose lenalidomide as initial therapy for older patients with acute myeloid leukemia. *Blood* **2011**, *117*, 1828–1833. [CrossRef] [PubMed]

75. Munn, D.H. Blocking IDO activity to enhance anti-tumor immunity. *Front. Biosci.* **2012**, *4*, 734–745. [CrossRef]

76. Mahoney, K.M.; Rennert, P.D.; Freeman, G.J. Combination cancer immunotherapy and new immunomodulatory targets. *Nat. Rev. Drug Discov.* **2015**, *14*, 561–584. [CrossRef] [PubMed]

77. Sun, P.; Ben, Q.; Tu, S.; Dong, W.; Qi, X.; Wu, Y. Serum interleukin-33 levels in patients with gastric cancer. *Dig. Dis. Sci.* **2011**, *56*, 3596–3601. [CrossRef] [PubMed]

78. Hu, L.A.; Fu, Y.; Zhang, D.N.; Zhang, J. Serum IL-33 as a diagnostic and prognostic marker in non-small cell lung cancer. *Asian Pac. J. Cancer Prev.* **2013**, *14*, 2563–2566. [CrossRef] [PubMed]

79. Gramatzki, D.; Frei, K.; Cathomas, G.; Moch, H.; Weller, M.; Mertz, K.D. Interleukin-33 in human gliomas: Expression and prognostic significance. *Oncol. Lett.* **2016**, *12*, 445–452. [CrossRef] [PubMed]

80. Levescot, A.; Flamant, S.; Basbous, S.; Jacomet, F.; Feraud, O.; Anne Bourgeois, E.; Bonnet, M.L.; Giraud, C.; Roy, L.; Barra, A.; et al. BCR-ABL-induced deregulation of the IL-33/ST2 pathway in CD34+ progenitors from chronic myeloid leukemia patients. *Cancer Res.* **2014**, *74*, 2669–2676. [CrossRef] [PubMed]

81. Mager, L.F.; Riether, C.; Schurch, C.M.; Banz, Y.; Wasmer, M.H.; Stuber, R.; Theocharides, A.P.; Li, X.; Xia, Y.; Saito, H.; et al. IL-33 signaling contributes to the pathogenesis of myeloproliferative neoplasms. *J. Clin. Investig.* **2015**, *125*, 2579–2591. [CrossRef] [PubMed]

82. Wasmer, M.H.; Krebs, P. The role of IL-33-dependent inflammation in the tumor microenvironment. *Front. Immunol.* **2016**, *7*, 682. [CrossRef] [PubMed]

83. Gabrilovich, D.I.; Ostrand-Rosenberg, S.; Bronte, V. Coordinated regulation of myeloid cells by tumours. *Nat. Rev. Immunol.* **2012**, *12*, 253–268. [CrossRef] [PubMed]

84. Wang, Z.; Zhang, L.; Wang, H.; Xiong, S.; Li, Y.; Tao, Q.; Xiao, W.; Qin, H.; Wang, Y.; Zhai, Z. Tumor-induced CD14+HLA-DR−/low myeloid-derived suppressor cells correlate with tumor progression and outcome of therapy in multiple myeloma patients. *Cancer Immunol. Immunother.* **2015**, *64*, 389–399. [CrossRef] [PubMed]

85. Chen, M.F.; Kuan, F.C.; Yen, T.C.; Lu, M.S.; Lin, P.Y.; Chung, Y.H.; Chen, W.C.; Lee, K.D. IL-6-stimulated CD11b+ CD14+ HLA-DR-myeloid-derived suppressor cells, are associated with progression and poor prognosis in squamous cell carcinoma of the esophagus. *Oncotarget* **2014**, *5*, 8716–8728. [CrossRef] [PubMed]

86. Laborde, R.R.; Lin, Y.; Gustafson, M.P.; Bulur, P.A.; Dietz, A.B. Cancer vaccines in the world of immune suppressive monocytes (CD14(+)HLA-DR(lo/neg) Cells): The gateway to improved responses. *Front. Immunol.* **2014**, *5*, 147. [CrossRef] [PubMed]

87. Movahedi, K.; Guilliams, M.; van den Bossche, J.; Van den Bergh, R.; Gysemans, C.; Beschin, A.; De Baetselier, P.; van Ginderachter, J.A. Identification of discrete tumor-induced myeloid-derived suppressor cell subpopulations with distinct T cell-suppressive activity. *Blood* **2008**, *111*, 4233–4244. [CrossRef] [PubMed]

88. Chen, X.; Eksioglu, E.A.; Zhou, J.; Zhang, L.; Djeu, J.; Fortenbery, N.; Epling-Burnette, P.; Van Bijnen, S.; Dolstra, H.; Cannon, J.; et al. Induction of myelodysplasia by myeloid-derived suppressor cells. *J. Clin. Investig.* **2013**, *123*, 4595–4611. [CrossRef] [PubMed]

89. Pyzer, A.R.; Stroopinsky, D.; Rajabi, H.; Washington, A.; Tagde, A.; Coll, M.; Fung, J.; Bryant, M.P.; Cole, L.; Palmer, K.; et al. MUC1 mediated induction of myeloid-derived suppressor cells in patients with acute myeloid leukemia. *Blood* **2017**, *129*, 1791–1801. [CrossRef] [PubMed]

90. Wesolowski, R.; Markowitz, J.; Carson, W.E., 3rd. Myeloid derived suppressor cells—A new therapeutic target in the treatment of cancer. *J. Immunother. Cancer* **2013**, *1*, 10. [CrossRef] [PubMed]

91. Qian, B.Z.; Pollard, J.W. Macrophage diversity enhances tumor progression and metastasis. *Cell* **2010**, *141*, 39–51. [CrossRef] [PubMed]

92. Biswas, S.K.; Mantovani, A. Macrophage plasticity and interaction with lymphocyte subsets: Cancer as a paradigm. *Nat. Immunol.* **2010**, *11*, 889–896. [CrossRef] [PubMed]

93. Bingle, L.; Brown, N.J.; Lewis, C.E. The role of tumour-associated macrophages in tumour progression: Implications for new anticancer therapies. *J. Pathol.* **2002**, *196*, 254–265. [CrossRef] [PubMed]

94. Al-Matary, Y.S.; Botezatu, L.; Opalka, B.; Hones, J.M.; Lams, R.F.; Thivakaran, A.; Schütte, J.; Köster, R.; Lennartz, K.; Schroeder, T.; et al. Acute myeloid leukemia cells polarize macrophages towards a leukemia supporting state in a growth factor independence 1 dependent manner. *Haematologica* **2016**, *101*, 1216–1227. [CrossRef] [PubMed]

95. Ries, C.H.; Cannarile, M.A.; Hoves, S.; Benz, J.; Wartha, K.; Runza, V.; Rey-Giraud, F.; Pradel, L.P.; Feuerhake, F.; Klaman, I.; et al. Targeting tumor-associated macrophages with anti-CSF-1R antibody reveals a strategy for cancer therapy. *Cancer Cell* **2014**, *25*, 846–859. [CrossRef] [PubMed]

96. Pekarek, L.A.; Starr, B.A.; Toledano, A.Y.; Schreiber, H. Inhibition of tumor growth by elimination of granulocytes. *J. Exp. Med.* **1995**, *181*, 435–440. [CrossRef] [PubMed]

97. Shojaei, F.; Singh, M.; Thompson, J.D.; Ferrara, N. Role of Bv8 in neutrophil-dependent angiogenesis in a transgenic model of cancer progression. *Proc. Natl. Acad. Sci. USA* **2008**, *105*, 2640–2645. [CrossRef] [PubMed]

98. Youn, J.I.; Gabrilovich, D.I. The biology of myeloid-derived suppressor cells: The blessing and the curse of morphological and functional heterogeneity. *Eur. J. Immunol.* **2010**, *40*, 2969–2975. [CrossRef] [PubMed]

99. Liu, C.; Workman, C.J.; Vignali, D.A. Targeting regulatory T cells in tumors. *FEBS J.* **2016**, *283*, 2731–2748. [CrossRef] [PubMed]

100. Turnis, M.E.; Sawant, D.V.; Szymczak-Workman, A.L.; Andrews, L.P.; Delgoffe, G.M.; Yano, H.; Beres, A.J.; Vogel, P.; Workman, C.J.; Vignali, D.A.A. Interleukin-35 limits anti-tumor immunity. *Immunity* **2016**, *44*, 316–329. [CrossRef] [PubMed]

101. Cao, X.; Cai, S.F.; Fehniger, T.A.; Song, J.; Collins, L.I.; Piwnica-Worms, D.R.; Ley, T.J. Granzyme B and perforin are important for regulatory T cell-mediated suppression of tumor clearance. *Immunity* **2007**, *27*, 635–646. [CrossRef] [PubMed]

102. Zhou, Q.; Bucher, C.; Munger, M.E.; Highfill, S.L.; Tolar, J.; Munn, D.H.; Levine, B.L.; Riddle, M.; June, C.H.; Vallera, D.A.; et al. Depletion of endogenous tumor-associated regulatory T cells improves the efficacy of adoptive cytotoxic T-cell immunotherapy in murine acute myeloid leukemia. *Blood* **2009**, *114*, 3793–3802. [CrossRef] [PubMed]

103. Ustun, C.; Miller, J.S.; Munn, D.H.; Weisdorf, D.J.; Blazar, B.R. Regulatory T cells in acute myelogenous leukemia: Is it time for immunomodulation? *Blood* **2011**, *118*, 5084–5095. [CrossRef] [PubMed]

104. Teague, R.M.; Kline, J. Immune evasion in acute myeloid leukemia: Current concepts and future directions. *J. Immunother. Cancer* **2013**, *1*, 1–11. [CrossRef] [PubMed]

105. Keir, M.E.; Butte, M.J.; Freeman, G.J.; Sharpe, A.H. PD-1 and its ligands in tolerance and immunity. *Annu. Rev. Immunol.* **2008**, *26*, 677–704. [CrossRef] [PubMed]

106. Pardoll, D.M. The blockade of immune checkpoints in cancer immunotherapy. *Nat. Rev. Cancer* **2012**, *12*, 252–264. [CrossRef] [PubMed]

107. Topalian, S.L.; Taube, J.M.; Anders, R.A.; Pardoll, D.M. Mechanism-driven biomarkers to guide immune checkpoint blockade in cancer therapy. *Nat. Rev. Cancer* **2016**, *16*, 275–287. [CrossRef] [PubMed]

108. Parry, R.V.; Chemnitz, J.M.; Frauwirth, K.A.; Lanfranco, A.R.; Braunstein, I.; Kobayashi, S.V.; Linsley, P.S.; Thompson, C.B.; Riley, J.L. CTLA-4 and PD-1 receptors inhibit T-cell activation by distinct mechanisms. *Mol. Cell. Biol.* **2005**, *25*, 9543–9553. [CrossRef] [PubMed]

109. Armand, P. Immune checkpoint blockade in hematologic malignancies. *Blood* **2015**, *125*, 3393–3400. [CrossRef] [PubMed]

110. Ansell, S.M.; Lesokhin, A.M.; Borrello, I.; Halwani, A.; Scott, E.C.; Gutierrez, M.; Schuster, S.J.; Millenson, M.M.; Cattry, D.; Freeman, G.J.; et al. PD-1 blockade with nivolumab in relapsed or refractory Hodgkin's lymphoma. *N. Engl. J. Med.* **2015**, *372*, 311–319. [CrossRef] [PubMed]

111. Kumar, A.; Casulo, C.; Yahalom, J.; Schoder, H.; Barr, P.M.; Caron, P.; Chiu, A.; Constine, L.S.; Drullinsky, P.; Friedberg, J.W.; et al. Brentuximab vedotin and AVD followed by involved-site radiotherapy in early stage, unfavorable risk Hodgkin lymphoma. *Blood* **2016**, *128*, 1458–1464. [CrossRef] [PubMed]

112. Lesokhin, A.M.; Ansell, S.M.; Armand, P.; Scott, E.C.; Halwani, A.; Gutierrez, M.; Millenson, M.M.; Cohen, A.D.; Schuster, S.J.; Lebovic, D.; et al. Nivolumab in patients with relapsed or refractory hematologic malignancy: Preliminary results of a phase Ib study. *J. Clin. Oncol.* **2016**, *34*, 2698–2704. [CrossRef] [PubMed]

113. Kondo, A.; Yamashita, T.; Tamura, H.; Zhao, W.; Tsuji, T.; Shimizu, M.; Shinya, E.; Takahashi, H.; Tamada, K.; Chen, L.; et al. Interferon-γ and tumor necrosis factor-α induce an immunoinhibitory molecule, B7-H1, via nuclear factor-κB activation in blasts in myelodysplastic syndromes. *Blood* **2010**, *116*, 1124–1131. [CrossRef] [PubMed]

114. Davids, M.S.; Kim, H.T.; Bachireddy, P.; Costello, C.; Liguori, R.; Savell, A.; Lukez, A.P.; Avigan, D.; Chen, Y.B.; McSweeney, P.; et al. Ipilimumab for patients with relapse after allogeneic transplantation. *N. Engl. J. Med.* **2016**, *375*, 143–153. [CrossRef] [PubMed]

115. Michot, J.M.; Bigenwald, C.; Champiat, S.; Collins, M.; Carbonnel, F.; Postel-Vinay, S.; Berdelou, A.; Varga, A.; Bahleda, R.; Hollebecque, A.; et al. Immune-related adverse events with immune checkpoint blockade: A comprehensive review. *Eur. J. Cancer* **2016**, *54*, 139–148. [CrossRef] [PubMed]

116. Mittal, D.; Gubin, M.M.; Schreiber, R.D.; Smyth, M.J. New insights into cancer immunoediting and its three component phases—Elimination, equilibrium and escape. *Curr. Opin. Immunol.* **2014**, *27*, 16–25. [CrossRef] [PubMed]

117. Daver, N.; Basu, S.; Garcia-Manero, G.; Cortes, J.E.; Ravandi, F.; Ning, J.; Xiao, L.; Juliana, L.; Kornblau, S.M.; Konopleva, M.; et al. Defining the immune checkpoint landscape in patients (pts) with acute myeloid leukemia (AML). *Blood* **2016**, *128*, 2900.

118. Garcia-Manero, G.; Daver, N.G.; Montalban-Bravo, G.; Jabbour, E.J.; DiNardo, C.D.; Kornblau, S.M.; Bose, P.; Alvarado, Y.; Ohanian, M.; Borthakur, G.; et al. A phase II study evaluating the combination of nivolumab (Nivo) or ipilimumab (Ipi) with azacitidine in pts with previously treated or untreated myelodysplastic syndromes (MDS). *Blood* **2016**, *128*, 344.

119. Mellman, I.; Coukos, G.; Dranoff, G. Cancer immunotherapy comes of age. *Nature* **2011**, *480*, 480–489. [CrossRef] [PubMed]

© 2017 by the authors. Licensee MDPI, Basel, Switzerland. This article is an open access article distributed under the terms and conditions of the Creative Commons Attribution (CC BY) license (http://creativecommons.org/licenses/by/4.0/).

International Journal of
Molecular Sciences

MDPI

Article

Clinical Outcomes and Co-Occurring Mutations in Patients with *RUNX1*-Mutated Acute Myeloid Leukemia

Maliha Khan [1], Jorge Cortes [1], Tapan Kadia [1], Kiran Naqvi [1], Mark Brandt [1], Sherry Pierce [1], Keyur P. Patel [2], Gautam Borthakur [1], Farhad Ravandi [1], Marina Konopleva [1], Steven Kornblau [1], Hagop Kantarjian [1], Kapil Bhalla [1] and Courtney D. DiNardo [1,*]

[1] Department of Leukemia, The University of Texas MD Anderson Cancer Center, Houston, TX 77030, USA; doc.maliha@gmail.com (M.K.); jcortes@mdanderson.org (J.C.); tkadia@mdanderson.org (T.K.); knaqvi@mdanderson.org (K.N.); mbrandt@mdanderson.org (M.B.); spierce@mdanderson.org (S.P.); gborthak@mdanderson.org (G.B.); fravandi@mdanderson.org (F.R.); mkonople@mdanderson.org (M.K.); skornblau@mdanderson.org (S.K.); hkantarjian@mdanderson.org (H.K.); kbhalla@mdanderson.org (K.B.)

[2] Department of Hematopathology, The University of Texas MD Anderson Cancer Center, Houston, TX 77030, USA; kppatel@mdanderson.org

*   Correspondence: cdinardo@mdanderson.org; Tel.: +1-713-794-1141

Received: 21 June 2017; Accepted: 14 July 2017; Published: 26 July 2017

Abstract: (1) Runt-related transcription factor 1 (*RUNX1*) mutations in acute myeloid leukemia (AML) are often associated with worse prognosis. We assessed co-occurring mutations, response to therapy, and clinical outcomes in patients with and without mutant *RUNX1* (*mRUNX1*); (2) We analyzed 328 AML patients, including 177 patients younger than 65 years who received intensive chemotherapy and 151 patients >65 years who received hypomethylating agents. *RUNX1* and co-existing mutations were identified using next-generation sequencing; (3) *RUNX1* mutations were identified in 5.1% of younger patients and 15.9% of older patients, and were significantly associated with increasing age (*p* = 0.01) as well as intermediate-risk cytogenetics including normal karyotype (*p* = 0.02) in the elderly cohort, and with lower lactate dehydrogenase (LDH; *p* = 0.02) and higher platelet count (*p* = 0.012) overall. Identified co-occurring mutations were primarily *ASXL1* mutations in older patients and *RAS* mutations in younger patients; FLT3-ITD and IDH1/2 co-mutations were also frequent. Younger *mRUNX1* AML patients treated with intensive chemotherapy experienced inferior treatment outcomes. In older patients with AML treated with hypomethylating agent (HMA) therapy, response and survival was independent of *RUNX1* status. Older *mRUNX1* patients with prior myelodysplastic syndrome or myeloproliferative neoplasms (MDS/MPN) had particularly dismal outcome. Future studies should focus on the prognostic implications of *RUNX1* mutations relative to other co-occurring mutations, and the potential role of hypomethylating agents for this molecularly-defined group.

Keywords: *RUNX1*; mutations; acute myeloid leukemia; hypomethylating agents; chemotherapy; prognosis

1. Introduction

The genetic diversity of acute myeloid leukemia (AML) and the complexity of its multi-step pathogenesis now allow the classification of AML based on molecular events. The evolution of AML has traditionally been proposed to follow the "two-hit model" [1], in which two classes of mutations are required for cancer development. Class I mutations are activating mutations that stimulate cell survival and proliferation, while class II mutations are inactivating mutations that interfere with hematopoietic differentiation [2]. Runt-related transcription factor 1 (*RUNX1*) is a key hematopoietic transcription factor that regulates genes involved in myeloid differentiation, and is generally considered to be a classical tumor suppressor (class II) mutation [2]. Mutations of *RUNX1* are reported in approximately 10–16% of AML patients [3,4] and 12–24% of myelodysplastic syndrome (MDS) patients [5,6]. *RUNX1* alterations predominate in the morphologically undifferentiated French-American-British (FAB) M0 subtype. Clinical features associated with this mutation include older age, male sex, and absence of cytogenetic abnormalities [4].

In patients with AML, mutant *RUNX1* often associates with certain class I mutations. Gene expression profiling has identified the co-occurrence of *RUNX1* with mutations of the chromatin remodeling gene *ASXL1* or partial tandem duplication of the transcription regulator *MLL* in all major cohorts to date, while *RUNX1* and *NPM1* (nucleophosmin) mutations are consistently negatively correlated [2,7]. *ASXL1* mutations appear to be the most frequent co-mutation with *RUNX1*, and *ASXL1/RUNX1* double mutants are associated with lower rates of therapeutic response [8,9]. Other significant associations, such as with alterations in the tyrosine kinase *FLT3*, splicing factor mutations, or isocitrate dehydrogenase (*IDH1* or *IDH2*) has not been firmly established [3,4,7,10]; however, concomitant *FLT3* mutations are thought to play a synergistic role with *RUNX1* mutations in the development of AML [11]. Upregulation of genes normally expressed in hematopoietic progenitor cells or lymphoid cells, and downregulation of promoters of myelopoeisis also ascribe a unique gene expression signature to *RUNX1*-mutated AML [3,12], implicating upregulation of oncogenic pathways such as BCR, TLR-4 and NOTCH1 to the pathogenicity of mutant *RUNX1*.

Although standard intensive chemotherapy has yielded high remission rates and superior long-term survival in younger AML patients [13], elderly patients continue to experience lower response rates and poor long-term outcomes [14], and hypomethylating agents (HMAs) are typically utilized in the older patient population [14]. Clinical outcomes with respect to the influence of *RUNX1* mutations based on age and treatment regimen have not been previously explored [4].

RUNX1 mutations are correlated with poor clinical outcomes. Gaidzik et al. [10] compared 53 mutant *RUNX1* and 831 wild-type *RUNX1* newly-diagnosed AML patients, and found inferior rates of event-free survival (EFS), relapse-free survival (RFS), and overall survival (OS) in *RUNX1*-mutated patients in patients 60 years of age or younger and treated with intensive chemotherapy (EFS, 8% vs. 30%; RFS, 26% vs. 44%; OS, 32% vs. 45%). In an analysis of AML patients treated with intensive chemotherapy by Tang et al. [7], the complete remission rate was lower in 62 newly-diagnosed patients with *RUNX1*-mutated AML compared with 408 without the mutation (56.8% vs. 77.5%). A statistically higher incidence of induction-related death in patients with mutant-*RUNX1* AML (10.8% vs. 6.5%) was also identified. Other studies have also correlated *RUNX1* mutations with resistance to chemotherapy and higher rates of refractory disease [2,7,10]. It is important to expand our understanding of the clinical outcomes of mutant and wild-type *RUNX1* in relation to specific treatment modalities, with attention to the older patient population treated with hypomethylating or lower-intensity approaches where *RUNX1* mutations are more often identified.

This study examined the frequency of *RUNX1* mutations in newly diagnosed patients with AML, and their effect on clinical outcomes and treatment response rates, including younger patients receiving chemotherapy and elderly patients receiving HMAs. Co-existing mutations and their effect on the clinical course of AML in conjunction with *RUNX1* were also examined.

2. Results

2.1. Frequency and Characteristics of RUNX1 Mutations

Overall, mutant *RUNX1* was identified in 33 (10.1%) patients with newly diagnosed AML. These rates were significantly higher in patients 65 years of age or older, 24 (15.9%), compared with the younger adult cohort, 9 (5.1%). None of the patients were identified as having more than one pathogenic *RUNX1* mutation at diagnosis. The majority (>90%) of *RUNX1* mutations described were either frameshift or nonsense in nature, resulting in early amino acid chain termination and truncated proteins. The specific characteristics of these mutations, including their allelic frequencies, types, and exon locations, have been elaborated in Appendix A Table A1.

2.2. Association of RUNX1 with Clinical Characteristics

Table 1 summarizes the clinicopathologic variables evaluated with respect to the impact of *RUNX1* mutational status. In the elderly cohort, a significant correlation was observed between the presence of mutated *RUNX1* and increasing age ($p = 0.01$); *mRUNX1* was rare in the younger patient cohort (5%). Along with an increasing age, *RUNX1* mutations occurred more frequently in older patients with intermediate-risk cytogenetics (particularly those with a normal karyotype), as compared to those with complex cytogenetic abnormalities ($p = 0.02$). Lower lactate dehydrogenase (LDH) levels and higher platelet counts were also found to be significant in *RUNX1*-mutated patients within the entire cohort overall ($p = 0.02$ and $p = 0.012$ respectively). *RUNX1* mutational status did not significantly correlate with WBC count, hemoglobin, or peripheral blood and bone marrow blasts, or history of prior myelodysplastic syndrome or myeloproliferative neoplasms (MDS/MPN) or therapy-related AML (t-AML). Among the mutant *RUNX1* patients, 15.2% ($n = 20$) of the patients classified as FAB M0 subtype.

Table 1. Clinical characteristics in the younger cohort receiving chemotherapy and the elderly cohort receiving hypomethylating agents according to *RUNX1* mutation status.

Clinical Characteristics	Younger Cohort (Chemotherapy)			Elderly Cohort (HMA)		
	Median (Range)		*p*-Value	Median (Range)		*p*-Value
	RUNX1^{mut}	*RUNX1^{WT}*		*RUNX1^{mut}*	*RUNX1^{WT}*	
	($n = 9$)	($n = 168$)		($n = 24$)	($n = 127$)	
Age, year	56 (31–63)	51 (17–64)	0.14	77 (65–92)	73 (65–91)	0.01 *
WBC, $\times 10^9$/L	4.95 (1.3–17.2)	6.9 (0.5–378.4)	0.43	3.2 (0.6–36.5)	3.2 (0.3–164.5)	0.98
Platelet count, $\times 10^9$/L	57.5 (28–213)	30 (1–584)	0.05	61 (7–416)	42 (3–324)	0.20
Hemoglobin, g/dL	8.95 (7.4–12.1)	9.3 (5.1–13.0)	0.38	9.4 (5.8–13.1)	9.4 (7.5–13.2)	0.37
Peripheral blood blasts, %	28 (1–86)	25.5 (0–97)	0.61	7 (0–90)	11 (0–95)	0.51
Neutrophils, %	14.5 (0–45)	12 (0–98)	0.93	21 (0–81)	21 (0–73.3)	0.95
Bone marrow blasts, %	51 (28–93)	54 (2–96)	0.66	50 (12–84)	44 (1–90)	0.80
Cytogenetic			0.94			0.02
Complex	2	37		2	47	
Diploid	3	65		14	50	
^ Intermediate	4	66		8	30	
LDH, U/L	668.5 (526–1649)	888 (310–12,489)	0.10	532 (276–2314)	650 (210–3921)	0.44
t-AML	0	10	0.45	4	18	0.75
Prior MDS/MPD	1	6	0.26	8	25	0.14

Abbreviations: LDH, lactate dehydrogenase; HMA, hypomethylating agent; Prior MDS/MPD: Prior myelodysplastic syndrome or myeloproliferative neoplasms; t-AML, therapy-related acute myeloid leukemia; WBC, white blood cell; * Boldface indicates statistical significance. ^ Intermediate also includes insufficient metaphases and not done.

2.3. Association of RUNX1 with Co-Occurring Mutations

In the older cohort, mutational analysis identified *ASXL1* mutations as the most frequent aberration in the presence of mutant *RUNX1*, co-occurring at a frequency of 37.5%—significantly higher compared to ASXL1 occurrence in wild-type *RUNX1* (14.2%) ($p = 0.007$). In comparison, the incidence of co-occurring *RAS* mutation was notable within the younger *mRUNX1* cohort, occurring in 4 of 9 patients (44.4%) ($p = 0.184$). The other most frequent co-occurring mutations in the younger cohort were those in *ASXL1*, *FLT3-ITD*, and *DNMT3A* genes at 33% each; compared to wild-type *RUNX1* (*ASXL1*, 2.4% $p < 0.005$; *FLT3-ITD*, 20.8% $p = 0.379$; *DNMT3A*, 20.2% $p = 0.384$). *IDH1* and *IDH2* mutations were also frequently identified in association with mutant *RUNX1*, with a frequency of mutant *IDH1* as 11.1% in *mRUNX1* compared to 4.8% wild-type ($p = 0.407$) in the younger cohort and 16.7% in *mRUNX1* compared to 12.6% in wild-type *RUNX1* in the elderly cohort ($p = 0.603$). *IDH2* mutations occurred at 22.2% (*mRUNX1*) and 15.5% (wild-type) ($p = 0.617$) and 20.8% (*mRUNX1*) compared to 16.5% (wild-type) ($p = 0.624$) in the younger and the elderly cohorts, respectively.

A number of other gene mutations were analyzed and showed varying association with mutated *RUNX1* (Figure 1) [15]. The frequencies of these co-mutations with mutant *RUNX1* in the younger and the elderly cohorts were, respectively, as follows: *FLT3-ITD*, 33.3% and 16.7%; *NPM1*, 0% and 4.2%; *RAS*, 44.4% and 8.3%; *IDH1*, 11.1% and 16.7%; *IDH2*, 22.2% and 20.8%; *TET2*, 0% and 12.5%; *PTPN11*, 11.1% and 8.3%; *JAK2*, 0% and 8.3%; *CEBPA*, 0% and 12.5%; *DNMT3A*, 33.3% and 16.7%; *EZH2*, 22.2% and 0%; *KIT*, 11.1% and 0%; *MPL*, 0% and 4.2%; *WT1*, 0% and 4.2%. Additionally, *MLL* translocation 11q23 was detected in four patients (1.2%) out of the total cohort of 328.

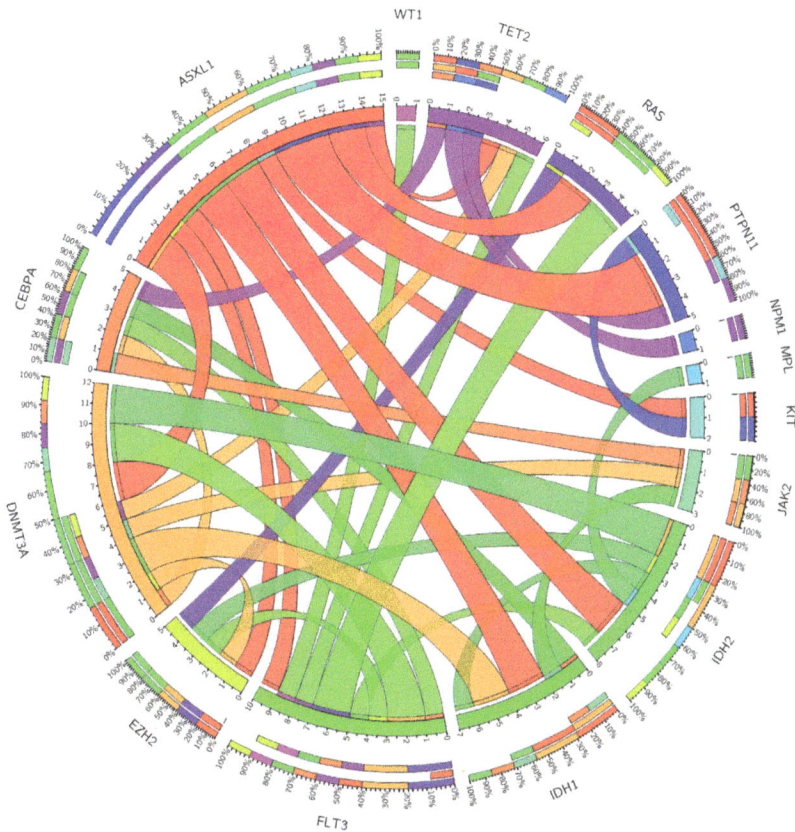

Figure 1. Co-occurring mutations among patients with *RUNX1* acute myeloid leukemia. The thickness of the connecting lines indicates the frequency with which the two mutations co-occur.

We additionally identified a number of mutations that were negatively associated with mutant *RUNX1*. Mutations in the *NPM1* gene were significantly lower in both the older (4.2%) and younger cohorts (0%) with *RUNX1* mutations, compared to those with wild-type *RUNX1* (older, 15.0%; younger, 26.2%). A similar negative correlation was identified between *RUNX1* and CEBPα mutations—particularly among the younger cohort, wherein CEBPα incidence with wild-type *RUNX1* was 16.1%, and was not identified (0%) in the setting of mutated *RUNX1*.

FLT3-ITD mutations were frequently identified in both cohorts, irrespective of *mRUNX1*: frequencies in the presence of mutated *RUNX1* versus wild-type *RUNX1* (older, 16.7% vs. 15.7%; younger, 30.0% vs. 20.8%). Of interest, co-occurring *FLT3-ITD* in the setting of *mRUNX1* appeared to confer a more favorable prognosis (median OS with mutated *FLT3-ITD* vs. wild-type *FLT3-ITD*, NR vs. 11.5 months; $p = 0.034$) with a notable improved OS in *FLT3-ITD* and *RUNX1* dual-mutated patients (Figure 2), although these differences were not significant when separated by age (younger, $p = 0.604$; older, $p = 0.126$) (Figure A1). This overall prognostic effect was additionally observed with improved EFS in dual-mutated patients (median EFS with mutated *FLT-ITD3* vs. wild-type *FLT3-ITD*, NR vs. 2.3 months; $p = 0.006$).

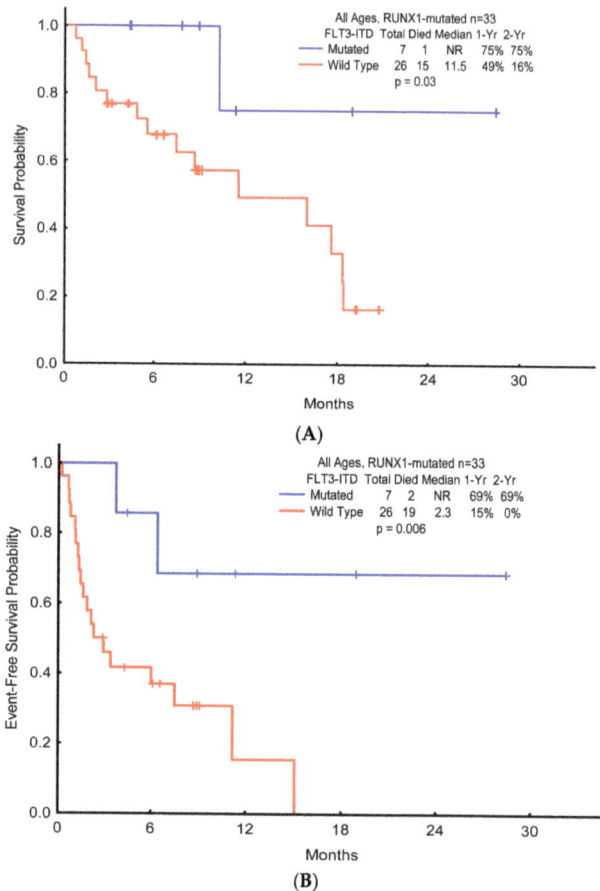

(A)

(B)

Figure 2. Survival probabilities of patients with *RUNX1* and *FLT3-ITD* mutations. (**A**) Overall survival; (**B**) Event-free survival.

2.4. Association of RUNX1 with Response to Frontline AML Therapy

In the elderly AML cohort, response to frontline therapy was generally unaffected by *RUNX1* mutational status (Table 2). There were no significant differences between patients with mutated and wild-type *RUNX1* for complete remission (CR) rates, overall response rate (ORR);(CR + complete remission with incomplete platelet recovery (CRp) + hematological improvement (HI) + partial remission (PR), HI, and/or early death (ED; death within 28 days of initiating treatment) in the cohort as a whole. Notably, a significant difference was observed for the subset of older AML patients with prior MDS/MPN (Table 3). Overall, ORR was obtained in 80% of MDS/MPN patients in the older cohort with wild-type *RUNX1*, but was achieved in only 25% of the patients identified with mutant *RUNX1* ($p = 0.004$).

Table 2. Clinical outcomes of all patients according to age, treatment, and *RUNX1* mutation status.

Response	Younger Cohort (Chemotherapy)			Elderly Cohort (HMA)		
	n (%)			n (%)		
	$RUNX1^{WT}$	$RUNX1^{mut}$	p-Value	$RUNX1^{WT}$	$RUNX1^{mut}$	p-Value
	(n = 168)	(n = 9)		(n = 127)	(n = 24)	
Complete remission	128 (76.2)	3 (33.3)	**0.004 ***	48 (37.8)	9 (37.5)	0.976
Complete remission with incomplete platelet recovery	9 (5.4)	3 (33.3)	0.271	11 (8.7)	1 (4.2)	0.667
Hematological improvement	4 (2.4)	1 (11.1)	0.631	17 (13.4)	3 (12.5)	0.603
Partial remission	2 (1.2)	0 (0.0)		4 (3.1)	0 (0.0)	
Overall response rate	143 (85.0)	7 (78.0)	0.549	80 (63.0)	13 (54.0)	0.418
No remission	19 (11.3)	2 (22.2)		22 (17.3)	7 (29.2)	
Death	5 (3.0)	0 (0.0)		22 (17.3)	3 (12.5)	
Early death **	1 (0.6)	0 (0.0)	0.818	3 (2.4)	1 (4.2)	0.617

* Boldface indicates statistical significance. Early death **: Death within 28 days of initiating treatment.

Table 3. Clinical outcomes of patients with prior MDS or MPD according to age, treatment, and *RUNX1* mutation status.

Response	Younger Cohort (Chemotherapy)			Elderly Cohort (HMA)		
	n (%)			n (%)		
	$RUNX1^{WT}$	$RUNX1^{mut}$	p-Value	$RUNX1^{WT}$	$RUNX1^{mut}$	p-Value
	(n = 6)	(n = 1)		(n = 25)	(n = 8)	
Complete remission	2 (33.3)	1 (100.0)	0.211	11 (44.0)	1 (12.5)	0.107
Complete remission with incomplete platelet recovery	1 (16.7)	0 (0.0)		4 (16.0)	0 (0.0)	
Hematological improvement	0 (0.0)	0 (0.0)		4 (16.0)	1 (12.5)	
Partial remission	1 (16.7)	0 (0.0)		1 (4.0)	0 (0.0)	
Overall response rate	4 (66.7)	1 (100.0)	0.497	20 (80.0)	2 (25.0)	**0.004 ***
No remission	1 (16.7)	0 (0.0)		1 (4.0)	3 (37.5)	
Death	1 (16.7)	0 (0.0)		4 (16.0)	2 (25.0)	
Early death **	0 (0.0)	0 (0.0)		0 (0.0)	1 (12.5)	

* Boldface indicates statistical significance. Early death **: Death within 28 days of initiating treatment.

In the younger cohort (Table 3), while CR rates were significantly reduced in patients with mutated versus wild-type *RUNX1* (33.3% vs. 76.2%; $p = 0.004$), no significant difference in overall response rate was observed.

2.5. Association of RUNX1 with Clinical Outcomes

OS, EFS, and RFS were not significantly different between mutant *RUNX1* compared with wild-type (Figures 3 and 4). In the younger cohort, the two-year OS with wild-type *RUNX1* and mutant *RUNX1*, respectively, was 60% and 50% (95% CI: ± 13% vs. ±98%); two-year RFS was 71% and 54% (95% CI: ±15% vs. ±85%); two-year EFS was 50% and 39% (95% CI: ±13% vs. ±85%). The clinical outcomes of the older cohort showed similar results, with no significant difference observed between the OS, EFS, and RFS of mutated-*RUNX1* patients versus that of wild-type-*RUNX1* patients. In a subgroup analysis of patients with a history of prior MDS/MPN ($n = 40$), those with mutant *RUNX1* had significantly shorter EFS compared with wild-type *RUNX1* (1.35 vs. 6.34 months; $p = 0.012$) (Figure 5) and trend to inferior OS ($p = 0.079$) in the older patient cohort.

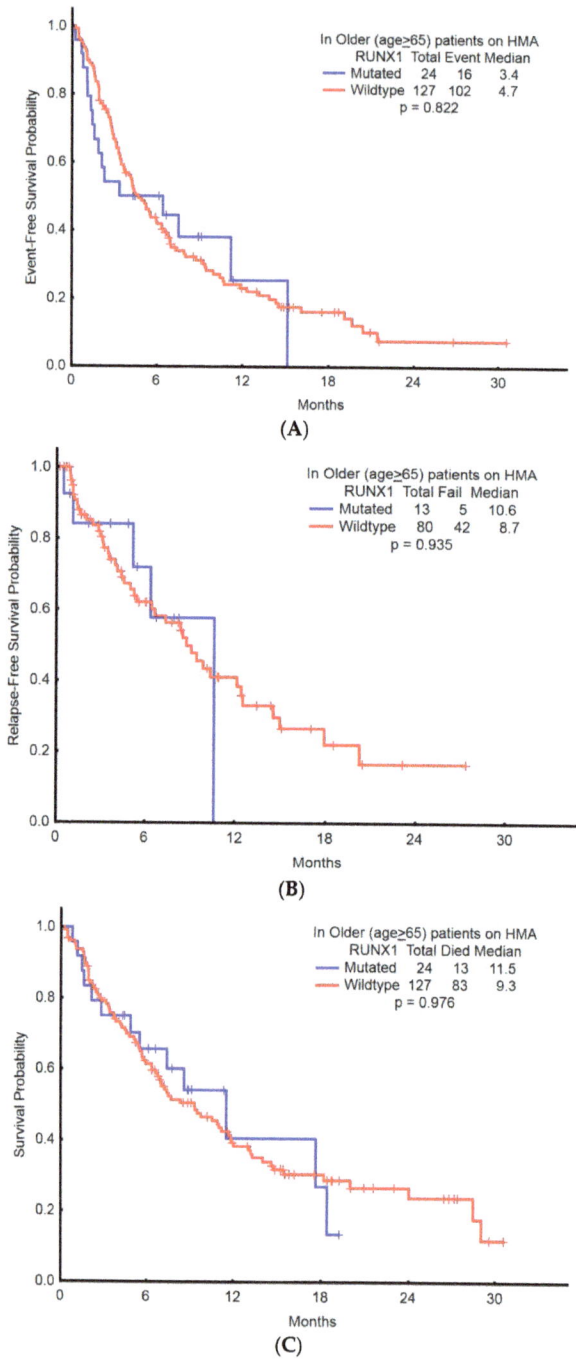

Figure 3. Survival probabilities among patients older than 65 years who received induction with hypomethylating agents, compared with the presence or absence of *RUNX1* mutation. (**A**) Overall survival; (**B**) Relapse-free survival; (**C**) Event-free survival.

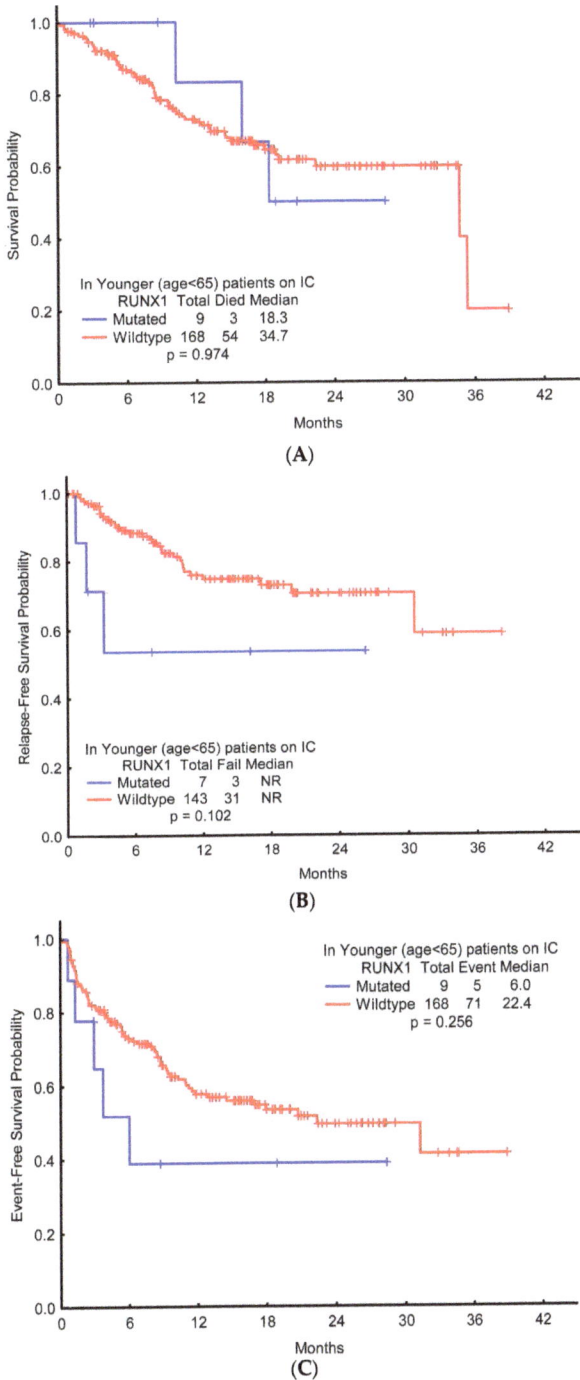

Figure 4. Survival probabilities among patients aged below 65 years who received induction with chemotherapy, compared with the presence or absence of *RUNX1* mutation. (**A**) Overall survival; (**B**) Relapse-free survival; (**C**) Event-free survival.

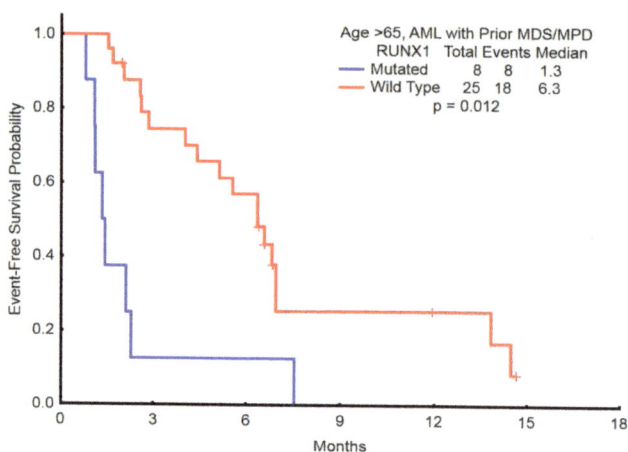

Figure 5. Event-free survival in patients with prior MDS/MPN who were older than 65 years and received HMA induction therapy.

3. Discussion

In our study, *RUNX1* mutations were detected at an overall incidence of 10.1%. This is consistent with previous studies that reported mutated *RUNX1* alleles in 5.6% to 16.1% of patients with AML [3,4,7,10,12].

Mutations in *RUNX1* associate with increasing age, also consistent with previous studies examining *RUNX1* mutation status and age [4,7,12,16]. Accumulation of cellular stress and impaired repair of double-stranded DNA breaks (e.g., after radiation exposure) may contribute to genomic instability [17] and increase the frequency of genetic mutations, including *RUNX1* [18–20]. Furthermore, *RUNX1* mutations in our elderly cohort clustered among patients identified with intermediate cytogenetic risk including cytogenetically-normal AML, consistent with previous studies [7,10]. Other clinical characteristics (white blood cell (WBC) count, hemoglobin, peripheral blood and bone marrow blasts, t-AML, and AML secondary to MDS/MPN) were not associated with *RUNX1* mutational status. Our analysis additionally confirmed a relationship between *RUNX1* mutation and platelet count and LDH level, as has been previously reported [4,7].

The most commonly co-occurring mutation overall and particularly prevalent in the older cohort was ASXL1. Of interest, clonal expansion of *ASXL1* is one of the most frequently identified genes in age-related clonal hematopoiesis, and subsequent *RUNX1* mutations may be implicated in malignant progression in these individuals [21,22]. Among younger patients, co-occurring *RAS* mutations (44%) and FLT3-ITD (33%) were also frequent. Their classification as class I mutations represents their capacity to promote stem cell proliferation and thereby cooperate with *RUNX1* mutations in AML pathogenesis [2,23].

Co-occurring *FLT3* and *RUNX1* mutations have been previously described, with a hypothesis that loss of function *RUNX1* mutations may predispose hematopoietic cells to overexpress activating mutations of *FLT3* [24]. When examining the clinical implications of this hypothesis in patients with mutant *RUNX1*, our analysis interestingly found superior OS and EFS in *mRUNX1* associated with *FLT3* mutations. Previous studies have identified that the adverse effect of *RUNX1* is independent of FLT3-ITD [25]. Evaluation in future studies will be important to see if this is a reproducible finding. It is also important to note that *NPM1* and biallelic *CEBPα* mutations confer a favorable prognosis in terms of disease-free survival and OS in the absence of *FLT3* mutations [2], and both are inversely related to the incidence of *RUNX1* mutations [7]. In our cohort, the incidence of *NPM1* mutations declined from 26.2% to 0% in younger patients and from 15.0% to 4.2% in older patients in the absence and presence of *RUNX1* mutations, respectively. Conversely, although partial tandem duplications in the *MLL* gene have been previously associated with mutated *RUNX1* and implicated in shorter survival [2,7], this association was not identified in our patient cohort, as only four patients were identified with the *MLL* rearrangement 11q23. An association of co-occurring *IDH1/IDH2* mutations with *RUNX1* mutations was additionally identified, consistent with Gaidzik et al. [10]. Furthermore, mutations in *RUNX1* and *IDH1/2* have been collectively described as the most frequent genetic lesions in M0-AML [26], the prognostic impact of the latter remaining controversial [27].

While *RUNX1* mutations are frequently associated with inferior responses to AML therapy, our analysis suggests that the poor outcomes seen in *RUNX1* mutations occur primarily within younger patients treated with intensive therapeutic strategies, as well as the subgroup of *RUNX1* mutant AML with prior MDS/MPN. The similar responses and outcome of the larger cohort of older *mRUNX1* treated with HMA approaches was a notable finding of this analysis.

A previous study in patients with MDS identified that treatment with HMAs may partially mitigate the adverse prognostic impact of gene mutations in patients with *RUNX1*, *ASXL1*, *EZH2*, or *ETV6* mutations [28]. Differences in outcomes between patient populations may also be linked with the distribution of co-existing mutations and cytogenetic abnormalities that may exacerbate or mitigate the effect of mutated *RUNX1*, [2] and our study is limited by a lack of complete co-mutational data, most notably for splicing factor mutations, as whole exome sequencing was not possible. The importance of *RUNX1* as a marker of MRD in responding patients was additionally not able to be assessed in our study.

4. Materials and Methods

4.1. Patients

A total of 328 patients with treatment-naïve AML treated at the University of Texas MD Anderson Cancer Center between August 2013 and July 2016 were analyzed in this study. The presence of de novo AML, or history of prior myelodysplastic syndrome (MDS) or myeloproliferative neoplasms (MPNs) was collected. As treatment intensity at our institution during this time period was strongly and near-universally correlated to patient age at diagnosis, patients were categorized into two cohorts according to their age and therapeutic regimen: patients younger than 65 years who received intensive chemotherapy (younger cohort; $n = 177$) and patients ≥ 65 years who received HMA therapy (elderly cohort; $n = 151$).

Variables in our study included patient demographics (age and sex), relevant diagnostic tests including white blood cell (WBC) count, hemoglobin (Hb) levels, platelet count, lactate dehydrogenase (LDH) levels, and percentage of bone marrow blasts, cytogenetics, FAB classification, *RUNX1* mutation, and co-occurrence of other mutations. Clinical outcomes were assessed using Kaplan–Meier estimates of OS, EFS, and RFS [29]. Response to therapy was evaluated using revised IWG criteria for AML [30].

All participating patients provided informed consent in compliance with the Declaration of Helsinki, and the study received ethical approval by the Institutional Review Board as part of a retrospective chart review protocol, PA11.0788 on 10 October 2011 within the Department of Leukemia at University of Texas MD Anderson Cancer Center.

4.2. Molecular Analysis

Diagnostic bone marrow samples were obtained for mutational analysis. Total genomic DNA was extracted from unenriched peripheral blood (PB) or bone marrow (BM) samples using ReliaPrep genomic DNA isolation kit (Promegal Corp, Madison, WI, USA) [31]. *FLT3* (internal tandem duplication and D835) was assessed by PCR followed by capillary electrophoresis on a Genetic Analyzer (Applied Biosystems, Foster City, CA, USA) [32]. Briefly, a total of 250 ng DNA was utilized to prepare sequencing libraries using Agilent HaloPlex custom Kit (Agilent Technologies, Santa Clara, CA, USA) [33]. The entire coding sequences of 28 genes (*ABL1*, *ASXL1*, *BRAF*, *DNMT3A*, *EZH2*,

FLT3, GATA1, GATA2, HRAS, IDH1, IDH2, IKZF2, JAK2, KIT, KRAS, MDM2, MLL, MPL, NPM1, NRAS, PTPN11, RUNX1, TET2, TP53, WT1) were interrogated on a custom-designed next-generation sequencing approach using the IlluminaMiSeq platform (Illumina; San Diego, CA, USA) [34]. The genomic reference sequence used was genome GRch37/hg19. The following software tools were utilized in the experimental setup and data analysis: Illumina Experiment Manager 1.6.0 (Illumina; San Diego, CA, USA), MiSeq Control Software 2.4 (Illumina; San Diego, CA, USA), Real Time Analysis 1.18.54 (Illumina; San Diego, CA, USA), Sequence Analysis Viewer 1.8.37 (Illumina; San Diego, CA, USA), MiSeq Reporter 2.5.1 (Illumina; San Diego, CA, USA), and SureCall 3.0.1.4 (Agilent Technologies; Santa Clara, CA, USA). A minimum of 80% reads at quality scores of AQ30 or higher were required to pass quality control. NM_001754 was utilized as the *RUNX1* reference sequence.

The lower limit of detection of this assay (analytical sensitivity) for single nucleotide variations was determined to be 5% (one mutant allele in the background of nineteen wild type alleles) to 10% (one mutant allele in the background of nine wild type alleles). Testing of patients with active hematologic malignancies was limited to somatic mutations only. Known germline polymorphisms identified in >20% of our in-house patient population were excluded. Additionally, germline polymorphisms previously reported in population databases such as dbSNP and ExAC were classified as "variants of probable germline origin" and were excluded from this analysis.

4.3. Statistical Analysis

Continuous variables were described as means, medians, standard deviations, and ranges, whereas the categorical variables were presented as frequencies and percentages. The categorical variables in patients with mutated *RUNX1* were compared with the wild-type using the Chi-squared test. Kaplan–Meier curves [26] were used to estimate unadjusted OS durations. OS was calculated from the start of treatment to the last follow up or death (censor = alive/dead). EFS was measured from the start of treatment to any adverse event or the last follow-up. Adverse events included either relapse (i.e., NR, ED) or death. RFS was determined from leukemia-free response (i.e., at the time of documented CR, CRp, CRi, PR) to relapse (loss), death, or last follow-up. For all analyses, $p < 0.05$ was considered statistically significant. All computations were carried out in Statistica 12.0 (StatSoftInc. Tulsa, OK, USA; 1984–2013).

5. Conclusions

In conclusion, *RUNX1* mutations are among the most frequent recurring genetic abnormalities in patients with AML, often associated with specific clinical features and poor outcomes. *ASXL1*, *DNMT3A*, *IDH1/2*, *FLT3-ITD*, and *RAS* mutations frequently co-occur with *RUNX1* mutations. *FLT3-ITD* mutations did not impart an inferior prognosis in *mRUNX1* AML, and of note appeared to confer an improved OS in this molecularly defined cohort. While younger *mRUNX1* patients treated with intensive chemotherapy and a subgroup of older *mRUNX1* patients with prior MDS/MPN exhibited inferior treatment responses, the majority of *mRUNX1* AML patients included older patients treated with HMA therapy, in whom treatment responses and clinical outcomes were not inferior compared to *RUNX1* wild-type. Further studies investigating the treatment response to HMAs—in particular in younger patient populations with *RUNX1*-mutated AML—are required to fully establish their therapeutic role in the individualized treatment of these patients.

Acknowledgments: This work was supported in part by the MD Anderson Cancer Center Support Grant P30 CA016672, Leukemia SPORE CA100632 and Charif Souki Cancer Research Fund.

Author Contributions: Maliha Khan Keyur P. Patel designed the study, collected and analyzed data, wrote the manuscript. Mark Brandt and Sherry Pierce provided statistical analysis. Jorge Cortes, Farhad Ravandi, Kapil Bhalla, Steven Kornblau, Gautam Borthakur, Marina Konopleva, Tapan Kadia, Kiran Naqvi, Keyur P. Patel and Hagop Kantarjian edited the manuscript. Courtney D. DiNardo conceived the study, designed the study, analyzed data, and wrote the manuscript.

Conflicts of Interest: The authors declare no conflict of interest.

Appendix A

Table A1. *RUNX1* mutations according to their minor allele frequency, exon location, and mutation type.

Mutation HGVS (MAF%)	Exon	Type
NM_001754.4 (*RUNX1*):c.497G>A p.R166Q [46.4]	5	Missense
NM_001754.4 (*RUNX1*):c.423_424insAAAGGGATTCT p.A142fs [5.9]	5	Frameshift
NM_001754.4 (*RUNX1*):c.891_894dupCCCA p.A299fs [8.7]	8	Frameshift
NM_001754.4 (*RUNX1*):c.238dupG p.E80fs [50.3]	4	Frameshift
NM_001754.4 (*RUNX1*):c.900_901insTCCG p.P301fs [17.0], NM_001754.4 (*RUNX1*):c.987_988dupGT p.F330fs [32.3], NM_001754.4(*RUNX1*):c.1281_1285del p.I428fs [13.6]	"8,9"	Frameshift
NM_001754.4 (*RUNX1*):c.292delG p.L98fs,NM_001754.4(*RUNX1*):c.492_493insAT p.G165fs [36.2]	"4,5"	Frameshift
NM_001754.4 (*RUNX1*):c.368A>C p.D123A [9.2], NM_001754.4(*RUNX1*):c.367G>C p.D123H [9.3], NM_001754.4 (*RUNX1*):c.361_364del p.L121fs [10.1]	5	Frameshift
NM_001754.4(*RUNX1*):c.422_423insA p.A142fs [26.2]	5	Frameshift
NM_001754.4 (*RUNX1*):c.418_422dupTACTC p.A142fs [21.0]	5	Frameshift
NM_001754.4 (*RUNX1*):c.422dupC p.A142fs [9.5], NM_001754.4(*RUNX1*):c.424_425insGGTGG p.A142fs [4.4]	5	Frameshift
NM_001754.4 (*RUNX1*):c.939_942dupCTCT p.A315fs [43.5]	8	Frameshift
NM_001754.4 (*RUNX1*):c.593A>G p.D198G [83.0]	6	Missense
NM_001754.4 (*RUNX1*):c.592G>A p.D198N [31.7]	6	Missense
NM_001754.4 (*RUNX1*):c.993_1005del p.D332fs	9	Frameshift
NM_001754.4 (*RUNX1*):c.402dupT p.G135fs [41.1]	5	Frameshift
NM_001754.4 (*RUNX1*):c.1023_1024insC p.I342fs	9	Frameshift
NM_001754.4 (*RUNX1*):c.1251_1252insC p.M418fs	9	Frameshift
NM_001754.4 (*RUNX1*):c.1244_1250dupAGTTCTC p.M418fs [19.5]	9	Frameshift
NM_001754.4 (*RUNX1*):c.548dupC p.P184fs [45.3]	6	Frameshift
NM_001754.4 (*RUNX1*):c.256_259delinsACCCGA p.P86fs	4	Missense
NM_001754.4 (*RUNX1*):c.484A>G p.R162G [70.4]	5	Missense
NM_001754.4 (*RUNX1*):c.485G>A p.R162K [42.2],NM_001754.4(*RUNX1*):c.388G>T p.V130F [44.7]	5	Missense
NM_001754.4 (*RUNX1*):c.496C>T p.R166* [39.1]	5	Nonsense
NM_001754.4 (*RUNX1*):c.601C>T p.R201* [40.7]	6	Nonsense
NM_001754.4 (*RUNX1*):c.955_958dupAGTC p.R320fs [49.7]	8	Frameshift
NM_001754.4 (*RUNX1*):c.952del p.S318fs [29.7]	8	Frameshift
NM_001754.4 (*RUNX1*):c.1352_1356dupACGTG p.V453fs [22.9], NM_001754.4 (*RUNX1*):c.611G>A p.R204Q [49.7]	"9,6"	Frameshift
NM_001754.4 (*RUNX1*):c.952T>G p.S318A [14.9], NM_001754.4 (*RUNX1*):c.954dupC p.S319fs [14.4], NM_001754.4(*RUNX1*) :c.950T>G p.L317S [15.0]	8	Frameshift
NM_001754.4 (*RUNX1*):c.977del p.D326fs [17.1], NM_001754.4 (*RUNX1*):c.352-1_352insTTTAG p.?	9	Frameshift, splice
NM_001754.4 (*RUNX1*):c.492_507dupCGGTCGAAGTGGAAGA p.G170fs [30.7], NM_001754.4 (*RUNX1*):c.484dupA p.R162fs [3.0]	5	Frameshift
NM_001754.4 (*RUNX1*):c.486dupG p.F163fs [51.2]	5	Frameshift
NM_001754.4 (*RUNX1*):c.1036dupC p.R346fs [2.8]	9	Frameshift
NM_001754.4 (*RUNX1*):c.167T>C p.L56S [51.5]	4	Missense

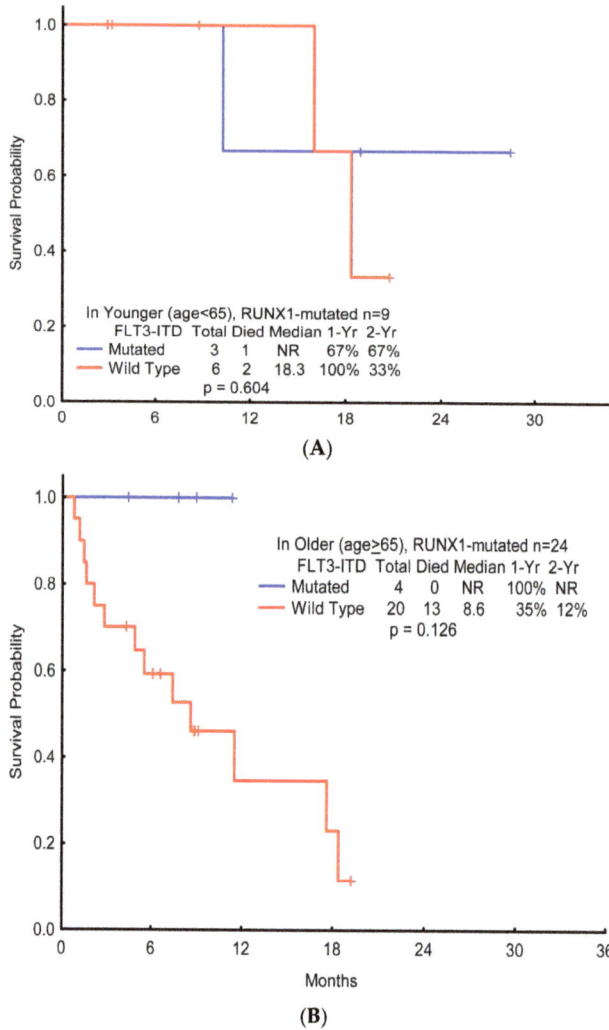

Figure A1. Survival probabilities of *FLT3-ITD* and *RUNX1* dual-mutated patients. (**A**) Patients aged below 65 years. (**B**) Patients older than 65 years.

References

1. Gilliland, D.G.; Griffin, J.D. The roles of FLT3 in hematopoiesis and leukemia. *Blood* **2002**, *100*, 1532–1542. [CrossRef] [PubMed]
2. Takahashi, S. Current findings for recurring mutations in acute myeloid leukemia. *J. Hematol. Oncol.* **2011**, *4*, 36. [CrossRef] [PubMed]
3. Greif, P.A.; Konstandin, N.P.; Metzeler, K.H.; Herold, T.; Pasalic, Z.; Ksienzyk, B.; Dufour, A.; Schneider, F.; Schneider, S.; Kakadia, P.M.; et al. RUNX1 mutations in cytogenetically normal acute myeloid leukemia are associated with a poor prognosis and up-regulation of lymphoid genes. *Haematologica* **2012**, *97*, 1909–1915. [CrossRef] [PubMed]

4. Gaidzik, V.I.; Teleanu, V.; Papaemmanuil, E.; Weber, D.; Paschka, P.; Hahn, J.; Wallrabenstein, T.; Kolbinger, B.; Köhne, C.H.; Horst, H.A.; et al. RUNX1 mutations in acute myeloid leukemia are associated with distinct clinico-pathologic and genetic features. *Leukemia* **2016**, *30*, 2282. [CrossRef] [PubMed]

5. Harada, H.; Harada, Y.; Niimi, H.; Kyo, T.; Kimura, A.; Inaba, T. High incidence of somatic mutations in the AML1/RUNX1 gene in myelodysplastic syndrome and low blast percentage myeloid leukemia with myelodysplasia. *Blood* **2004**, *103*, 2316–2324. [CrossRef] [PubMed]

6. Chen, C.Y.; Lin, L.I.; Tang, J.L.; Ko, B.S.; Tsay, W.; Chou, W.C.; Yao, M.; Wu, S.J.; Tseng, M.H.; Tien, H.F.; et al. RUNX1 gene mutation in primary myelodysplastic syndrome—The mutation can be detected early at diagnosis or acquired during disease progression and is associated with poor outcome. *Br. J. Haematol.* **2007**, *139*, 405–414. [CrossRef] [PubMed]

7. Tang, J.L.; Hou, H.A.; Chen, C.Y.; Liu, C.Y.; Chou, W.C.; Tseng, M.H.; Huang, C.F.; Lee, F.Y.; Liu, M.C.; Yao, M.; et al. AML1/RUNX1 mutations in 470 adult patients with de novo acute myeloid leukemia: Prognostic implication and interaction with other gene alterations. *Blood* **2009**, *114*, 5352–5361. [CrossRef] [PubMed]

8. Chou, W.C.; Huang, H.H.; Hou, H.A.; Chen, C.Y.; Tang, J.L.; Yao, M.; Tsay, W.; Ko, B.S.; Wu, S.J.; Huang, S.Y.; et al. Distinct clinical and biological features of de novo acute myeloid leukemia with additional sex comb-like 1 (ASXL1) mutations. *Blood* **2010**, *116*, 4086–4094. [CrossRef] [PubMed]

9. Schnittger, S.; Eder, C.; Alpermann, T.; Fasan, A.; Grossmann, V.; Kohlmann, A.; Kern, W.; Haferlach, C.; Haferlach, T. ASXL1 exon 12 mutations are frequent in AML with intermediate risk karyotype and are independently associated with an adverse outcome. *Leukemia* **2013**, *27*, 82–91. [CrossRef] [PubMed]

10. Gaidzik, V.I.; Bullinger, L.; Schlenk, R.F.; Zimmermann, A.S.; Rock, J.; Paschka, P.; Corbacioglu, A.; Krauter, J.; Schlegelberger, B.; Ganser, A.; et al. RUNX1 mutations in acute myeloid leukemia: Results from a comprehensive genetic and clinical analysis from the AML study group. *J. Clin. Oncol.* **2011**, *29*, 364–372. [CrossRef] [PubMed]

11. Osato, M. Point mutations in the RUNX1/AML1 gene: Another actor in RUNX leukemia. *Oncogene* **2004**, *23*, 4284–4296. [CrossRef] [PubMed]

12. Mendler, J.H.; Maharry, K.; Radmacher, M.D.; Mrozek, K.; Becker, H.; Metzeler, K.H.; Schwind, S.; Whitman, S.P.; Khalife, J.; Kohlschmidt, J.; et al. RUNX1 mutations are associated with poor outcome in younger and older patients with cytogenetically normal acute myeloid leukemia and with distinct gene and MicroRNA expression signatures. *J. Clin. Oncol.* **2012**, *30*, 3109–3118. [CrossRef] [PubMed]

13. Hann, I.M.; Stevens, R.F.; Goldstone, A.H.; Rees, J.K.; Wheatley, K.; Gray, R.G.; Burnett, A.K. Randomized comparison of DAT versus ADE as induction chemotherapy in children and younger adults with acute myeloid leukemia. Results of the Medical Research Council's 10th AML trial (MRC AML10). Adult and Childhood Leukaemia Working Parties of the Medical Research Council. *Blood* **1997**, *89*, 2311–2318. [PubMed]

14. Al-Ali, H.K.; Jaekel, N.; Niederwieser, D. The role of hypomethylating agents in the treatment of elderly patients with AML. *J. Geriatr. Oncol.* **2014**, *5*, 89–105. [CrossRef] [PubMed]

15. Krzywinski, M.; Schein, J.; Birol, I.; Connors, J.; Gascoyne, R.; Horsman, D.; Jones, S.J.; Marra, M.A. Circos: An information aesthetic for comparative genomics. *Genome Res.* **2009**, *19*, 1639–1645. [CrossRef] [PubMed]

16. Schneider, F.; Hoster, E.; Schneider, S.; Dufour, A.; Benthaus, T.; Kakadia, P.M.; Bohlander, S.K.; Braess, J.; Heinecke, A.; Sauerland, M.C.; et al. Age-dependent frequencies of NPM1 mutations and FLT3-ITD in patients with normal karyotype AML (NK-AML). *Ann Hematol.* **2012**, *91*, 9–18. [CrossRef] [PubMed]

17. Li, Z.; Zhang, W.; Chen, Y.; Guo, W.; Zhang, J.; Tang, H.; Xu, Z.; Zhang, H.; Tao, Y.; Wang, F.; et al. Impaired DNA double-strand break repair contributes to the age-associated rise of genomic instability in humans. *Cell Death Differ.* **2016**, *23*, 1765–1777. [CrossRef] [PubMed]

18. Papamichos-Chronakis, M.; Peterson, C.L. Chromatin and the genome integrity network. *Nat. Rev. Genet.* **2013**, *14*, 62–75. [CrossRef] [PubMed]

19. Harada, H.; Harada, Y.; Tanaka, H.; Kimura, A.; Inaba, T. Implications of somatic mutations in the AML1 gene in radiation-associated and therapy-related myelodysplastic syndrome/acute myeloid leukemia. *Blood* **2003**, *101*, 673–680. [CrossRef] [PubMed]

20. Zharlyganova, D.; Harada, H.; Harada, Y.; Shinkarev, S.; Zhumadilov, Z.; Zhunusova, A.; Tchaizhunusova, N.J.; Apsalikov, K.N.; Kemaikin, V.; Zhumadilov, K.; et al. High frequency of AML1/RUNX1 point mutations in radiation-associated myelodysplastic syndrome around Semipalatinsk nuclear test site. *J. Radiat. Res.* **2008**, *49*, 549–555. [CrossRef] [PubMed]

21. Genovese, G.; Kähler, A.K.; Handsaker, R.E.; Lindberg, J.; Rose, S.A.; Bakhoum, S.F.; Chambert, K.; Mick, E.; Neale, B.M.; Fromer, M.; et al. Clonal hematopoiesis and blood-cancer risk inferred from blood DNA sequence. *NEJM* **2014**, *371*, 2477–2487. [CrossRef] [PubMed]

22. Xie, M.; Lu, C.; Wang, J.; McLellan, M.D.; Johnson, K.J.; Wendl, M.C.; McMichael, J.F.; Schmidt, H.K.; Yellapantula, V.; Miller, C.A.; et al. Age-related mutations associated with clonal hematopoietic expansion and malignancies. *Nat. Med.* **2014**, *20*, 1472–1478. [CrossRef] [PubMed]

23. Klein, K.; Kaspers, G.; Harrison, C.J.; Beverloo, H.B.; Reedijk, A.; Bongers, M.; Cloos, J.; Pession, A.; Reinhardt, D.; Zimmerman, M.; et al. Clinical Impact of Additional Cytogenetic Aberrations, cKIT and RAS Mutations, and Treatment Elements in Pediatric t(8;21)-AML: Results From an International Retrospective Study by the International Berlin-Frankfurt-Münster Study Group. *J. Clin. Oncol.* **2015**, *33*, 4247–4258. [CrossRef] [PubMed]

24. Matsuno, N.; Osato, M.; Nanri, T.; Shigesada, K.; Fukushima, T.; Motoji, T.; Kusumoto, S.; Naoe, T.; Ohno, R.; Mitsuya, H.; et al. Dual mutations in the AML1 and FLT3 genes are associated with leukemogenesis in acute myeloblastic leukemia of the M0 type. *Leukemia* **2003**, *17*, 2492–2499. [CrossRef] [PubMed]

25. Schnittger, S.; Dicker, F.; Kern, W.; Wendland, N.; Sundermann, J.; Alpermann, T.; Haferlach, C.; Haferlach, T. RUNX1 mutations are frequent in de novo AML with noncomplex karyotype and confer an unfavorable prognosis. *Blood* **2011**, *117*, 2348–2357. [CrossRef] [PubMed]

26. Kao, H.W.; Liang, D.C.; Wu, J.H.; Kuo, M.C.; Wang, P.N.; Yang, C.P.; Shih, Y.S.; Lin, T.H.; Huang, Y.H.; Shih, L.Y.; et al. Gene mutation patterns in patients with minimally differentiated acute myeloid leukemia. *Neoplasia* **2014**, *16*, 481–488. [CrossRef] [PubMed]

27. Green, C.L.; Evans, C.M.; Zhao, L.; Hills, R.K.; Burnett, A.K.; Linch, D.C.; Gale, R.E. The prognostic significance of IDH2 mutations in AML depends on the location of the mutation. *Blood* **2011**, *118*, 409–412. [CrossRef] [PubMed]

28. Bejar, R.; Lord, A.; Stevenson, K.; Bar-Natan, M.; Pérez-Ladaga, A.; Zaneveld, J.; Wang, H.; Caughey, B.; Stojanov, P.; Getz, G.; et al. TET2 mutations predict response to hypomethylating agents in myelodysplastic syndrome patients. *Blood* **2014**, *124*, 2705–2712. [CrossRef] [PubMed]

29. Kaplan, E.L.; Meier, P. Nonparametric estimation from incomplete observations. *J. Am. Stat. Assoc.* **1958**, *53*, 457–481. [CrossRef]

30. Cheson, B.D.; Bennett, J.M.; Kopecky, K.J.; Büchner, T.; Willman, C.L.; Estey, E.H.; Schiffer, C.A.; Doehner, H.; Tallman, M.S.; Lister, T.A.; et al. Revised recommendations of the International Working Group for Diagnosis, Standardization of Response Criteria, Treatment Outcomes, and Reporting Standards for Therapeutic Trials in Acute Myeloid Leukemia. *J. Clin. Oncol.* **2003**, *21*, 4642–4649. [CrossRef] [PubMed]

31. Casaril, A.E.; de Oliveira, L.P.; Alonso, D.P.; de Oliveira, E.F.; Barrios, S.P.G.; Infran, J.D.O.M.; de Souza Fernandes, W.; Oshiro, E.T.; Ferreira, A.M.T.; Ribolla, P.E.M.; et al. Standardization of DNA extraction from sand flies: Application to genotyping by next generation sequencing. *Exp. Parasitol.* **2017**, *177*, 66–72. [CrossRef] [PubMed]

32. Warren, M.; Luthra, R.; Yin, C.C.; Ravandi, F.; Cortes, J.E.; Kantarjian, H.M.; Medeiros, L.J.; Zuo, Z. Clinical impact of change of FLT3 mutation status in acute myeloid leukemia patients. *Mod. Pathol.* **2012**, *25*, 1405–1412. [CrossRef] [PubMed]

33. Wendt, F.R.; Zeng, X.; Churchill, J.D.; King, J.L.; Budowle, B. Analysis of Short Tandem Repeat and Single Nucleotide Polymorphism Loci From Single-Source Samples Using a Custom HaloPlex Target Enrichment System Panel. *Am. J. Forensic Med. Pathol.* **2016**, *37*, 99–107. [CrossRef] [PubMed]

34. Patel, K.P.; Newberry, K.J.; Luthra, R.; Jabbour, E.; Pierce, S.; Cortes, J.; Singh, R.; Mehrotra, M.; Routbort, M.J.; Luthra, M.; et al. Correlation of mutation profile and response in patients with myelofibrosis treated with ruxolitinib. *Blood* **2015**, *126*, 790–797. [CrossRef] [PubMed]

© 2017 by the authors. Licensee MDPI, Basel, Switzerland. This article is an open access article distributed under the terms and conditions of the Creative Commons Attribution (CC BY) license (http://creativecommons.org/licenses/by/4.0/).

International Journal of
Molecular Sciences

MDPI

Article

Azacitidine for Front-Line Therapy of Patients with AML: Reproducible Efficacy Established by Direct Comparison of International Phase 3 Trial Data with Registry Data from the Austrian Azacitidine Registry of the AGMT Study Group

Lisa Pleyer [1,3,*], Hartmut Döhner [4], Hervé Dombret [5], John F. Seymour [6], Andre C. Schuh [7], CL Beach [8], Arlene S. Swern [8], Sonja Burgstaller [9], Reinhard Stauder [10], Michael Girschikofsky [11], Heinz Sill [12], Konstantin Schlick [1], Josef Thaler [9], Britta Halter [10], Sigrid Machherndl Spandl [11], Armin Zebisch [12], Angelika Pichler [13], Michael Pfeilstöcker [14], Eva M. Autzinger [15], Alois Lang [16], Klaus Geissler [17], Daniela Voskova [18], Wolfgang R. Sperr [19], Sabine Hojas [20], Inga M. Rogulj [21], Johannes Andel [22] and Richard Greil [1,2,3]

[1]   3rd Medical Department with Hematology and Medical Oncology, Hemostaseology,
      Rheumatology and Infectious Diseases, Laboratory for Immunological and Molecular Cancer Research,
      Oncologic Center, Paracelsus Medical University, Salzburg 5020, Austria; k.schlick@salk.at (K.S.);
      r.greil@mac.com (R.G.)
[2]   Salzburg Cancer Research Institute—Center for Clinical Cancer and Immunology Trials,
      Salzburg 5020, Austria
[3]   Cancer Cluster, Salzburg 5020, Austria
[4]   Department of Medicine and Internal Medicine III, Universitätsklinikum Ulm, Ulm D-89081, Germany;
      hartmut.doehner@uniklinik-ulm.de
[5]   Institut Universitaire d'Hématologie, Hôpital Saint Louis, University Paris Diderot, Paris 75010, France;
      herve.dombret@mac.com
[6]   Peter MacCallum Cancer Centre, East Melbourne, Australia, and University of Melbourne, Parkville 3000,
      Australia; john.seymour@petermac.org
[7]   Princess Margaret Cancer Centre, Toronto, ON M5G 2M9, Canada; andre.schuh@uhn.ca
[8]   Celgene Corporation, Summit, NJ 07901, USA; clbeach@celgene.com (C.L.B.), aswern@celgene.com (A.S.S.)
[9]   Department of Internal Medicine IV, Klinikum WelsGrieskirchen, Wels 4600, Austria;
      sonja.burgstaller@klinikum-wegr.at (S.B.); josef.thaler@klinikum-wegr.at (J.T.)
[10] Department of Internal Medicine V, Innsbruck Medical University, Innsbruck 6020, Austria;
      reinhard.stauder@i-med.ac.at (R.S.), britta.halter@i-med.ac.at (B.H.)
[11] Department of Hematology and Oncology, Elisabethinen Hospital, Linz 4020, Austria;
      michael.girschikofsky@elisabethinen.or.at (M.G.), sigrid.machherndl-spandl@elisabethinen.or.at (S.M.S.)
[12] Department of Hematology, Medical University of Graz, Graz 8036, Austria;
      heinz.sill@medunigraz.at (H.S.), armin.zebisch@medunigraz.at (A.Z.)
[13] Department for Hematology and Oncology, LKH Leoben, Leoben 8700, Austria; angelika.pichler@kages.at
[14] 3rd Medical Department for Hematology and Oncology, Hanusch Hospital, Vienna 1140, Austria;
      m.pfeilstoecker@aon.at
[15] First Medical Department, Center for Oncology, Hematology and Palliative Care, Wilhelminenspital,
      Vienna 1160, Austria; eva-maria.autzinger@wienkav.at
[16] Department of Internal Medicine, Landeskrankenhaus Feldkirch (LKH) Feldkirch, Feldkirch 6800, Austria;
      alois.lang@lkhf.at
[17] 5th Medical Department, Hospital Hietzing, Vienna 1130, Austria; klaus.geissler@wienkav.at
[18] Department of Internal Medicine III, General Hospital, Linz 4020, Austria; daniela.voskova@gmail.com
[19] Department of Internal Medicine I, Division of Hematology and Hemostaseology,
      Medical University of Vienna, Vienna 1090, Austria; wolfgang.r.sperr@meduniwien.ac.at
[20] Department of Internal Medicine, LKH Fürstenfeld, Fürstenfeld 8280, Austria;
      sabineelisabeth.hojas@kages.at
[21] Department of Hematology, Clinical Hospital Merkur, Zagreb 10000, Croatia; imandac@yahoo.com
[22] Department of Internal Medicine II, LKH Steyr, Steyr 4400, Austria; johannes.andel@gespag.at

* Correspondence: l.pleyer@salk.at; Tel.: +43-57255-58271

Academic Editors: Geoffrey Brown and Ewa Marcinkowska
Received: 8 December 2016; Accepted: 8 February 2017; Published: 15 February 2017

Abstract: We recently published a clinically-meaningful improvement in median overall survival (OS) for patients with acute myeloid leukaemia (AML), >30% bone marrow (BM) blasts and white blood cell (WBC) count $\leq$15 G/L, treated with front-line azacitidine versus conventional care regimens within a phase 3 clinical trial (AZA-AML-001; NCT01074047; registered: February 2010). As results obtained in clinical trials are facing increased pressure to be confirmed by real-world data, we aimed to test whether data obtained in the AZA-AML-001 trial accurately represent observations made in routine clinical practice by analysing additional AML patients treated with azacitidine front-line within the Austrian Azacitidine Registry (AAR; NCT01595295; registered: May 2012) and directly comparing patient-level data of both cohorts. We assessed the efficacy of front-line azacitidine in a total of 407 patients with newly-diagnosed AML. Firstly, we compared data from AML patients with WBC $\leq$ 15 G/L and >30% BM blasts included within the AZA-AML-001 trial treated with azacitidine ("AML-001" cohort; n = 214) with AAR patients meeting the same inclusion criteria ("AAR (001-like)" cohort; n = 95). The current analysis thus represents a new sub-analysis of the AML-001 trial, which is directly compared with a new sub-analysis of the AAR. Baseline characteristics, azacitidine application, response rates and OS were comparable between all patient cohorts within the trial or registry setting. Median OS was 9.9 versus 10.8 months (p = 0.616) for "AML-001" versus "AAR (001-like)" cohorts, respectively. Secondly, we pooled data from both cohorts (n = 309) and assessed the outcome. Median OS of the pooled cohorts was 10.3 (95% confidence interval: 8.7, 12.6) months, and the one-year survival rate was 45.8%. Thirdly, we compared data from AAR patients meeting AZA-AML-001 trial inclusion criteria (n = 95) versus all AAR patients with World Health Organization (WHO)-defined AML ("AAR (WHO-AML)" cohort; n = 193). Within the registry population, median OS for AAR patients meeting trial inclusion criteria versus all WHO-AML patients was 10.8 versus 11.8 months (p = 0.599), respectively. We thus tested and confirmed the efficacy of azacitidine as a front-line agent in patients with AML, >30% BM blasts and WBC $\leq$ 15 G/L in a routine clinical practice setting. We further show that the efficacy of azacitidine does not appear to be limited to AML patients who meet stringent clinical trial inclusion criteria, but instead appears efficacious as front-line treatment in all patients with WHO-AML.

Keywords: acute myeloid leukaemia (AML); AZA-AML-001 trial; Austrian Azacitidine Registry (AAR); real-world data; azacitidine

---

1. Introduction

Acute myeloid leukaemia (AML) is predominantly a disease of the elderly, with a median age at diagnosis of roughly 70 years [1]. Older patients with AML are often precluded from intensive treatment, owing to multiple poor-risk prognostic factors, including a higher proportion of adverse cytogenetics, myelodysplasia-related changes (MRC), poor Eastern Cooperative Oncology Group performance status (ECOG-PS) and/or significant comorbidities [1–5]. In 2008, azacitidine was approved by the European Medicines Agency (EMA) for the treatment of AML patients with 20%–30% bone marrow (BM) blasts, older than 64 years and who are ineligible for HSCT. For AML patients with more than 30% BM blasts, azacitidine remained an off-label indication until 30 October 2015 and was not reimbursed in most countries. These patients were either treated with other options or not treated at all, which is substantiated by a large population-based study published in 2012, showing that only 38% of 5480 AML patients older than 65 years received leukaemia therapy, whereas 62% received BSC [6]. In Austria, off-label drug use is permitted if a dire clinical need, lack of alternative substances and presumed efficacy can be substantiated. Haematologists at specialized centres started treating AML patients with >30% BM blasts with azacitidine as early as 2007 in Austria, indicating that the physicians were convinced they were doing the best for their patients. This assumption was based on the significant improvement of overall survival (OS) obtained in the AZA-MDS-001 trial [7] and the Cancer and Leukemia Group B (CALGB) protocols, in which 32% [8] and 38% [9,10] of the trial population had AML with 20%–30% BM blasts, respectively. In anticipation of a need to test and potentially confirm the efficacy of azacitidine in a real-world population, the Austrian Azacitidine Registry (AAR) of the Arbeitsgemeinschaft Medikamentöse Tumortherapie (AGMT) Study Group was founded in February 2009 to establish a platform to document the off-label use of azacitidine [11–20].

In 2010, the international phase 3 randomised AZA-AML-001 clinical trial testing azacitidine versus conventional care regimens (intensive chemotherapy, low-dose cytarabine or BSC as preselected by the treating physician) in AML patients older than 65 years with newly-diagnosed AML, >30% BM blasts and ≤15 G/L white blood cell (WBC) count was initiated. In this trial, a clinically-meaningful improvement in OS for azacitidine versus conventional care regimens (10.4 vs. 6.5 months; $p = 0.1009$) was reported by part of our group on 16 July 2015 [21], and EMA approval of azacitidine was expanded on 30 October 2015 to include AML patients with >30% BM blasts.

While well-designed randomised clinical trials are the gold-standard for investigating treatment options for (AML) patients, they are considered to include extremely selected, skewed and/or not fully-representative patient populations [22]. Results obtained may therefore under-represent some patient groups [22], for example those with comorbidities or anticipated intolerance to treatment [23]. Thus, caution needs to be exercised when using newly-approved drugs in patients who would not have met clinical trial inclusion criteria, and conclusions drawn from clinical trials cannot eo ipso be generalized to all patients with AML. In fact, experts in the field have recently ascertained that all trial results should be extrapolated with caution, and population-based studies of real-world patients have a prominent role in examining the prognosis, as well as the management, efficacy and toxicity of new agents after regulatory approval and outside of clinical trials of higher-risk MDS [24,25] and AML [26].

Therefore, a cooperative data analysis between the sponsors and data owners of the AZA-AML-001 trial and the AGMT-AAR-registry was agreed upon, with the aim to test whether the use of azacitidine within the routine clinical practice setting can recapitulate the median OS time of 10.4 months observed in AML patients with >30% BM blasts and ≤15 G/L treated with azacitidine front line within the AZA-AML-001 trial. Of note, 27 of the 241 patients (11.2%) treated with azacitidine within the AZA-AML-001 trial were excluded from the current analysis, as they had ≤30% BM blasts after central pathology review and, therefore, did not fulfil the inclusion criteria of this study (Figure 1). The current analysis thus represents a new sub-analysis of the AML-001 trial, which is directly compared with a new sub-analysis of the AAR. Secondly, we wanted to expand on the data from the AZA-AML-001 trial regarding the efficacy of azacitidine as front-line therapy for AML patients [21,27,28], by analysing additional patients from the AAR and directly comparing the patient-level data of both cohorts. Thirdly, we aimed to analyse whether the efficacy and safety of azacitidine as a front-line agent in AML is limited to patients meeting strict clinical trial inclusion criteria by analysing and directly comparing patient-level data from a wider patient population of the AAR that includes all patients with AML as defined by the World Health Organization (WHO).

2. Results

2.1. Clinical Trial versus Registry Subsets

Our predefined criteria (>30% BM blasts and ≤15 G/L WBC) for the direct comparative analysis of the clinical trial with registry data were met by 214 patients of the AZA-AML-001 trial ("AML-001" subset) and 95 patients from the AAR ("AAR (001-like)" subset) (Figure 1). This represents a novel and thus far unpublished subset of the AML-001-trial, as 27 of the previously published cohort of 241 patients [21] were excluded, as they had <30% BM blasts after central pathology review.

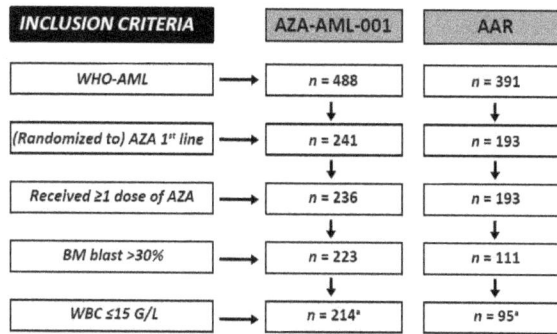

Figure 1. CONSORT diagram. [a] Subset of patients included in pooled analysis (AML-001-like).

Baseline characteristics were similar between the "AML-001" subset and the "AAR (001-like)" subset, including gender, median age, median BM blast count, WBC count, haemoglobin, absolute neutrophil count (ANC), platelets (PLT), prior myelodysplastic syndromes (MDS), transfusion dependence (TD) and National Comprehensive Cancer Network (NCCN) cytogenetic risk (Table 1). The only notable differences were the presence of patients with ECOG-PS 3–4 (9.5%; nine patients) and the presence of patients with good-risk cytogenetics (2.1%; two patients) within the "AAR (001-like)" subset, both of which were exclusion criteria in the "AML-001" subset (Table 1).

Table 1. Baseline characteristics of acute myeloid leukaemia (AML) patients treated with azacitidine front-line per patient subset.

Baseline Characteristics	AML-001 Trial Subset ($n = 214$)	AAR (001-Like) Subset ($n = 95$)	AAR (WHO-AML) Subset ($n = 193$)
Age, median (mean) [SD], years	76 (75.5) [5.6]	77 (75.2) [11.5]	77 (75.6) [10.2]
Age ≥75 years, %	58.4	56.8	58.5
Male, %	57.5	54.7	58.6
ECOG-PS, %			
0–1	76.6	67.3	67.9
2	23.4	23.2	24.4
3–4 [a]	0	9.5	1.6
AML classification, %			
AML-MRC [b]	57.0	66.3	70.8
AML-NOS	37.3	24.2	18.8
AML-RCA	2.3	4.2	3.7
t-AML	3.3	5.3	6.8
Antecedent haematological disease, %	19.6	25.3	31.6
Prior MDS, %	19.6	21.1	24.9
NCCN cytogenetic risk, %			
Good [c]	0	2.1	2.4
Intermediate	65.0	58.9	66.9
Normal	47.7	42.1	47.3
Poor	34.6	27.4	30.8
BM blasts, median (range), %	73.0 (31–100)	55.5 (31–96)	40.0 (20–100)
BM blasts ≥50%, %	78.0	58.9	38.9

Table 1. *Cont.*

Baseline Characteristics	AML-001 Trial	AAR (001-Like)	AAR (WHO-AML)
	Subset	Subset	Subset
	(*n* = 214)	(*n* = 95)	(*n* = 193)
Number of cytopenias, %			
0–1	13.1	14.7	16.6
2–3	86.9	85.3	83.4
RBC-TD, %	70.1	60.0	52.9
PLT-TD, %	40.2	30.5	24.9
WBC, median (range), G/L	3.0 (0.3–14.7)	2.1 (0.6–14.4)	2.5 (0.6–74.1)
Hb, median (range), g/dL	9.5 (5.0–13.4)	9.1 (5.8–13.6)	9.1 (5.8–14.2)
ANC, median (range), G/L	0.3 (0–5.3)	0.5 (0–7.7)	0.6 (0–37.2)
PLT, median (range), G/L	54.0 (3–585)	49.0 (7–1.270)	52.0 (6–1,270)

[a] ECOG-PS > 2 was an exclusion criterion in the AZA-AML-001 trial. Nine patients included within the AAR had an ECOG-PS > 2; [b] The definition of MRC in the AZA-AML-001 trial was based on the presence of NCCN poor-risk cytogenetics. This means that the following cytogenetic aberrations (included in the WHO definition of myelodysplasia-related changes) were not accounted for: -9q, -12p, -13q, -13, t(12p), t(2;11), t(3;5), t(3;21), t(5;7), t(5;10), t(5;17), t(11;16), isochromosome(17q), idic(X)(q13), possibly resulting in a lower number of patients within this subgroup. In addition, the presence of prior chronic myelomonocytic leukaemia and prior myeloproliferative neoplasia was an exclusion criterion in the AML-001 trial, possibly resulting in a slightly lower number of patients within this subgroup; and [c] NCCN cytogenetic good risk was an exclusion criterion in the AZA-AML-001 trial. Two patients were included within the AAR had NCCN good risk cytogenetics.

Treatment-related characteristics, such as median number of azacitidine cycles (six vs. five), percentage of patients receiving ≥6 cycles (50% vs. 46%) or ≥12 cycles of azacitidine (29% vs. 24%), median total days of azacitidine application (42 vs. 34 days), daily dose (130 vs. 132 mg), patient status at data cut-off and reasons for azacitidine discontinuation were similar between the clinical trial and registry cohorts, respectively (Table 2).

Table 2. Treatment characteristics of AML patients treated with azacitidine front-line per patient subset.

Treatment Characteristics	AML-001	AAR	AAR
	Trial	(001-Like)	(WHO-AML)
	(*n* = 214)	(*n* = 95)	(*n* = 193)
AZA cycles, median, *n*	6	5	6
(Mean) [SD]	(8.4) [7.1]	(8.5) [9.1]	(8.4) [6.0]
AZA cycles ≥6, %	50.0	46.3	51.3
AZA cycles ≥12, %	28.5	24.2	24.9
Days of AZA application, median, days	42	34	39
(Mean) [SD]	(58.0) [49.8]	(55.8) [61.1]	(57.1) [57.3]
Daily of AZA dose, median, mg	130.1	131.6	132.0
(Mean) [SD]	(129.4) [17.8]	(128.7) [26.5]	(126.4) [33.3]
Reasons for AZA discontinuation, %			
AE/no response/relapse/PD/death	66.8	74.8	73.1
Withdrew consent/patient's wish	11.7	9.5	7.3
Others	11.7	9.5	11.4
Still on AZA at study closure	9.8	6.3	8.3

Median OS (9.9 vs. 10.7 months, $p = 0.9553$), median relapse-free survival (RFS; 16.3 vs. 13.8 months, $p = 0.6817$), median event-free survival (EFS; 6.9 vs. 8.3 months, $p = 0.2909$), median complete response (CR)/CR with incomplete blood count recovery (CRi) duration (8.6 vs. 11.1 months, $p = 0.1740$), one-year survival rates (54.2% vs. 53.7%, $p = 0.924$) and 30-day mortality rates (7.0% vs. 8.4%, $p = 0.843$) were comparable between the clinical trial and the registry cohorts (Table 3; Figure 2A). In addition, median OS and one-year survival rates were comparable between the clinical trial and registry cohorts for all subgroups analysed: (a) AML with MRC (AML-MRC); (b) NCCN poor-risk cytogenetics; (c) NCCN intermediate-risk cytogenetics; and (d) normal karyotype (Figure 2B–E). Regarding response, achievement of red blood cell- (RBC-) and PLT-transfusion independence (TI) was also similar between trial and registry cohorts, respectively ($p = 0.7522$, $p = 1.0000$; Table 3). However, the overall response rate (ORR; CR/CRi/partial response (PR)) was significantly higher in the AZA-AML-001 trial, as compared to the AAR (30.4% vs. 18.9%, $p = 0.0379$), which was due to a lower rate of CR/CRi in the registry population.

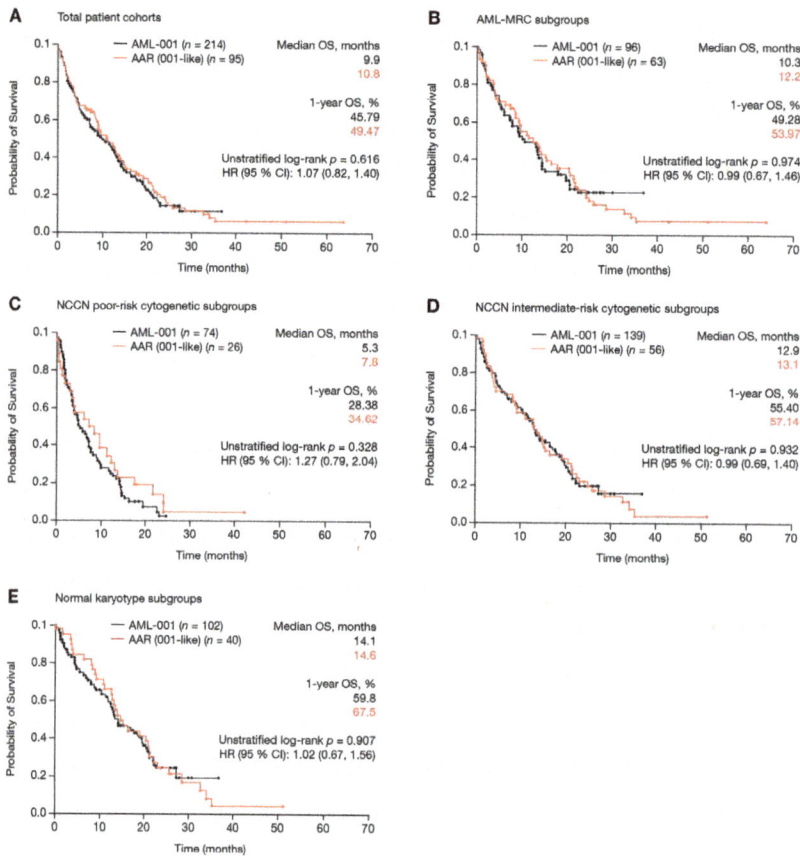

Figure 2. Overall survival (OS) in AML patients with >30% BM blasts and <15 G/L WBC treated with azacitidine front-line within the AML-001 trial and the AAR-AML-001-like cohorts. (**A**) Total patient cohorts; (**B**) AML-MRC; (**C**) AML with NCCN poor-risk cytogenetics; (**D**) AML with NCCN intermediate-risk cytogenetics; and (**E**) Normal karyotype.

Univariate and multivariate Cox regression analyses of OS were performed to evaluate the similarity of the two groups adjusted for baseline covariates (Table 4). The baseline variables age as a continuous variable (hazards ratio (HR): 1.02; 95% confidence interval (CI): 1.00, 1.04, $p = 0.0182$), age < versus ≥75 years (HR: 0.70; 95% CI: 0.54, 0.90, $p = 0.0053$), PLT-TD (HR: 0.68; 95% CI: 0.53, 0.88, $p = 0.0028$), ECOG-PS (HR: 0.54; 95% CI: 0.41, 0.71, $p < 0.001$) and NCCN cytogenetic risk (HR:

0.51; 95% CI: 0.39, 0.66, $p < 0.001$) had a significant impact on survival in univariate analysis (Table 4). Of note, patient affiliation with the clinical trial or the registry cohort did not have an impact on OS (HR: 1.02; 95% CI: 0.78, 1.32, $p = 0.8998$).

Table 3. Outcome of AML patients treated with azacitidine front-line per patient subset.

Outcome	AML-001 Trial ($n = 214$)	AAR (001-Like) ($n = 95$)	*p*-Value	AAR (001-Like) ($n = 95$)	AAR (WHO-AML) ($n = 193$)	*p*-Value
Median OS, mo	9.9	10.7	0.9553 [a]	10.7	11.8	0.599 [a]
Median RFS (CR/CRi), mo	16.3	13.8	0.6817 [a]	13.8	13.3	0.621 [a]
Median EFS (all pts), mo	6.9	8.3	0.2909 [a]	8.3	8.1	0.941 [a]
Median CR/CRi duration, mo	8.6	11.1	0.1740 [a]	11.1	11.5	0.818 [a]
1-Year survival, %	54.2	53.7	0.843 [b]	53.7	50.8	0.476 [b]
30-Day mortality, %	7.0	8.4	0.924 [b]	8.4	7.8	0.848 [b]
ORR (CR, CRi, PR), %	30.4	18.9	0.0379 [b]	18.9	23.1	0.685 [b]
RBC-TI, %	39.3	42.1	0.7522 [b]	42.1	42.2	0.517 [b]
PLT-TI, %	37.2	35.7	1.0000 [b]	35.7	41.7	0.688 [b]

[a] Median times for OS, RFS, EFS and CR/CRi duration were estimated by the KM method, and the *p*-value was based on the log-rank test; and [b] Calculated according to the χ-squared test for categorical variables and the *t*-test for continuous variable.

Table 4. Univariate and multivariate analysis of the effects of baseline covariates on the OS of AML patients treated with azacitidine front-line with >30% BM blasts and <15 G/L WBC within the AZA-AML-001 trial and the AAR.

Univariate Analysis		
Baseline Parameter	HR (95% CI)	*p*-Value
Study group (AML-001 vs. AAR)	1.02 (0.78, 1.32)	0.8998
Age (as a continuous variable)	1.02 (1.00, 1.04)	0.0182
Age (<75 vs. ≥75 years)	0.70 (0.54, 0.90)	0.0053
Gender (female vs. male)	0.82 (0.64, 1.05)	0.1243
RBC-TD (No vs. Yes)	0.89 (0.69, 1.16)	0.3857
PLT-TD (No vs. Yes)	0.68 (0.53, 0.88)	0.0028
ECOG-PS (0–1 vs. ≥2)	0.54 (0.41, 0.71)	<0.001
MDS-related changes present (Yes vs. No)	0.90 (0.70, 1.16)	0.4326
Prior MDS (No vs. Yes)	1.01 (0.74, 1.38)	0.9366
No. of cytopenias at baseline (0–1 vs. 2–3)	0.83 (0.58, 1.19)	0.3082
NCCN cytogenetic risk (Intermediate vs. Poor)	0.51 (0.39, 0.66)	<0.001
BM blasts (30%–49% vs. ≥50%)	0.90 (0.69, 1.18)	0.4511
WBC (as a continuous variable)	1.00 (0.97, 1.04)	0.8400
ANC (as a continuous variable)	1.05 (0.93, 1.17)	0.4514
PLT count (as a continuous variable)	1.00 (1.00, 1.00)	0.1487
Hb (as a continuous variable)	0.96 (0.88, 1.05)	0.3344
Multivariate Analysis		
Baseline Covariate	HR (95% CI)	*p*-Value
Age (<75 vs. ≥75 years)	0.76 (0.58, 0.98)	0.0366
PLT-TD (No vs. Yes)	0.69 (0.53, 0.90)	0.0057
ECOG-PS (0–1 vs. ≥2)	0.65 (0.48, 0.87)	0.0041
NCCN cytogenetic risk (Intermediate vs. Poor)	0.51 (0.39, 0.67)	<0.001
AML-001 vs. AAR [a]	1.11 (0.84, 1.47)	0.4509

[a] As this covariate was the variable most critical to the intended analysis, it was kept in the final multivariate analysis model.

In multivariate analysis, PLT-TD (HR: 0.69; 95% CI: 0.53, 0.90, $p = 0.0057$), ECOG-PS (HR: 0.65; 95% CI: 0.48, 0.87, $p = 0.0041$) and NCCN cytogenetic status (HR: 0.51; 95% CI: 0.39, 0.67, $p < 0.001$) remained significant. As the baseline covariate "AML-001" versus "AAR (001-like)" was the variable most critical to this publication, this variable was kept in the final multivariate model, and the two groups were comparable even after adjustment for significant baseline covariates (HR: 1.11; 95% CI: 0.84, 1.47, p = 0.4509), indicating no relevant difference between the outcomes of patients from either cohort (Table 4).

Having determined no relevant differences in baseline variables, treatment-related characteristics or various measures of outcome between the AZA-AML-001 trial and the "real-world" cohorts of the AAR, both cohorts were pooled (n = 309) for outcome analyses: median OS was 10.3 (95% CI: 8.65, 12.56) months, and the Kaplan–Meier (KM) estimate of one-year survival rate was 45.8%.

2.2. Registry Subsets Meeting Clinical Trial Inclusion Criteria versus All WHO-AML Patients

Baseline characteristics were similar between the registry subset meeting clinical trial inclusion criteria ("AAR (001 like)") and all WHO-AML patients included within the registry ("AAR (WHO-AML)") treated with azacitidine front-line, with the exception of BM blast count, which was expectedly lower for the "WHO-AML" patient subgroup as these also included patients with 20%–30% BM blasts (Table 1). Treatment-related characteristics were similar between the two groups (Table 2). The occurrence of Grade 3–4 treatment-emergent adverse events (TEAEs) per total applied azacitidine cycle was similar for the "001-like" and "WHO-AML" subsets (0.24 vs. 0.22; Table S1). In all WHO-AML patients treated with azacitidine front-line within the AAR, 35% of all TEAEs and 20% of all Grade 3–4 TEAEs were deemed as azacitidine-related; 9% of Grade 3–4 azacitidine-related AEs resulted in hospitalization, 6% in dose interruption, 9% in dose reduction and 3% in termination of treatment with azacitidine (Table S1).

Response, including ORR, TI and median CR/CRi duration, was similar between the "001-like" and "WHO-AML" subsets (Table 3). Other outcome measures including one-year survival rate and 30-day mortality rate, as well as median OS, median RFS duration and median EFS duration were also similar (Table 3; Figure 3A). The same held true in the subgroups of patients with AML-MRC, NCCN poor-risk cytogenetics, NCCN intermediate-risk cytogenetics and normal karyotype (Table 3; Figure 3B–E).

3. Discussion

Currently, both physicians and regulatory agencies are placing a stronger focus of attention on the performance of new drugs in the routine clinical practice setting, rather than merely relying on results obtained in clinical trials, and pressure is rising to recapitulate such results in unselected patients in clinical practice [22,23,29–34].

The EMA has reacted to this pressure by publishing its intention to "expand the use of patient registries" [34].

In anticipation of a need to test and potentially confirm the efficacy of azacitidine in a real-world population, the AAR established a platform to document the off-label use of azacitidine even before azacitidine was approved for the treatment of AML in the European Union and before the AZA-AML-001 trial was initiated. This indicates the lack of alternative treatments, an unmatched medical need and the high expectation of physicians that azacitidine would provide clinical benefit for their patients. Physicians' anticipation and conviction that they were acting in the best interest of their patients, independent of the current EMA label, were based on previously published results in patients with higher-risk MDS and AML [7–10]. The AAR of the AGMT Study Group is a national registry with Ethics Committee approval, which collects data on MDS, CMML and AML patients specifically treated with hypomethylating agents. Data entry into a study-specific eCRF is performed by clinical trial personnel and/or the treating physicians, after the collection of patient written informed consent from all patients alive at the time of data entry. In contrast, other study groups have collected data from all patients diagnosed with MDS/AML [24] within registries or compassionate use programs approved by internal review boards [35], with informed patient consent. Some have also used anonymised minimal datasets of the entire country population obtained via medical claims from healthcare insurance companies and/or national cancer and/or leukaemia registries to assess treatment of patients with MDS and/or AML [5,36–38].

In a collaboration effort with the sponsors, data owners and authors of the AZA-AML-001 trial, this study compared data of AML patients treated with azacitidine front-line within the setting of the AZA-AML-001 clinical trial [21] (and subsets thereof [39–41]) versus the clinical practice registry setting of the AAR. To our best knowledge, such a comparison of patient-level data between a randomised phase 3 clinical trial and a nationwide registry has not been published in AML before.

For the first analysis, patients were selected according to the most relevant inclusion criteria of the AZA-AML-001 trial, namely the presence of WHO-AML, front-line treatment with azacitidine, BM blasts >30% and a WBC count of ≤15 G/L (Figure 1). Other clinical trial inclusion/exclusion criteria were not used in order to maintain the additional information gained by a "less-controlled, wider-ranging, naturalistic (registry) setting" [32]. Baseline and treatment characteristics, as well as safety and survival data were similar between the two groups, despite the fact that the ORR, mainly due to higher CR/CRi rates, was significantly higher in the clinical trial than in the registry setting (Table 3). While achievement of CR/CRi remains the primary treatment goal for all AML patients irrespective of age, we and others have previously shown that achievement of CR/CRi is not necessarily a prerequisite for OS benefit in AML patients treated with non-intensive therapeutic options [42–46]. Interestingly, a recent analysis of a Danish population-based cohort reports that AML patients treated with intensive chemotherapeutic regimens within clinical trials not only had higher CR rates compared with patients treated off-trial (80.2% vs. 68.5%), but also had superior one-year survival (61% vs. 45% for patients older than 60 years). A possible explanation for the worse survival observed in AML patients treated with standard intensive therapy regimens off-trial in Denmark might be that they had a less favourable profile than patients treated on-trial [26].

We thus confirm the safety and efficacy of front-line azacitidine in these patients, as well as for all subgroups analysed. This included stratification according to cytogenetic risk category and AML-MRC. These results also indicate that there was no bias towards improved patient outcome within the AZA-AML-001 trial through the use of more stringent exclusion criteria as compared to the registry population. Pooled data from the clinical trial and the registry cohorts revealed a median OS of 10.3 months and a one-year survival rate of 45.8% for 309 AML patients. We thus expand on recently published data from the AZA-AML-001 trial [21] and confirm the efficacy of azacitidine in this patient group.

We further aimed to evaluate whether the efficacy of azacitidine might be limited to patient subsets fulfilling the AZA-AML-001 clinical trial inclusion criteria (i.e., front-line treatment with azacitidine, BM blasts >30% and a WBC count of ≤15 G/L). In this regard, no relevant difference in baseline parameters, treatment-related parameters, TEAEs, response and survival outcomes existed between patients of the AAR that met clinical trial inclusion criteria versus all WHO-AML patients (Tables 1–3; Table S1; Figure 3A). These results further confirm that the inclusion criteria used by the AZA-AML-001 trial did not select for a patient population with particularly good or particularly bad features and/or treatment outcomes. These data expand on the clinical trial data by demonstrating the safety and efficacy of front-line azacitidine in all patients with WHO-AML, irrespective of BM blast or WBC counts. This remained true when analysing patient subgroups stratified according to NCCN cytogenetic risk status and the presence of MRC separately (Figure 3B–E). In this regard, it seems noteworthy that azacitidine may overcome baseline factors generally considered to be indicators of adverse prognosis in AML, such as adverse cytogenetics [46], elevated WBC count [47] and MRC [48]. For example, azacitidine showed significant survival benefit over conventional care regimens in the NCCN poor-risk cytogenetic subgroup analysis of the AZA-AML-001 trial (median OS 6.4 vs. 3.2 months, p = 0.019; one-year survival rate 30.9% vs. 14.0%) [39], and we have previously reported that neither WBC < versus ≥15 G/L (12.8 vs. 13.5 months, p = 0.250) [14], nor the presence of MRC adversely affect OS of AML patients treated with azacitidine front-line (13.2 vs. 8.9 months, p = 0.104) [13].

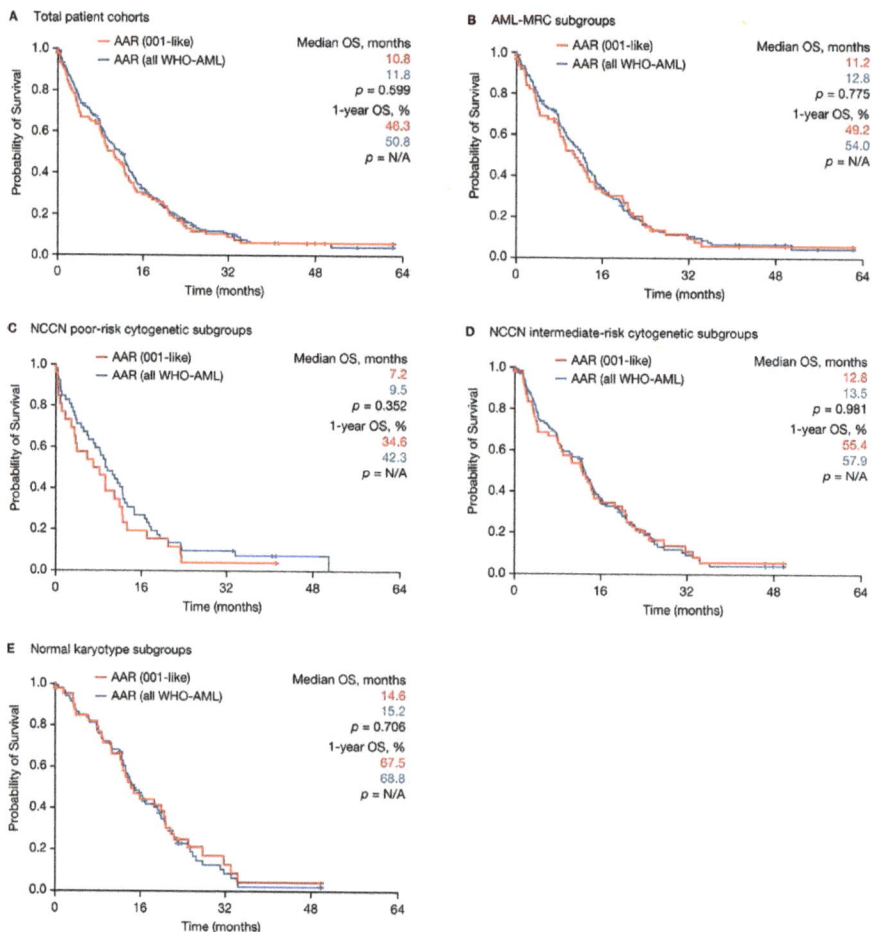

Figure 3. Median OS and one-year survival rates in AML patients with >30% BM blasts and <15 G/L WBC and WHO-AML patients treated with azacitidine front-line within the Austrian Azacitidine Registry. (**A**) Total patient cohorts; (**B**) AML-MRC; (**C**) AML with NCCN poor-risk cytogenetics; (**D**) AML with NCCN intermediate-risk cytogenetics; and (**E**) Normal karyotype.

It is clear that observational studies cannot replace clinical trials, but they do play a well-accepted complementary role. One might argue that the retrospective, uncontrolled nature of a registry setting is a drawback of the current analyses. However, precisely this fact represents one of the acknowledged advantages of registry settings. The necessity to assess the effect of new drugs in patients that often escape inclusion in clinical trials, in order to gain additional insights into a drug's ability to achieve its intended use, safety and outcome in the routine clinical practice setting, in contrast to the confined and controlled environment of a clinical trial, is supported by several groups, including the EMA [22,23,29–34,37,49–51].

Another recurrent concern about registries is the potential presence of a bias regarding the inclusion of patients. In the AAR, 20% of all patients (and 40% of patients aged ≥70 years) diagnosed with AML across Austria were included in the registry between 2008 and 2012 (Table S2). In comparison, a large retrospective analysis of 4416 patients with MDS from the SEER database revealed that only 7.7% of patients diagnosed before May 2007 and 13.4% of patients diagnosed thereafter were treated with hypomethylating agents in the U.S. [52]. Similarly, a more recent and larger analysis (*n* = 8580) from the Medicare database (cut-off date: 31 December 2012) published similar findings, namely that only 14.7% of MDS patients aged ≥66 years had received hypomethylating agents [53]. The Medicare database used the International Classification of Diseases for Oncology 3rd edition (ICD-O-3), which included patients with AML and 20%–30% BM blasts. No data on the use of hypomethylating agents has been published for all types of WHO-AML, including AML with >30% BM blasts. In comparison, seven AML patients were randomised in the U.S. during the recruitment period of 30 months (2.5 years) in the international AZA-AML-001 trial, which accounts for approximately 0.06% of all newly-diagnosed AML patients in the U.S. during this time period (4.889 new AML cases per year in 2013 [54]). In light of all of the above, an inclusion rate of 20% of AML patients of all ages and 40% of AML patients >70 years in the AAR (cut-off date: 31 December 2012) seems to be a very high nationwide coverage of AML patients treated with azacitidine, indicating limited selection bias. In addition, 20% of patients had received only one cycle of azacitidine, 64% ≤3 treatment cycles, 26% ≥3 comorbidities and 26% ECOG-PS ≥2, further indicating limited selection bias.

4. Materials and Methods

4.1. Design and Aim of the Study

This analysis selected patients with AML receiving azacitidine as front-line therapy within the AZA-AML-001 trial and the AAR. In total, 429 patients were identified that met this criterion, 241 from the AZA-AML-001 trial and 193 from the AAR (Figure 1). In the first analyses, we further selected for the presence of >30% BM blasts and ≤15 G/L WBC and directly compared baseline and treatment characteristics, as well as treatment outcomes and OS of patients included within the AZA-AML-001 trial (*n* = 214; "AML-001" subset) versus patients included within the AAR (*n* = 95; "AAR (001-like)" subset), using patient-level data from both cohorts (Figure 1). Comparative analyses of patient subsets from the AZA-AML-001 trial and the AAR grouped according to NCCN cytogenetic risk categories, as well as the presence of MRC were also performed.

Secondly, we wanted to explore whether the strict inclusion criteria of the AZA-AML-001 trial hampered the generalizability of the outcome data, as compared to the registry population. We next pooled patient data from the "AML-001" subset and the "AAR (001-like)" subset (*n* = 309), for outcome analyses (Figure 1).

Thirdly, we further aimed to evaluate whether there was a difference in baseline factors, treatment-related factors, TEAEs and the efficacy of azacitidine as front-line therapy for AML patients of the AAR who met clinical trial inclusion criteria ("AAR (001-like)" subset) versus all comers ("AAR (WHO-AML)" subset) (Figure 1).

4.2. Setting of the Study

The multicentre, randomised, open-label, parallel-group AZA-AML-001 trial (NCT01074047; registered February 2010) evaluated the efficacy and safety of azacitidine versus conventional care regimens in 488 patients aged ≥65 years with newly diagnosed AML with >30% BM blasts and ≤15 G/L WBC and was conducted in 18 countries. Details regarding the study design and the randomization process have recently been published [21] (study start date June 2010; last patient randomised 5 November 2012; final data collection date 22 January 2014). The AAR (NCT01595295; registered May 2012) is a multicentre database that includes 900 patients with AML, MDS or chronic myelomonocytic leukaemia (CMML) who were treated with azacitidine during the course of their disease. The AAR was initiated by the AGMT Study Group in 2009 to provide representative insight into the clinical management of patients treated with azacitidine in a clinical practice setting in Austria [11–20,55,56]. The AAR adheres to published quality guidelines of the U.S. Department of Health and Human Services Agency for Healthcare Research and Quality (AHRQ) [57]. The final data collection date for this analysis was 31 May 2016.

4.3. Definitions and Endpoints

For all patients included in these analyses (n = 407), AML was defined according to WHO 2008 criteria [58]. Front-line therapy with azacitidine was defined as the absence of prior disease-modifying therapy (in the AZA-AML-001 trial, hydroxyurea was allowed up to 2 weeks before the screening haematology sample was taken). Cytogenetic risk was classified according to NCCN 2009 cytogenetic risk categories [59].

Median OS was defined as time from Day 1 (date of randomization for the AML-001 group; Day 1 of Cycle 1 azacitidine treatment for the AAR group) to death from any cause. Event-free survival (EFS; for all patients) and relapse-free survival (RFS; for patients achieving CR/CRi) were assessed from Day 1 to the occurrence of an event (events included treatment failure, progressive disease, relapse after CR/CRi, death from any cause or lost to follow-up, whichever occurred first). Participants who were still alive without any of these events were censored at the date of their last assessment.

Response was determined based on International Working Group (IWG) Response Criteria for AML [60]. Morphologic CR was defined as <5% blasts in the BM (histology and/or aspirate with marrow spicules ($\geq$200 nucleated cells, absence of blasts with Auer rods) and an ANC of $\geq$1.0 G/L, a PLT count $\geq$100 G/L in the peripheral blood, as well as TI (no transfusions for 1–4 weeks prior to each assessment)). Morphologic CRi was defined as BM blasts <5% in histology and/or cytology, not meeting $\leq$1 of the peripheral blood CR criteria. PR was defined as an at least 50% reduction of baseline BM blast count in histology and/or aspirate (BM blasts must be >5% and <25%) and an ANC of $\geq$1.0 G/L, a PLT count $\geq$ 100 G/L in the peripheral blood, as well as TI (no transfusions for 1–4 weeks prior to each assessment)). No duration of these findings was required for confirmation of this response. The ORR included CR, CRi and PR [60]. RBC- and PLT-TI was defined as achievement of TI for at least 56 consecutive days and was assessed for patients who were transfusion dependent at baseline ($\geq$1 transfusion within the 56 days prior to Day 1). Duration of CR/CRi was defined as the time from the date CR/CRi was first documented until the date of documented relapse/disease progression from CR/CRi. Participants who were lost to follow-up without documented relapse or were alive at last follow-up without documented relapse were censored at the date of their last response assessment.

AEs were assessed according to the Common Terminology Criteria for Adverse Events (CTCAEv.4; [61]). The safety population included all patients who had received at least one dose of azacitidine and had at least one safety assessment thereafter. TEAEs were defined as new or worsening AEs between the time of first azacitidine dose and the end of the safety follow-up period, which was set at 28 days after the last dose of azacitidine. Treatment-emergent haematological toxicity was calculated from differential blood counts and transfusion status at Day 1 of each respective azacitidine treatment cycle.

4.4. Characteristics of Participants

Inclusion criteria of the AZA-AML-001 trial comprised newly-diagnosed patients with AML, ineligibility for HSCT, age $\geq$65 years, BM blasts >30%, WBC $\leq$15 G/L, ECOG-PS $\leq$2 and intermediate- or poor-risk cytogenetics (NCCN 2009 criteria); exclusion criteria included antecedent chronic CMML or other myeloproliferative neoplasm (MPN) and concurrent malignancies [21]. The only inclusion criteria of the AAR were a diagnosis of AML, MDS or CMML and treatment with at least one dose of azacitidine. No formal exclusion criteria existed, as the aim was to document the use and efficacy of azacitidine irrespective of age, comorbidities and/or previous lines of treatment. Further details have recently been published in this journal [8,13].

4.5. Processes, Interventions and Comparisons

This is a non-interventional retrospective comparative analysis of patient-level data of the above-defined patient populations with the above-defined aims and using the below-defined methods.

4.6. Statistical Analyses

Patient-level data for both studies were analysed by the sponsor of the AZA-AML-001 trial (in contrast to recently-presented data of an indirect comparison [15] based on previously-reported results [21,39,40]). No patients were excluded from efficacy and safety analyses. Survival distribution functions were estimated by the KM method, and survival curves were compared using the log-rank method. Univariate and multivariate analyses using Cox-regression of OS adjusted for significant baseline covariates were performed. Covariates in the univariate analysis with a p-value < 0.10 were included in the multivariate model. A backward selection method with a stay-level of 0.1 was used as the criterion to select covariates for the final model. The study covariate (i.e., AML-001 vs. AAR) was kept in the final multivariate model.

5. Conclusions

We tested and confirmed the efficacy of azacitidine as a front-line agent in patients with AML treated within the AGMT-AAR. Data obtained within the AZA-AML-001 trial could be recapitulated both in an AAR cohort of patients meeting AZA-AML-001 trial inclusion criteria (AML, >30% BM blasts and WBC ≤15 G/L), as well as in a wider AAR patient population that included all patients with WHO-AML. We thus not only confirm the value of azacitidine as reported by the clinical trial in a real-world scenario, but also expand on these data by showing that the safety and efficacy of azacitidine as a front-line agent is not limited to patients meeting stringent clinical trial inclusion criteria. Instead, azacitidine appears efficacious as a front-line treatment in a wider patient population that includes all patients with WHO-AML.

Supplementary Materials: Supplementary materials can be found at www.mdpi.com/1422-0067/18/2/415/s1.

Acknowledgments: The authors would like to acknowledge and thank Lucinda Huxley from FireKite, an Ashfield company, part of UDG Healthcare plc, for proofreading as an English native speaker, as well as editing of the references and figures to the required journal format. She had no influence on planning the study, interpreting the data, the writing or content of the manuscript, nor the decision to submit. Editorial assistance was funded by Celgene. The Austrian Azacitidine Registry (AAR) is a Registry of the Arbeitsgemeinschaft Medikamentöse Tumortherapie (AGMT) Study Group, which served as the responsible sponsor and holds the full and exclusive rights to data. Financial support for the AGMT was provided by Celgene. Celgene had no role in the study design, data collection, data analysis, data interpretation, writing of the manuscript, nor the decision to submit the manuscript for publication. The Austrian Azacitidine Registry is a registry of the AGMT Study Group, which served as the responsible sponsor and holds the full and exclusive rights to data. Financial support for the AGMT was provided by Celgene. The AZA-AML-001 trial was supported by research funding from Celgene Corporation, Summit, NJ, USA. Editorial assistance (proofreading by an English native speaker and editing of the references and figures to the required journal format) was funded by Celgene. The funding bodies had no role in the study design, data collection, data analysis, data interpretation, writing of the manuscript, nor the decision to submit the manuscript for publication.

Author Contributions: Lisa Pleyer: wrote the manuscript, contributed patients, provided substantial contributions to the conception and design, as well as the acquisition, analysis and interpretation of data, provided intellectual input, critically revised the manuscript and gave final approval of the version to be published. Lisa Pleyer, Hartmut Döhner, Hervé Dombret, John F. Seymour, Andre C. Schuh, CL Beach, Arlene S. Swern, Sonja Burgstaller, Reinhard Stauder, Michael Girschikofsky, Heinz Sill, Konstantin Schlick, Josef Thaler, Britta Halter, Sigrid Machherndl Spandl, Angelika Pichler, Michael Pfeilstöcker, Eva M. Autzinger, Alois Lang, Klaus Geissler, Daniela Voskova, Wolfgang R. Sperr, Sabine Hojas, Inga M. Rogulj, Johannes Andel and Richard Greil provided patient data and/or entered data into the eCRF. All co-authors provided substantial contributions to the conception and design, as well as the acquisition, analysis and interpretation of data, provided intellectual input, critically revised the manuscript and gave final approval of the version to be published. All authors agreed to be accountable for all aspects of the work.

Conflicts of Interest: Lisa Pleyer has been a consultant for Agios, Celgene, Bristol-Myers Squibb and Novartis and reports receiving honoraria and travel support from Agios, Celgene, Bristol-Myers Squibb, Novartis and AOP Orphan Pharmaceuticals. Hartmut Döhner has been a consultant for Celgene and reports having received honoraria from Celgene. Hervé Dombret has been a consultant for Celgene, Amgen and Erytech Pharma, reports receiving research funding from Amgen, Roche-Genentech, Novartis, Ariad, Kite Pharma, Oncoethix and Ambit and reports receiving honoraria from Celgene, Amgen, Roche-Genentech, Novartis, Ariad, Kite Pharma, Daiichi Sankyo, Pfizer, Servier, Sanofi, Astellas, Janssen, Sunesis, Agios, Seattle Genetics, Cellectis, Boehringer-Ingelheim, Karyopharm and Lilly. John F. Seymour has been a consultant to Celgene and received honoraria and travel support. Andre C. Schuh has been a consultant for Celgene, Amgen and Lundbeck and reports receiving honoraria from Celgene, Amgen and Lundbeck. CL Beach and Arlene S. Swern are full time employees and stockholders of Celgene Corp. Sonja Burgstaller has been a consultant for Celgene and Novartis, a member on the Board of Directors or advisory committees for Celgene and Novartis and reports receiving research funding from Celgene and honoraria from AOP Orphan Pharmaceuticals, Celgene, Mundipharma and Novartis. Reinhard Stauder has been a consultant and a member on the Board of Directors or advisory committees for Celgene and reports receiving research funding and honoraria from Celgene, Teva (Ratiopharm) and Novartis. Michael Girschikofsky has been a consultant for Mundipharma and reports receiving honoraria from Mundipharma and Pfizer and research funding from Pfizer. Heinz Sill reports receiving research funding from Celgene and has been an advisory board member for Celgene. Sigrid Machherndl Spandl has been a member of an advisory board for Celgene. Armin Zebisch reports receiving honoraria from Celgene. Michael Pfeilstöcker has been a consultant for Celgene and Novartis and reports receiving honoraria from Celgene, Novartis and Janssen-Cilag. Alois Lang has been a consultant for Celgene. Klaus Geissler has been a member of advisory boards for Celgene. Wolfgang R. Sperr has been a consultant for Celgene. Richard Greil reports receiving honoraria from Bristol-Myers-Squibb, Cephalon, Amgen, Eisai, Mundipharma, Merck, Janssen-Cilag, Genentech, Novartis, AstraZeneca, Boehringer Ingelheim, Pfizer, Roche and Sanofi Aventis, research funding from Cephalon, Celgene, Amgen, Mundipharma, Genentech, Pfizer, GSK and Ratiopharm and has been a consultant for Bristol-Myers-Squibb, Cephalon and Celgene. The other authors declare that no competing interests exist.

Abbreviations

AAR	Austrian Azacitidine Registry
AE	Adverse event
AGMT	Arbeitsgemeinschaft Medikamentöse Tumortherapie
AHRQ	Agency for Healthcare Research and Quality
AML	Acute myeloid leukaemia
ANC	Absolute neutrophil count
AZA	Azacitidine
BM	Bone marrow
BSC	Best supportive care
CI	Confidence interval
CMML	Chronic myelomonocytic leukaemia
CR	Complete response
CRi	Morphologic CR with incomplete blood count recovery
ECOG-PS	Eastern Cooperative Oncology Group Performance Status
eCRF	Electronic case report form
EFS	Event-free survival
EMA	European Medicines Agency
HSCT	Hematopoietic stem cell transplantation
Hb	Haemoglobin
HR	Hazards ratio
IWG	International Working Group
KM	Kaplan–Meier
MDS	Myelodysplastic syndromes
MPN	Myeloproliferative neoplasm
MRC	Myelodysplasia-related changes
NCCN	National Comprehensive Cancer Network
NOS	Not otherwise specified
ORR	Overall response rate
OS	Overall survival
PLT	Platelet
PR	Partial response
pts	Patients
RBC	Red blood cell
RCA	Recurrent cytogenetic abnormalities
RFS	Relapse-free survival
SD	Standard deviation
t-AML	Therapy-related AML

TD	Transfusion dependence
TEAE	Treatment-emergent adverse event
TI	Transfusion independence
vs.	Versus
WBC	White blood cell
WHO	World Health Organisation

References

1. Klepin, H.D.; Balducci, L. Acute myelogenous leukemia in older adults. *Oncologist* **2009**, *14*, 222–232. [CrossRef] [PubMed]
2. Appelbaum, F.R.; Gundacker, H.; Head, D.R.; Slovak, M.L.; Willman, C.L.; Godwin, J.E.; Anderson, J.E.; Petersdorf, S.H. Age and acute myeloid leukemia. *Blood* **2006**, *107*, 3481–3485. [CrossRef] [PubMed]
3. Sorror, M.L.; Maris, M.B.; Storb, R.; Baron, F.; Sandmaier, B.M.; Maloney, D.G.; Storer, B. Hematopoietic cell transplantation (HCT)-specific comorbidity index: A new tool for risk assessment before allogeneic HCT. *Blood* **2005**, *106*, 2912–2919. [CrossRef] [PubMed]
4. Giles, F.J.; Borthakur, G.; Ravandi, F.; Faderl, S.; Verstovsek, S.; Thomas, D.; Wierda, W.; Ferrajoli, A.; Kornblau, S.; Pierce, S.; et al. The haematopoietic cell transplantation comorbidity index score is predictive of early death and survival in patients over 60 years of age receiving induction therapy for acute myeloid leukaemia. *Br. J. Haematol.* **2007**, *136*, 624–627. [CrossRef] [PubMed]
5. Lazarevic, V.; Hörstedt, A.S.; Johansson, B.; Antunovic, P.; Billström, R.; Derolf, Å.; Hulegårdh, E.; Lehmann, S.; Möllgård, L.; Nilsson, C.; et al. Incidence and prognostic significance of karyotypic subgroups in older patients with acute myeloid leukemia: The swedish population-based experience. *Blood Cancer J.* **2014**, *4*, e188. [CrossRef] [PubMed]
6. Oran, B.; Weisdorf, D.J. Survival for older patients with acute myeloid leukemia: A population-based study. *Haematologica* **2012**, *97*, 1916–1924. [CrossRef] [PubMed]
7. Fenaux, P.; Mufti, G.J.; Hellstrom-Lindberg, E.; Santini, V.; Finelli, C.; Giagounidis, A.; Schoch, R.; Gattermann, N.; Sanz, G.; List, A.; et al. Efficacy of azacitidine compared with that of conventional care regimens in the treatment of higher-risk myelodysplastic syndromes: A randomised, open-label, phase III study. *Lancet Oncol.* **2009**, *10*, 223–232. [CrossRef]
8. Fenaux, P.; Mufti, G.J.; Hellstrom-Lindberg, E.; Santini, V.; Gattermann, N.; Germing, U.; Sanz, G.; List, A.F.; Gore, S.; Seymour, J.F.; et al. Azacitidine prolongs overall survival compared with conventional care regimens in elderly patients with low bone marrow blast count acute myeloid leukemia. *J. Clin. Oncol.* **2010**, *28*, 562–569. [CrossRef] [PubMed]
9. Silverman, L.R.; Demakos, E.P.; Peterson, B.L.; Kornblith, A.B.; Holland, J.C.; Odchimar-Reissig, R.; Stone, R.M.; Nelson, D.; Powell, B.L.; DeCastro, C.M.; et al. Randomized controlled trial of azacitidine in patients with the myelodysplastic syndrome: A study of the cancer and leukemia group B. *J. Clin. Oncol.* **2002**, *20*, 2429–2440. [CrossRef] [PubMed]
10. Silverman, L.R.; McKenzie, D.R.; Peterson, B.L.; Holland, J.F.; Backstrom, J.T.; Beach, C.L.; Larson, R.A. Further analysis of trials with azacitidine in patients with myelodysplastic syndrome: Studies 8421, 8921, and 9221 by the cancer and leukemia group B. *J. Clin. Oncol.* **2006**, *24*, 3895–3903. [CrossRef] [PubMed]
11. Pleyer, L.; Stauder, R.; Burgstaller, S.; Schreder, M.; Tinchon, C.; Pfeilstocker, M.; Steinkirchner, S.; Melchardt, T.; Mitrovic, M.; Girschikofsky, M.; et al. Azacitidine in patients with who-defined AML—Results of 155 patients from the Austrian azacitidine registry of the AGMT-study group. *J. Hematol. Oncol.* **2013**, *6*, 32. [CrossRef] [PubMed]
12. Pleyer, L.; Burgstaller, S.; Girschikofsky, M.; Linkesch, W.; Stauder, R.; Pfeilstöcker, M.; Schreder, M.; Tinchon, C.; Sliwa, T.; Lang, A.; et al. Azacitidine in 302 patients with who-defined acute myeloid leukemia: Results from the Austrian azacitidine registry of the AGMT-study group. *Ann. Hematol.* **2014**, *93*, 1825–1838. [PubMed]
13. Pleyer, L.; Burgstaller, S.; Stauder, R.; Girschikofsky, M.; Sill, H.; Schlick, K.; Thaler, J.; Halter, B.; Hherndl-Spandl, S.; Zebisch, A.; et al. Azacitidine front-line in 339 patients with myelodysplastic syndromes and acute myeloid leukaemia: Comparison of French-American-British and World Health Organization classifications. *J. Hematol. Oncol.* **2016**, *9*, 39. [CrossRef] [PubMed]

14. Pleyer, L.; Burgstaller, S.; Stauder, R.; Girschikofsky, M.; Linkesch, W.; Pfeilstöcker, M.; Autzinger, E.M.; Tinchon, C.; Sliwa, T.; Lang, A.; et al. Azacitidine in patients with acute myeloid leukemia: Assessing the potential negative impact of elevated baseline white blood cell count on outcome. *Blood* **2014**, *124*, abstract 3683.

15. Pleyer, L.; Burgstaller, S.; Stauder, R.; Girschikofsky, M.; Sill, H.; Schlick, K.; Thaler, J.; Halter, B.; Machherndl-Spandl, S.; Zebisch, A.; et al. Azacitidine in acute myeloid leukemia with >30% bone marrow blasts and <15 g/L white blood cell count: Results from the Austrian Azacitidine Registry of the AGMT study group versus randomized controlled phase III clinical trial data. *Blood* **2015**, *126*, abstract 2515.

16. Pleyer, L.; Burgstaller, S.; Stauder, R.; Girschikofsky, M.; Linkesch, W.; Pfeilstöcker, M.; Autzinger, E.M.; Tinchon, C.; Sliwa, T.; Lang, A.; et al. Azacitidine in patients with treatment-related acute myeloid leukemia: Retrospective analysis of the Austrian Azacitidine Registry. *Blood* **2014**, *124*, abstract 2284.

17. Pleyer, L.; Burgstaller, S.; Stauder, R.; Girschikofsky, M.; Linkesch, W.; Pfeilstöcker, M.; Autzinger, E.M.; Tinchon, C.; Sliwa, T.; Lang, A.; et al. Azacitidine in acute myeloid leukemia: Comparison of patients with AML-MRF vs. AML-NOS enrolled in the Austrian Azacitidine Registry. *Blood* **2014**, *124*, abstract 3681.

18. Pleyer, L.; Burgstaller, S.; Stauder, R.; Girschikofsky, M.; Linkesch, W.; Pfeilstöcker, M.; Autzinger, E.M.; Tinchon, C.; Sliwa, T.; Lang, A.; et al. Azacitidine in patients with acute myeloid leukemia: Impact of intermediate-risk vs. high-risk cytogenetics on patient outcomes. *Blood* **2014**, *124*, abstract 955.

19. Pleyer, L.; Germing, U.; Sperr, W.R.; Linkesch, W.; Burgstaller, S.; Stauder, R.; Girschikofsky, M.; Schreder, M.; Pfeilstocker, M.; Lang, A.; et al. Azacitidine in CMML: Matched-pair analyses of daily-life patients reveal modest effects on clinical course and survival. *Leuk. Res.* **2014**, *38*, 475–483. [CrossRef] [PubMed]

20. Pleyer, L.; Burgstaller, S.; Stauder, R.; Girschikofsky, M.; Linkesch, W.; Pfeilstöcker, M.; Autzinger, E.M.; Tinchon, C.; Sliwa, T.; Lang, A.; et al. Azacitidine in patients with relapsed/refractory acute myeloid leukemia: Retrospective analysis of the Austrian Azacitidine Registry. *Blood* **2014**, *124*, abstract 943.

21. Dombret, H.; Seymour, J.F.; Butrym, A.; Wierzbowska, A.; Selleslag, D.; Jang, J.H.; Kumar, R.; Cavenagh, J.; Schuh, A.C.; Candoni, A.; et al. International phase 3 study of azacitidine vs. conventional care regimens in older patients with newly diagnosed AML with >30% blasts. *Blood* **2015**, *126*, 291–299. [CrossRef] [PubMed]

22. Mosenifar, Z. Population issues in clinical trials. *Proc. Am. Thorac. Soc.* **2007**, *4*, 185–187. [CrossRef] [PubMed]

23. Denson, A.C.; Mahipal, A. Participation of the elderly population in clinical trials: Barriers and solutions. *Cancer Control* **2014**, *21*, 209–214. [PubMed]

24. Bernal, T.; Martinez-Camblor, P.; Sanchez-Garcia, J.; de Paz, R.; Luno, E.; Nomdedeu, B.; Ardanaz, M.T.; Pedro, C.; Amigo, M.L.; Xicoy, B.; et al. Effectiveness of azacitidine in unselected high-risk myelodysplastic syndromes: Results from the Spanish registry. *Leukemia* **2015**, *29*, 1875–1881. [CrossRef] [PubMed]

25. Dinmohamed, A.G.; van Norden, Y.; Visser, O.; Posthuma, E.F.M.; Huijgens, P.C.; Sonneveld, P.; van de Loosdrecht, A.A.; Jongen-Lavrencic, M. Effectiveness of azacitidine for the treatment of higher-risk myelodysplastic syndromes in daily practice: Results from the Dutch population-based PHAROS MDS registry. *Leukemia* **2015**, *29*, 2449–2451. [CrossRef] [PubMed]

26. Ostgard, L.S.; Norgaard, M.; Sengelov, H.; Medeiros, B.C.; Kjeldsen, L.; Overgaard, U.M.; Severinsen, M.T.; Marcher, C.W.; Jensen, M.K.; Norgaard, J.M. Improved outcome in acute myeloid leukemia patients enrolled in clinical trials: A national population-based cohort study of danish intensive chemotherapy patients. *Oncotarget* **2016**, *7*, 72044–72056. [PubMed]

27. Gahn, B.; Haase, D.; Unterhalt, M.; Drescher, M.; Schoch, C.; Fonatsch, C.; Terstappen, L.W.; Hiddemann, W.; Buchner, T.; Bennett, J.M.; et al. De novo aml with dysplastic hematopoiesis: Cytogenetic and prognostic significance. *Leukemia* **1996**, *10*, 946–951. [PubMed]

28. Miesner, M.; Haferlach, C.; Bacher, U.; Weiss, T.; Macijewski, K.; Kohlmann, A.; Klein, H.U.; Dugas, M.; Kern, W.; Schnittger, S.; et al. Multilineage dysplasia (MLD) in acute myeloid leukemia (AML) correlates with MDS-related cytogenetic abnormalities and a prior history of MDS or MDS/MPN but has no independent prognostic relevance: A comparison of 408 cases classified as "AML not otherwise specified" (AML-NOS) or "aml with myelodysplasia-related changes" (AML-MRC). *Blood* **2010**, *116*, 2742–2751. [PubMed]

29. Food and Drug Administration. Guidance for Industry and Food and Drug Administration Staff—Postmarket Surveillance under Section 522 of the Federal Food, Drug and Cosmetic Act. 2016. Available online: http://www.fda.gov/downloads/medicaldevices/deviceregulationandguidance/guidancedocuments/ucm268141.pdf (accessed on 10 February 2017).

30. Juliusson, G.; Lazarevic, V.; Horstedt, A.S.; Hagberg, O.; Hoglund, M. Acute myeloid leukemia in the real world: Why population-based registries are needed. *Blood* **2012**, *119*, 3890–3899. [CrossRef] [PubMed]

31. Larsson, S.; Lawyer, P. Improving Health Care Value: The Case for Disease Registries. (The Boston Consulting Group). 2011. Available online: http://2eic.com/sites/default/files/bcg_-_registries_can_add_health_care_value.pdf (accessed on 11 January 2017).

32. Noe, L.; Larson, L.; Trotter, J. Utilizing Patient Registries to Support Health Economics Research: Integrating Observational Data with Economic Analyses, Models, and Other Applications. Available online: https://www.ispor.org/news/articles/oct05/patient_registr.asp (accessed on 11 January 2017).

33. Stark, N.J. Registry Studies: Why and How? Available online: http://clinicaldevice.typepad.com/cdg_whitepapers/2011/07/registry-studies-why-and-how.html (accessed on 11 January 2017).

34. European Medicines Agency. Patient Registries. Available online: http://www.ema.europa.eu/ema/index.jsp?curl=pages/regulation/general/general_content_000658.jsp (accessed on 21 September 2016).

35. Thepot, S.; Itzykson, R.; Seegers, V.; Recher, C.; Raffoux, E.; Quesnel, B.; Delaunay, J.; Cluzeau, T.; Marfaing Koka, A.; Stamatoullas, A.; et al. Azacitidine in untreated acute myeloid leukemia. A report on 149 patients. *Am. J. Hematol.* **2014**, *89*, 410–416. [CrossRef] [PubMed]

36. Dinmohamed, A.G.; van Norden, Y.; Visser, O.; Posthuma, E.F.; Huijgens, P.C.; Sonneveld, P.; van de Loosdrecht, A.A.; Jongen-Lavrencic, M. The use of medical claims to assess incidence, diagnostic procedures and initial treatment of myelodysplastic syndromes and chronic myelomonocytic leukemia in The Netherlands. *Leuk. Res.* **2015**, *39*, 177–182. [CrossRef] [PubMed]

37. Ostgard, L.S.; Norgaard, J.M.; Severinsen, M.T.; Sengelov, H.; Friis, L.; Jensen, M.K.; Nielsen, O.J.; Norgaard, M. Data quality in the Danish national acute leukemia registry: A hematological data resource. *Clin. Epidemiol.* **2013**, *5*, 335–344. [CrossRef] [PubMed]

38. Dinmohamed, A.G.; Brink, M.; Visser, O.; Jongen-Lavrencic, M. Population-based analyses among 184 patients diagnosed with large granular lymphocyte leukemia in The Netherlands between 2001 and 2013. *Leukemia* **2016**, *30*, 1449–1451. [CrossRef] [PubMed]

39. Döhner, H.; Seymour, J.F.; Butrym, A.; Willemze, R.; Selleslag, D.; Jang, J.H.; Cavenagh, J.; Kumar, R.; Schuh, A.C.; Candoni, A.; et al. Overall survival in older patients with newly diagnosed acute myeloid leukemia (AML) with >30% bone marrow blasts treated with azacitidine by cytogenetic risk status: Results of the AZA-AML-001 study. *Blood* **2014**, *124*, abstract 621.

40. Seymour, J.F.; Döhner, H.; Butrym, A.; Wierzbowska, A.; Selleslag, D.; Jang, J.H.; Cavenagh, J.D.; Kumar, R.; Schuh, A.C.; Candoni, A.; et al. Azacitidine (AZA) versus conventional care regimens (CCR) in older patients with newly diagnosed acute myeloid leukemia (>30% bone marrow blasts) with morphologic dysplastic changes: A subgroup analysis of the AZA-AML-001 trial. *Blood* **2014**, *124*, 10.

41. Seymour, J.F.; Döhner, H.; Schuh, A.C.; Stone, R.M.; Minden, M.; Weaver, J.; Songer, S.; Beach, C.L.; Dombret, H. Azacitidine (AZA) vs. Conventional Care Regimens (CCR) in Patients with Acute Myeloid Leukemia (AML) with Myelodyspasia-Related Changes (MRC) in AZA-AML-001 per Central Review. Available online: http://learningcenter.ehaweb.org/eha/2016/21st/133461/john.f.seymour.azacitidine.28aza29.vs.conventional.care.regimens.28ccr29.in.html?f=p16m3l9759 (accessed on 9 February 2017).

42. Schuh, A.; Dombret, H.; Sandhu, I.; Seymour, J.F.; Stone, R.M.; Kathrin Al-Ali, H.; Alimena, G.; Lewis, I.; Kyun, S.S.; Geddes, M.; et al. Overall Survival (OS) without Complete Remission (CR) in Older Patients with Acute Myeloid Leukemia (AML): Azacitidine (aza) vs. Conventional Care Regimens (CCR) in the AZAAML001 Study. Available online: http://learningcenter.ehaweb.org/eha/2015/20th/100716/%5B%5B\protect\T1\textdollaritem.link%5D%5D (accessed on 9 February 2017).

43. Ramos, F.; Thepot, S.; Pleyer, L.; Maurillo, L.; Itzykson, R.; Bargay, J.; Stauder, R.; Venditti, A.; Seegers, V.; Martinez-Robles, V.; et al. Azacitidine frontline therapy for unfit acute myeloid leukemia patients: Clinical use and outcome prediction. *Leuk. Res.* **2015**, *39*, 296–306. [CrossRef] [PubMed]

44. Tombak, A.; Ucar, M.A.; Akdeniz, A.; Tiftik, E.N.; Şahin, D.G.; Akay, O.M.; Yıldırım, M.; Nevruz, O.; Kış, C.; Gürkan, E.; et al. The role of azacitidine in the treatment of elderly patients with AML—Results of a retrospective multicenter study. *Turk. J. Haematol.* **2016**, *33*, 273–280. [CrossRef] [PubMed]

45. Sudan, N.; Rossetti, J.M.; Shadduck, R.K.; Latsko, J.; Lech, J.A.; Kaplan, R.B.; Kennedy, M.; Gryn, J.F.; Faroun, Y.; Lister, J. Treatment of acute myelogenous leukemia with outpatient azacitidine. *Cancer* **2006**, *107*, 1839–1843. [CrossRef] [PubMed]

46. Papaemmanuil, E.; Gerstung, M.; Bullinger, L.; Gaidzik, V.I.; Paschka, P.; Roberts, N.D.; Potter, N.E.; Heuser, M.; Thol, F.; Bolli, N.; et al. Genomic classification and prognosis in acute myeloid leukemia. *N. Engl. J. Med.* **2016**, *374*, 2209–2221. [CrossRef] [PubMed]

47. Greenwood, M.J.; Seftel, M.D.; Richardson, C.; Barbaric, D.; Barnett, M.J.; Bruyere, H.; Forrest, D.L.; Horsman, D.E.; Smith, C.; Song, K.; et al. Leukocyte count as a predictor of death during remission induction in acute myeloid leukemia. *Leuk. Lymph.* **2006**, *47*, 1245–1252. [CrossRef] [PubMed]

48. Stone, R.M.; Mazzola, E.; Neuberg, D.; Allen, S.L.; Pigneux, A.; Stuart, R.K.; Wetzler, M.; Rizzieri, D.; Erba, H.P.; Damon, L.; et al. Phase III open-label randomized study of cytarabine in combination with amonafide L-malate or daunorubicin as induction therapy for patients with secondary acute myeloid leukemia. *J. Clin. Oncol.* **2015**, *33*, 1252–1257. [CrossRef] [PubMed]

49. Sorensen, H.T.; Lash, T.L.; Rothman, K.J. Beyond randomized controlled trials: A critical comparison of trials with nonrandomized studies. *Hepatology* **2006**, *44*, 1075–1082. [CrossRef] [PubMed]

50. Juliusson, G.; Antunovic, P.; Derolf, A.; Lehmann, S.; Mollgard, L.; Stockelberg, D.; Tidefelt, U.; Wahlin, A.; Hoglund, M. Age and acute myeloid leukemia: Real world data on decision to treat and outcomes from the Swedish acute leukemia registry. *Blood* **2009**, *113*, 4179–4187. [CrossRef] [PubMed]

51. Hulegardh, E.; Nilsson, C.; Lazarevic, V.; Garelius, H.; Antunovic, P.; Rangert, D.A.; Mollgard, L.; Uggla, B.; Wennstrom, L.; Wahlin, A.; et al. Characterization and prognostic features of secondary acute myeloid leukemia in a population-based setting: A report from the Swedish acute leukemia registry. *Am. J. Hematol.* **2015**, *90*, 208–214. [CrossRef] [PubMed]

52. Wang, R.; Gross, C.P.; Maggiore, R.J.; Halene, S.; Soulos, P.R.; Raza, A.; Galili, N.; Ma, X. Pattern of hypomethylating agents use among elderly patients with myelodysplastic syndromes. *Leuk. Res.* **2011**, *35*, 904–908. [CrossRef] [PubMed]

53. Zeidan, A.M.; Wang, R.; Davidoff, A.J.; Ma, S.; Zhao, Y.; Gore, S.D.; Gross, C.P.; Ma, X. Disease-related costs of care and survival among medicare-enrolled patients with myelodysplastic syndromes. *Cancer* **2016**, *122*, 1598–1607. [CrossRef] [PubMed]

54. NCI. Seer Cancer Statistics Review 1975–2013. Available from: http://seer.Cancer.Gov/csr/1975_2013/browse_csr.Php?Sectionsel=13&pagesel=sect_13_table.13.html (accessed on 27 July 2016).

55. Pleyer, L.; Stauder, R.; Thaler, J.; Ludwig, H.; Pfeilstöcker, M.; Steinkirchner, T.; Melchardt, T.; Weltermann, A.; Lang, A.; Linkesch, W.; et al. Overall survival data of patients with MDS, AML and CMML from the Austrian Azacitidine Registry of 184 consecutive patients. *Leuk. Res.* **2011**, *35*, 100. [CrossRef]

56. Pleyer, L.; Stauder, R.; Thaler, J.; Ludwig, H.; Pfeilstocker, M.; Steinkirchner, S.; Melchardt, T.; Weltermann, A.; Lang, A.; Linkesch, W.; et al. Age- and comorbidity-specific evaluation of azacitidine treatment, response and overall survival in 184 patients in the Austrian Azacitidine Registry. *Leuk. Res.* **2011**, *35*, 101. [CrossRef]

57. Gliklich, R.E.; Dreyer, N.A.; Leavy, M.B. Registries for Evaluating Patient Outcomes: A User's Guide. Agency for Healthcare research and Quality (AHRQ): Rockville, MD, USA, 2014.

58. Wandt, H.; Schakel, U.; Kroschinsky, F.; Prange-Krex, G.; Mohr, B.; Thiede, C.; Pascheberg, U.; Soucek, S.; Schaich, M.; Ehninger, G. MLD according to the WHO classification in AML has no correlation with age and no independent prognostic relevance as analyzed in 1766 patients. *Blood* **2008**, *111*, 1855–1861. [CrossRef] [PubMed]

59. National Comprehensive Cancer Network. NCCN Clinical Practice Guidelines in Oncology for Acute Myeloid Leukemia. Available online: https://www.nccn.org/professionals/physician_gls/f_guidelines.asp (accessed on 10 February 2017).

60. Cheson, B.D.; Bennett, J.M.; Kopecky, K.J.; Buchner, T.; Willman, C.L.; Estey, E.H.; Schiffer, C.A.; Doehner, H.; Tallman, M.S.; Lister, T.A.; et al. Revised recommendations of the international working group for diagnosis, standardization of response criteria, treatment outcomes, and reporting standards for therapeutic trials in acute myeloid leukemia. *J. Clin. Oncol.* **2003**, *21*, 4642–4649. [CrossRef] [PubMed]

61. NCI Common Terminology Criteria for Adverse Events (CTCAE). Available online: http://evs.nci.nih.gov/ftp1/CTCAE/About.html (accessed on 9 February 2017).

© 2017 by the authors. Licensee MDPI, Basel, Switzerland. This article is an open access article distributed under the terms and conditions of the Creative Commons Attribution (CC BY) license (http://creativecommons.org/licenses/by/4.0/).

International Journal of
Molecular Sciences

MDPI

Article

Clinical Outcomes of 217 Patients with Acute Erythroleukemia According to Treatment Type and Line: A Retrospective Multinational Study

Antonio M. Almeida [1,*], Thomas Prebet [2], Raphael Itzykson [3], Fernando Ramos [4], Haifa Al-Ali [5], Jamile Shammo [6], Ricardo Pinto [7], Luca Maurillo [8], Jaime Wetzel [9], Pellegrino Musto [10], Arjan A. Van De Loosdrecht [11], Maria Joao Costa [12], Susana Esteves [1], Sonja Burgstaller [13], Reinhard Stauder [14], Eva M. Autzinger [15], Alois Lang [16], Peter Krippl [17], Dietmar Geissler [18], Jose Francisco Falantes [19], Carmen Pedro [20], Joan Bargay [21], Guillermo Deben [22], Ana Garrido [23], Santiago Bonanad [24], Maria Diez-Campelo [25], Sylvain Thepot [26], Lionel Ades [3], Wolfgang R. Sperr [27], Peter Valent [27], Pierre Fenaux [3], Mikkael A. Sekeres [9], Richard Greil [28,29,30,31] and Lisa Pleyer [28,29,30,31,*]

1   Instituto Português de Oncologia de Lisboa (IPOL), 1200-795 Lisbon, Portugal; sesteves@ipolisboa.min-saude.pt
2   Institut Paoli Calmettes, Marseille, France and Yale New Haven Hospital, New Haven, CT 06512, USA; thomas.prebet@yale.edu
3   Hopital Saint-Louis, Assistance Publique-Hôpitaux de Paris (AP-HP), Paris Diderot University, 75010 Paris, France; itzykson@gmail.com (R.I.); lionel.ades@sls.aphp.fr (L.A.); pierre.fenaux@sls.aphp.fr (P.F.)
4   Hospital Universitario de Leon, 24071 Leon, Spain; mail@fernandoramosmd.es
5   University Hospital of Halle, 06120 Halle, Germany; alah@medizin.uni-leipzig.de
6   Rush University Medical Center, Chicago, IN 60612, USA; Jamile_Shammo@rush.edu
7   Hospital Sao Joao, 4200-319 Porto, Portugal; rjsmpinto@gmail.com
8   University Tor Vergata, 00173 Rome, Italy; luca.maurillo@uniroma2.it
9   Cleveland Clinic Taussig Cancer Institute, Cleveland, OH 44195, USA; FENSTEJ@ccf.org (J.W.); SEKEREM@ccf.org (M.A.S.)
10 RCCS-CROB, Referral Cancer Center of Basilicata, 85028 Rionero in Vulture (Pz), Italy; p.musto@tin.it
11 Department of Hematology VU University Medical Center, 1081 HV Amsterdam, The Netherlands; a.vandeloosdrecht@vumc.nl
12 Centro Hospitalar Lisboa Norte Hospital Santa Maria, 1649-035 Lisbon, Portugal; mjoaocosta@gmail.com
13 Department of Internal Medicine IV, Hospital Wels-Grieskirchen, 4600 Wels, Austria; sonja.burgstaller@klinikum-wegr.at
14 Department of Internal Medicine V (Haematology and Oncology), Innsbruck Medical University, 6020 Innsbruck, Austria; reinhard.stauder@i-med.ac.at
15 1st Department of Internal Medicine, Center for Oncology and Hematology, Wilhelminenspital, 1160 Vienna, Austria; eva-maria.autzinger@wienkav.at
16 Internal Medicine, Hospital Feldkirch,6800 Feldkirch, Austria; alois.lang@lkhf.at
17 Department of Internal Medicine, Hospital Fürstenfeld, 8280 Fürstenfeld, Austria; peter.krippl@lkh-fuerstenfeld.at
18 Department for Internal Medicine, Klinikum Klagenfurt am Wörthersee, 9020 Pörtschach am Wörthersee, Austria; dietmar.geissler@kabeg.at
19 Hospital Universitario Virgen del Rocio, 41013 Sevilla, Spain; jfalantes@gmail.com
20 Hospital del Mar, 08003 Barcelona, Spain; MPedro@parcdesalutmar.cat
21 Hospital Son Llatzer, 07198 Palma de Mallorca, Spain; jbargay@hsll.es
22 Hospital Universitario, 15006 A Coruña, Spain; gdebari@canalejo.org
23 Hospital de la Santa Creu i Sant Pau, 08026 Barcelona, Spain; AGarridoD@santpau.cat
24 Hospital Universitario de la Ribera, 46600 Alzira, Spain; sbonanad@gmail.com
25 Hospital Universitario de Salamanca, 37007 Salamanca, Spain; mdiezcampelo@usal.es
26 Centre Hospitalier Universitaire, 49100 Angers, France; Sylvain.Thepot@chu-angers.fr
27 Department of Internal Medicine I, Division of Hematology & Hemostaseology and Ludwig Boltzmann Cluster Oncology, Medical University of Vienna, 1090 Vienna, Austria; wolfgang.r.sperr@meduniwien.ac.at (W.R.S.); peter.valent@meduniwien.ac.at (P.V.)
28 3rd Med. Department, Paracelsus Medical University, 5020 Salzburg, Austria; r.greil@salk.at

Int. J. Mol. Sci. **2017**, *18*, 837

29   Salzburg Cancer Research Institute, 5020 Salzburg, Austria
30   Cancer Cluster Salzburg, 5020 Salzburg, Austria
31   Arbeitsgemeinschaft Medikamentöse Tumortherapie (AGMT), 5020 Salzburg, Austria
*   Correspondence: amalmeida@ipolisboa.min-saude.pt (A.M.A.); l.pleyer@salk.at (L.P.);
     Tel.: +351-21-724-9036 (A.M.A.); +43-(0)5-7255-58271 (L.P.)

Academic Editors: Geoffrey Brown and Ewa Marcinkowska
Received: 10 February 2017; Accepted: 6 April 2017; Published: 14 April 2017

Abstract: Acute erythroleukemia (AEL) is a rare disease typically associated with a poor prognosis. The median survival ranges between 3–9 months from initial diagnosis. Hypomethylating agents (HMAs) have been shown to prolong survival in patients with myelodysplastic syndromes (MDS) and AML, but there is limited data of their efficacy in AEL. We collected data from 210 AEL patients treated at 28 international sites. Overall survival (OS) and PFS were estimated using the Kaplan-Meier method and the log-rank test was used for subgroup comparisons. Survival between treatment groups was compared using the Cox proportional hazards regression model. Eighty-eight patients were treated with HMAs, 44 front line, and 122 with intensive chemotherapy (ICT). ICT led to a higher overall response rate (complete or partial) compared to first-line HMA (72% vs. 46.2%, respectively; $p \leq 0.001$), but similar progression-free survival (8.0 vs. 9.4 months; $p = 0.342$). Overall survival was similar for ICT vs. HMAs (10.5 vs. 13.7 months; $p = 0.564$), but patients with high-risk cytogenetics treated with HMA first-line lived longer (7.5 for ICT vs. 13.3 months; $p = 0.039$). Our results support the therapeutic value of HMA in AEL.

Keywords: acute erythroleukemia; azacitidine; decitabine

1. Introduction

Acute erythroleukemia (AEL) is a rare subtype of acute myeloid leukemia (AML), accounting for 3–5% of all AML cases [1]. It is characterized by an expanded erythroid component with a variable, but increased, percentage of blasts [1]. Although recognized as a distinct entity by most classification systems, the diagnostic criteria have changed from system to system, which has been subject to discussion [2–5]. The recently-published WHO 2016 classification [6] advocates the use of blast percentage on the total cell population rather than that of the non-erythroid component. This reclassifies almost all cases of AEL into myelodysplasia (MDS) or AML subtypes [7,8].

Typical laboratory features include pancytopenia, few peripheral blood blasts, the presence of dysplasia in BM and peripheral blood, especially with dysplastic PAS-positive erythroblasts overexpression of the multidrug resistance (MDR) gene product P-glycoprotein, frequent occurrence of high-risk karyotypes, and a high frequency of mutations, especially of *TP53* [2,9–14]. In addition, AEL is frequently secondary to previous myelodysplastic syndrome (MDS) [15]. Consequently, it is associated with a poor prognosis, with a median overall survival (OS) of 3–14 months from diagnosis [1,2,10,14]. The only recurring molecular alteration reported has been translocation t(1;16) generating the fusion gene NFIA/CBFA2T3 [16]. Furthermore, a high proerythroblast/myeloblast ratio correlates with significant increases in cytogenetic aberrations, proliferation markers, and worse outcomes [1,17–19], although this is not consensual [10,14]. In fact, several authors believe that the association of AEL with adverse prognostic factors imparts the adverse prognosis, rather than the diagnosis of AEL itself [14,20].

Due to the rarity of the disease (2–5% of all leukemias), few publications focus on this entity alone, with single cases or case series predominating [9,10,15,21] and patients with AEL are usually treated similarly to patients with other types of AML [1,3]. When treated with intensive chemotherapy (ICT), the median OS of AEL patients range between 7.6 and nine months [14,22]. The poor results achieved with ICT in AEL are likely due to the adverse prognostic factors described above.

Hypomethylating agents (HMAs; azacitidine and decitabine) have become the first-line therapy of choice for patients with MDS [23,24], CMML [25–27], and AML [28–36] who are not candidates for, or decline, intensive chemotherapy (ICT) and/or allo-SCT. HMAs have demonstrated improved outcomes for patients with AML when compared to conventional care regimens, including ICT, low-dose cytarabine, or best supportive care (BSC) [28–30,37,38]. Despite some limitations, several studies indicate that the OS of older AML patients treated with HMAs may not be inferior to those treated with ICT [28,29,39–44].

The few existing studies of HMA in AEL report favourable response rates and survival times [21,45–48]. Larger patient series or randomized clinical trials are lacking. In this international effort, we report on the largest cohort of AEL patients in whom we describe baseline characteristics, overall response rates (ORR), and OS in those treated with HMAs or ICT. In an exploratory analysis, we also compare the treatment outcomes of patients receiving first-line HMA line with those treated with ICT.

2. Results

2.1. Total Acute Erythroleukemia (AEL) Cohort (n = 217)

The overall sample comprised 210 patients with AEL. Of these, 88 (41%) received treatment with HMA in the first or subsequent lines of therapy (82 were treated with azacitidine, six with decitabine) and 122 (56%) received ICT alone. Median age at diagnosis was 69 years (range: 28–88) for the HMA group, and 60 years (range: 20–86) for the ICT group. Poor cytogenetic risk was found in 51% of the HMA and 43% of the ICT groups. Baseline patient characteristics according to treatment group and line of therapy are detailed in Table 1. In the whole AEL cohort, 135 deaths were documented, 79 (59%) due to disease progression, 21 (15%) due to infection, 12 (9%) due to other causes, and in 23 cases (17%) the cause of death was unknown. The median follow-up of all patients was 7.7 (range, 0.2–148.5) months. One patient from the ICT group and four patients from the HMA group were not evaluable for PFS or OS (data regarding time to treatment start and/or death were missing) and, thus, were excluded from the survival analysis. For the total treated cohort (first-line HMA, second-line or later HMA, ICT), the median PFS was 7.1 (range: 6.3–9.4) months, the median OS was 11.1 (range: 9.8–14.3) months and the one-year survival rate was 49% (range: 42–57%) (Tables S1 and S2).

2.2. AEL Treated with HMA (n = 88)

In the cohort treated with HMAs, 41 patients (47%) received HMA as a front-line treatment, 45 as a second-line or later treatment, and two patients were excluded from the analysis as no data were provided regarding the treatment line of HMA. Prior disease-modifying treatments in patients receiving HMA as a second-line or later therapy included allo-SCT (5/45), ICT (40/45), low-dose cytarabine (5/45), and/or IMiDs (immunomodulatory agents, e.g., Lenalidomide) (4/45); four patients received concomitant growth factors, one patient received growth factors without prior disease-modifying treatment. The median time from initial diagnosis to treatment was 0.72 (range, 0.03–18.43) months in patients treated with first-line HMA, and 7.6 (range, 0.07–85.27) months in the group receiving HMA as a second-line or later treatment (*n* = 45). In patients treated with HMA, the median number of cycles in patients for whom data were available (*n* = 72) was five (range, 1–37); those treated with first-line HMA received a median of seven cycles (range, 1–37), and those treated in the second-line received a median of three cycles (range, 1–22). Those treated with azacitidine (*n* = 82) were treated with 28 day cycles: 35% received the schedule 5-2-2 (75 mg/m^2 days 1–5, rest days 6–7, administer days 8–9), 32% received the schedule 1–7 (75 mg/m^2 days 1–7), 26% received the schedule 1–5 (100 mg/m^2 days 1–5), and 7% received other schedules. Those treated with decitabine (*n* = 6) received 15 mg/m^2 for three days every six weeks. At the time of data assessment, 66 patients (76%) had died, of which seven died of subsequent allo-SCT complications. Twenty-two patients (24%) were alive; of these, nine had stopped treatment with HMA, nine were still on treatment with HMA (eight with azacitidine and one with decitabine), and four patients were alive at follow-up, but it was unknown whether they were still receiving HMA or not. The main reason for treatment discontinuation was disease progression (*n* = 39, 62%). Other reasons included infection/toxicity (*n* = 8, 12%), death (*n* = 8, 12%), allo-SCT (*n* = 5, 8%), and others (*n* = 4, 6%). Causes for death were similarly distributed between HMA and ICT treatment groups (Table S3).

Table 1. Baseline clinical and demographic characteristics according to treatment group and line.

Parameter	HMA All Lines (n = 88)	First-Line HMA (n = 41)	First-Line ICT (n = 122)	p-Value First-Line HMA vs. ICT
Median age at diagnosis, years	69	73	60	0.1698
(min–max)	(28–88)	(44–88)	(20–86)	
Male gender, n (%)	54 (61)	26 (63)	88 (72)	0.3919
BM blasts at start of treatment				
Median	22	22	24	
Mean (Standard Deviation)	25.8 (17.2)	25.8 (15.9)	27.1 (15.8)	0.8576
Unknown, n (%)	12 (14)	2 (5)	6 (5)	
Hemoglobin at start of treatment, n (%)				
$\leq$10 g/dL	55 (63)	32 (78)	32/64 (50)	1.00
Pts. with unknown hemoglobin	3 (3)	0 (0)	71 (58)	
Median WBC count at start of treatment, $\times 10^9$/L	2.35	2.42	1.81	0.7294
(min–max)	(0.1–32.3)	(0.6–24.0)	(0.2–23.9)	
Neutrophil count at start of treatment, n (%)				
$\leq 0.5 \times 10^9$/L	34 (39)	18 (44)	18/57 (31)	0.7326
Pts. with unknown neutrophil count	5 (6)	1 (2)	79 (65)	
Platelet count at start of treatment, n (%)				
$\leq 50 \times 10^9$/L	54 (61)	24 (69)	62 (51)	0.8673
Unknown	3 (3)	0 (0)	10 (8)	
AML subtype, n (%)				
Primary	66 (75)	35 (85)	81 (66)	0.4373
Secondary	11 (13)	4 (10)	17 (14)	
Unknown	11 (13)	2 (5)	24 (20)	
MRC cytogenetic risk group, n (%)				
Good risk	1 (1)	0 (0)	0 (0)	0.6943
Intermediate risk	39 (44)	17 (42)	51 (42)	
Poor risk	45 (51)	22 (54)	53 (43)	
Unknown	3 (3)	2 (5)	18 (15)	

MRC = Medical research council.

Table 2. Responses of AEL patients treated with HMA or ICT.

	HMA All Lines (n = 75)[1]	HMA 1st Line (n = 39)[2]	HMA ≥ 2nd Line (n = 34)[3]	ICT 1st Line (n = 119)[4]
Overall response acc. to ELN, *n* (%)	30 (40.0)	18 (46.2)	10 (29.4)	86 (72.3)
Complete	20 (26.7)	12 (30.8)	7 (20.6)	79 (66.4)
Partial	10 (13.3)	6 (15.4)	3 (8.8)	7 (5.9)
Overall response including HI, *n* (%)	44 (58.7)	25 (64.1)	17 (50.0)	ND
HI without marrow response	14 (18.7)	7 (17.9)	7 (20.6)	ND
ANC	9 (12.0)	6 (15.4)	3 (8.8)	
RBC	7 (9.3)	5 (12.8)	2 (5.9)	
PLT	9 (12.0)	5 (12.8)	4 (11.8)	
Transfusion independence, *n/n* (%)[5]				
RBC-TI	19/55 (35)	13/32 (40.6)	6/21 (28.6)	ND
PLT-TI	8/28 (29)	3/14 (21.4)	4/12 (33.3)	
Stable disease	26 (34.7)	11 (28.2)	15 (44.1)	16 (13.4)
Primary disease progression	5 (6.7)	3 (7.7)	2 (5.9)	17 (14.3)
Time to first response, days[6]				ND
Median (min–max)	79 (18–822)[7]	66 (18–233)	85 (30–822)	
Time to best response, days[8]				ND
Median (min–max)	120 (20–1150)[7]	143 (20–353)	89.5 (30–1150)	

[1] Data available for 75 patients; [2] Data available for 39 patients; [3] Data available for 34 patients; [4] Data on HI was not assessed in this subgroup of patients, as this response form is considered irrelevant for AML-patients treated with ICT; [5] Evaluated in the subset of patients who were transfusion dependent at the start of HMA therapy; [6] Data available for 51 patients; [7] The longest time (822 days to fist response and 1150 days to best response) is a single patient. Other late responders are all ~200 days (6.6 months); and [8] Data available for 52 patients; and ND: not detected.

Response data for patients treated with HMA were available for 75 patients. Among these, best overall response rate (ORR) according to the ELN criteria (complete, CR, or partial, PR) of patients treated with HMA was 40%; when including hematological improvement (HI), ORR rose to 59%; 27% had CR, 13% had PR, and 19% had HI; 35% of patients who were initially dependent on red blood cell transfusion achieved transfusion independence, and 29% of patients who were initially platelet transfusion dependent achieved transfusion independence (Table 2). Of those with an abnormal karyotype at the start of treatment, 11 (21%) of 51 HMA patients reached cytogenetic remission and 40 (53%) of 75 ICT-treated patients achieved cytogenetic remission. The median time to first response was 2.6 months (range, 0.6–27.4) and the median time to best response was 3.9 months (range, 0.66–38.3), respectively.

After a median follow-up of 12.3 (range, 0.03–35.2) and 4.8 (range, 0.0–68.8) months for patients treated with first-line HMA and second line or later HMA treatment, respectively, the median (range) PFS was longer for those treated with HMA in first-line treatment compared to second-line or later (9.4 (range 4.2–14.5) vs. 3.4 (2.0–6.3) months, respectively; Table 1). The median OS (range) was also longer for those treated with first-line HMA compared to second-line or later (13.7 (12.3–20.5) vs. 9.8 (4.6–13.5), respectively; Table S2).

The median OS for AEL patients treated with HMA (all treatment lines) was superior for patients with intermediate- compared to high-risk cytogenetics (13.5 vs. 12.3 months; $p = 0.0376$) (Figure 1A). AEL patients treated with first-line HMA with intermediate-risk cytogenetics had a median OS of 29.3, whereas those with high-risk cytogenetics had a median OS of 13.3 months (Table 3).

Ten (11.3%) patients had an allogeneic bone marrow transplant following treatment with HMA. The median OS in this subgroup was 9.66 months (range, 2.8–25).

In univariate analysis, response to HMA had a significant impact on OS (Figure 1B). The median survival in patients with CR was 18.2 months, 12.7 months in patients with PR or HI, and 4.5 months in patients with no response (stable disease, SD, or primary progressive disease, PD; $p < 0.001$).

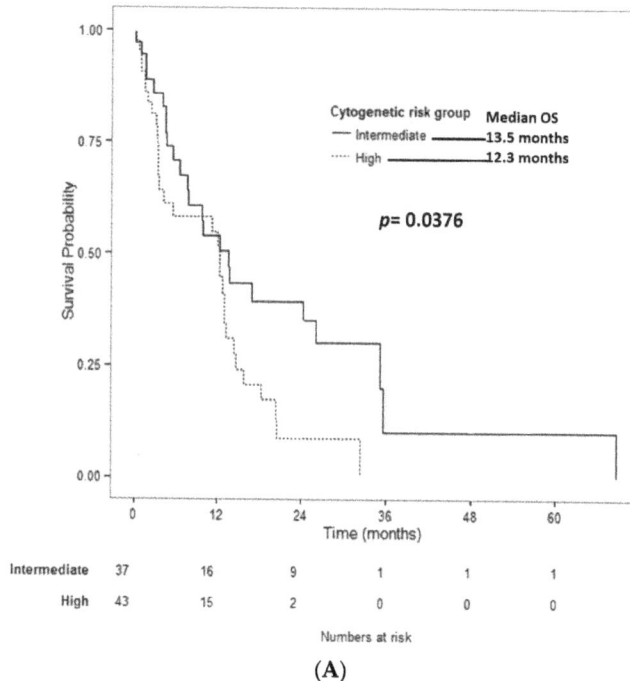

(A)

Figure 1. *Cont.*

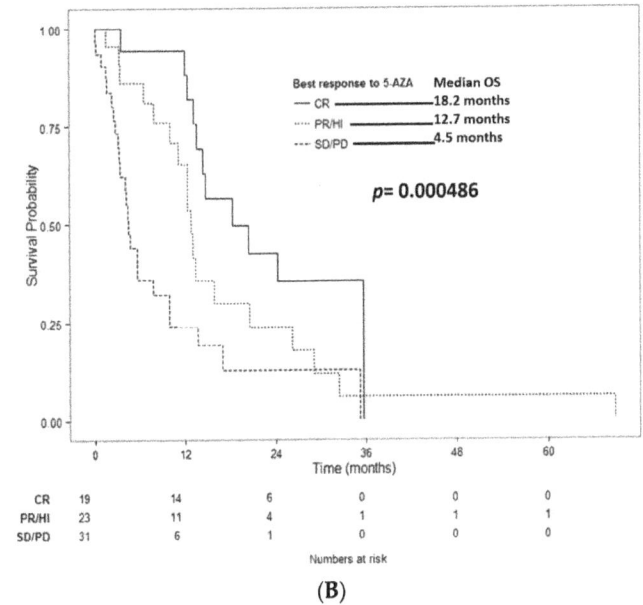

Figure 1. (A) Overall survival of HMA-treated patients stratified by cytogenetic risk group (total HMA cohort): the median OS for patients treated with HMA was superior for patients with intermediate-compared to high-risk cytogenetics (13.5 months vs. 12.3 months; $p = 0.0376$); and (B) the overall survival by response to HMA: the median survival in patients with CR was 18.2 months, 12.7 months in patients with PR or HI, and 4.5 months in patients with no response (SD or primary PD; $p < 0.001$).

Table 3. Comparison of AEL patient characteristics and outcomes according to front-line treatment with ICT or HMA in univariate analysis.

Outcomes	First-Line ICT	First-Line HMA	*p*-Value
Overall response acc. to ELN, %	72.3	46.2	0.016
Complete response	64.4	30.8	<0.001
Partial response	5.9	15.4	0.101
Stable disease, %	13.4	28.2	0.001
Primary disease progression, %	14.3	7.7	0.004
Median time to best response, months	NA [1]	89.5	NA [1]
Median PFS, months	8.0	9.4	0.107
MRC intermediate cytogenetic risk	22.7	5.9	0.004
MRC high cytogenetic risk	6.5	11.3	0.279
1-year PFS, %	41.8	40.6	0.896
Median OS total cohort, months	10.5	13.7	0.564
MRC intermediate cytogenetic risk	16.9	29.3	0.277
MRC high cytogenetic risk	7.5	13.3	0.039
1-year OS total cohort, %	46.7	65.8	0.072

[1] NA = not available.

2.3. AEL Treated with ICT (n = 122)

In the group of 122 patients receiving front-line ICT treatment, response data were available for 119 patients. The most frequently used (*n* = 81; 66%) induction regimen was Daunorubicin

(45 or 60 mg/m^2 × 3 days) with Cytarabine (100 mg/m^2 bid × 7 days). Similar 3 + 7 regimens using Idarubicin 12 mg/m^2 or Mitoxantrone 12 mg/m^2 for three days, instead of Daunorubicin, were used in 25 (20%) and eight (7%) patients, respectively. Information regarding induction regimen was not available in eight (7%) patients.

ORR according to the ELN criteria was 72%; CR in 79 patients was 66%; PR in seven patients was 6%; SD in 16 patients was 13%; PPD in 17 patients was 14% (see Table 2). Data on HI was not assessed in this subgroup of patients, as this response form is considered irrelevant for AML-patients treated with ICT.

At the time of data assessment, 84 patients (69%) had died, and 37 (31%) were alive. The main cause of death was disease progression (65%) (Table S3).

Median follow-up was 7.8 (range, 0.03–148.5) months for patients treated with ICT. Median PFS was 8.0 months (range, 6.8–14.5) for AEL-patients treated with ICT (Table S1). Median OS for patients treated with ICT was 10.5 (range, 9.1–20.0) months (Table S2). Median OS for AEL-patients treated with ICT was not significantly superior for patients with intermediate- vs. high-risk cytogenetics (16.9 vs. 7.5 months; p = 0.277) (Table 3). For AEL-patients treated with ICT, the median OS of intermediate- vs. high-risk cytogenetics was 29.3 vs. 13.3 months, p = 0.0.039 (Table 3). In univariate analysis, the response to ICT had a significant impact on overall survival. The median OS in patients with CR was 23.17 months, as compared to 4.07 months in patients with PR, and 5.63 months in patients with no response (SD or primary PD; p < 0.001).

Twenty-three (18.8%) patients had an allogeneic bone marrow transplant following treatment with ICT. Median OS in this subgroup was 5.9 months (range, 2.0–17.9).

2.4. Comparison of AEL Treated with ICT vs. HMA

There were no significant differences in baseline characteristics (Table 1) or causes of death in the HMA vs. ICT group (Table 3).

AEL-patients treated with ICT had a higher rate of CR (66% vs. 30.8%; p < 0.001), and ORR according to the ELN criteria (CR + PR) (72% vs. 46.2%, p = 0.016) compared to patients treated with first-line HMA, respectively. Notably, there were significantly more progressions in the ICT group compared to the HMA group (14.3% vs. 7.7%, p = 0.004) and more disease stability in the HMA group (28.2% vs. 13.4%, p = 0.001).

Despite this higher response rate, there was no significant difference in median PFS (8.0 vs. 9.4 months; p = 0.342) or 1-year PFS rates (42% vs. 41%; p = 0.896) (Table S2). In multivariate analyses controlling for cytogenetic risk group and age, treatment with ICT was not superior to treatment with first-line HMA in prolonging PFS (p = 0.6907) (Table 4).

Table 4. PFS and OS comparison for first-line treatment with HMA vs. ICT, controlling for cytogenetic risk group and age.

PFS Comparison	Hazard Ratio	95% CI	*p*-Value
First line AZA vs. ICT	0.90	0.54–1.51	0.6907
Cytogenetic risk group: High vs. Intermediate	1.86	1.19–2.90	0.0064
Age Per additional year	1.03	1.01–1.05	0.0118
OS Comparison	Hazard Ratio	95% CI	*p*-Value
First line AZA vs. ICT	0.75	0.45–1.23	0.2489
Cytogenetic risk group High vs. Intermediate	2.40	1.54–3.69	<0.0001
Age Per additional year	1.03	1.01–1.05	0.0032

A likelihood ratio test was used to compare models with and without interaction between first line treatment and cytogenetic risk group: *p*-value = 0.0994.

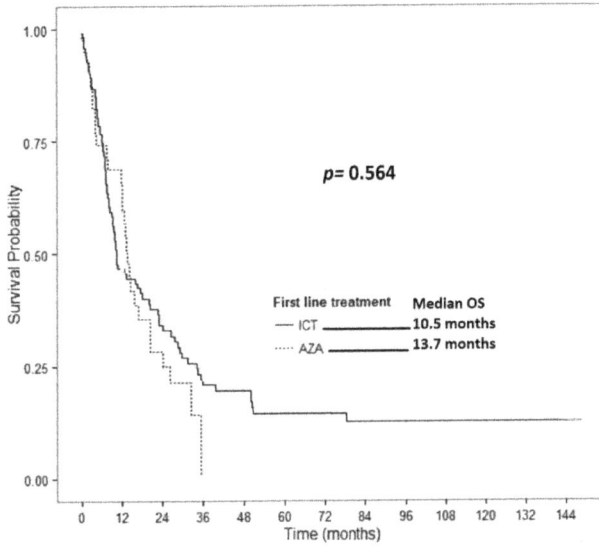

(A)

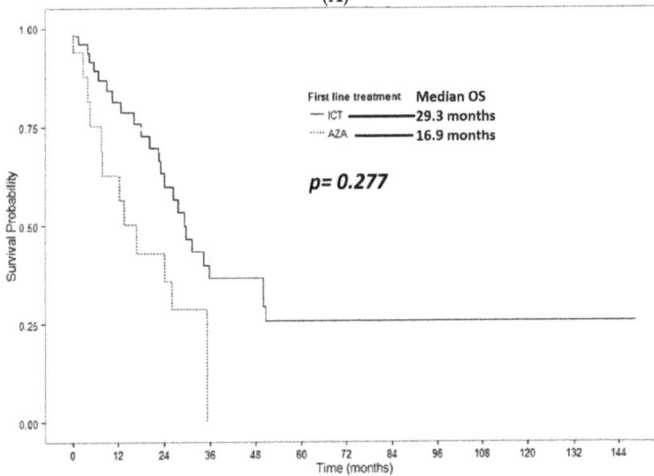

(B)

Figure 2. *Cont.*

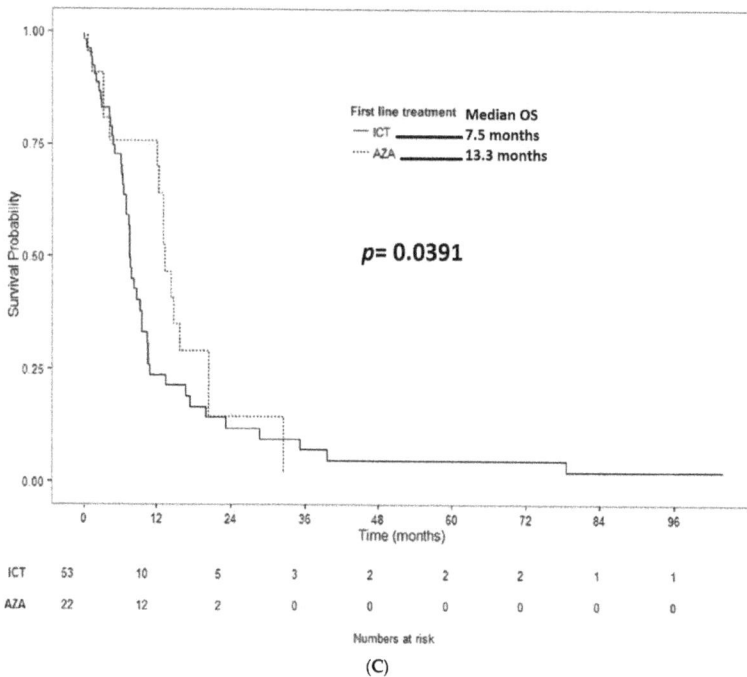

Figure 2. Overall survival of AEL patients stratified by type of first line treatment. (**A**) Total cohorts: median OS for patients treated with first-line HMA was similar to that of those treated with first-line ICT (13.7 months vs. 10.5 months; *p* = 0.564); (**B**) stratified by MRC intermediate cytogenetic risk: AEL-patients with intermediate-risk cytogenetics treated with first-line HMA did not have a significantly different median survival as compared to AEL-patients treated with first-line ICT (16.9 months vs. 29.3 months; *p* = 0.277); and (**C**) stratified by MRC high cytogenetic risk: AEL-patients with high-risk cytogenetics treated with first-line HMA had a significantly longer median survival as compared to AEL-patients treated with first-line ICT (13.3 months vs. 7.5 months; *p* = 0.0391).

Comparing AEL-patients treated with ICT vs. first-line HMA, no significant differences in 1-year survival rates (47% vs. 66%; *p* = 0.072) or median OS times could be detected (10.5 vs. 13.7 months; *p* = 0.564), respectively, though absolute numbers favored HMAs (Figure 2A). When stratified by the cytogenetic risk group, there was no significant difference in the median survival of AEL-patients with intermediate cytogenetic risk treated with ICT vs. first-line HMA (16.9 vs. 29.3 months; *p* = 0.277; Figure 2B). However, a shorter survival was detected for AEL-patients with high risk cytogenetics treated with ICT, as compared to those treated with first-line HMA (7.5 vs. 13.3 months; *p* = 0.039; Figure 2C). In multivariate analysis, controlling for age and cytogenetic risk, treatment with ICT was not superior to treatment with first-line HMA in prolonging OS (*p* = 0.2489), whereas both the MRC cytogenetic risk group (*p* < 0.0001) and age per additional year (*p* = 0.0032) did (Table 4).

3. Discussion

No prospective clinical trial has been conducted exclusively in patients with AEL. Little is known about the responses to specific drugs in AEL. Case reports and small series indicate possible efficacy of azacitidine [21,45,49], interferon-α [50], and even high dose erythropoietin combined with granulocyte colony-stimulating factor [51].

It was demonstrated several decades ago that HMA can induce erythroid differentiation and increase the synthesis of hemoglobin in both murine and human erythroleukemia cell lines in vitro [52–54]. In addition, the HMA decitabine was shown to induce down-regulation of the multidrug resistance (MDR) gene phospho-glycoprotein in a human erythroleukemia cell line, which coincided with modulation of response to cytostatic drugs [55,56].

We report here the largest series to date of patients with AEL treated with HMA. The overall response rate of 46% in the front-line setting, with a CR rate of 30% and an additional HI rate of 18%, in our cohort are encouraging and similar to those reported in other smaller studies [21,46,49]. Our study reinforces that, when treating AEL with HMA, any type of response, including hematological improvement, is beneficial. The observation that, despite a significantly lower ORR rate than ICT, PFS similarly suggests that the significantly higher SD rate also has an impact on survival. This highlights the importance of maintaining treatment in all patients who do not progress, even in the absence of marrow responses.

It is also noteworthy that initial responses were seen after a median of 79 days, but the best responses were documented after a median of 120 days, confirming that responses improve with continued treatment and reinforcing the importance of not interrupting treatment too early due to a lack of response.

When compared to AEL patients who were treated with ICT alone, those treated with HMA in as a first-line had similar progression-free and overall survival. This is significant considering the more advanced age of the HMA group. Older patients tolerate intensive chemotherapy poorly. In addition, aggressive treatment options are associated with long hospital admissions and poor quality of life, which may not be justified in an elderly patient group with a disease that is unlikely to be cured. HMA are administered in an outpatient setting and associated with reduced hospital admissions. Given the lack of a curative option for most patients and similar survival, the toxicity profile of HMA makes this option more attractive [57]. In addition, our data shows that adverse karyotype patients have better outcomes when treated with HMA compared to ICT. This suggests that HMA may be the preferred treatment option for older individuals with a poor prognosis karyotype, as is often seen in AEL.

Nevertheless, it is important to note that HMAs do not preclude the option of a bone marrow transplant. The therapeutic goal in younger patients with a donor should be to cure the disease and allo-SCT is the only option. Reduced intensity conditioning regimens have opened the option of allo-SCT to more elderly and frail patients but the toxicities associated with conventional intensive AML induction chemotherapy can increase the risk of death or compromise allo-SCT. It has been shown that Azacitidine before SCT does not significantly affect rates of remission, relapse, acute and chronic GVHD, and survival after transplant, and may actually be an alternative for inducing remission in patients with higher risk MDS [58], and eventually AEL [10]. Despite only 10 patients treated with HMA in our cohort having a subsequent SCT, their median survival is encouraging.

Our series analyses patients with AEL but we now know that there is great genetic heterogeneity in myeloid disorders, with a large variety of mutations having been described which have differing impacts on the natural history of the disease. Very recent analyses of the mutational profiles have significantly increased our understanding and prognostication of acute leukemias [59,60]. Future studies in this regard are needed in order to identify those patients with AEL who are most likely to respond to HMA.

4. Methods

4.1. Patient Population

Patient data were collected retrospectively and pooled from registries or patient files from 28 different Institutions representing eight different countries (Austria, France, Germany, Italy, Netherlands, Portugal, Spain, and USA). All subjects gave their informed consent for inclusion before they participated in the study. The study was conducted in accordance with the Declaration of Helsinki, and the protocol was approved by the Medical Ethics Committee of each individual centre.

AEL diagnosis by WHO 2008 criteria was the only entry criterion and this was confirmed by local diagnostic laboratories. MRC cytogenetic risk stratification was applied to all patients.

Patients were included in the HMA group if they had received HMA at any stage of their treatment, whether first, or subsequent, lines. Patients in the ICT group must have been treated with ICT in first-line and never received HMA. Patients diagnosed between March 1998 and November 2014 were included. Treatment choice was made by the treating physician according to personal practice and local protocols. Seven AEL patients treated only with supportive care were proposed for the study, but their outcomes were not included in the analysis.

4.2. Definition of Endpoints

Response was defined according to European Leukemia Network (ELN) criteria for AML, and included complete remission (CR) and partial remission (PR) [61]. In addition, hematologic improvement (HI) was assessed according to the modified International Working Group (IWG) criteria 2006 [62]. OS was defined as time from start of treatment with HMA (either first- or second-line or later) or ICT to death from any cause, or last follow-up. Patients who underwent allogeneic stem cell transplantation (allo-SCT) after treatment with HMA or after ICT were censored at the date of allo-SCT. Progression-free survival (PFS) was defined as the time from the start of treatment until disease relapse/progression, or death from any cause.

4.3. Statistics

Descriptive statistics were used to describe the baseline patient characteristics. OS and PFS were estimated using the Kaplan-Meier method and the log-rank test was used for subgroup comparisons. The comparison of baseline features between the subgroups HMA and ICT was performed using the Pearson's χ-squared test for categorical baseline variables and the Wilcoxon rank sum test for quantitative variables. In the exploratory analysis to evaluate the impact of treatment (HMA first-line vs. ICT) on OS and PFS we used the Cox proportional hazards regression model. Adjusted hazard ratios were calculated controlling for the potential confounding factors age and cytogenetic risk group. The likelihood ratio test was used to test the interaction between treatment and cytogenetic risk groups. All tests were two-tailed and p-values less than 0.05 were considered to be statistically significant. No adjustment was made for multiple comparisons. All analyses were performed using R [63].

5. Conclusions

Our data reinforces the utility of HMA in patients with acute erythroleukamia, especially those with poorest prognosis. Future studies in this regard are needed in order to identify those patients with AEL who are most likely to respond to HMA.

Supplementary Materials: Supplementary materials can be found at www.mdpi.com/1422-0067/18/4/837/s1.

Acknowledgments: Publication costs were supported by the University of Salzburg.

Author Contributions: Antonio M. Almeida and Lisa Pleyer conceived and designed the database; Susana Esteves performed the statistical analysis; Antonio M. Almeida wrote the paper; all authors contributed with patient data and critical analysis of the manuscript.

Conflicts of Interest: Antonio M. Almeida: speaker and advisory board for Celgene; Arjan A. Van De Loosdrecht: speaker and advisory board Celgene, advisory board Novartis; Jamile Shammo: Received research funding and honoraria for speaking engagements and consultancy from Celgene; Peter Valent received a research grant and speaker's honoraria from Celgene and served as an advisory board member for Celgene. Fernando Ramos: Honoraria/Consultation fees for Celgene, Janssen, Amgen, Novatis, Pfizer, Glaxo-Smith-Kline, Merck-Sharp & Dohme. Maria Diez Campelo: speaker, research founding and advisory boards for Celgene.

References

1. Santos, F.P.; Bueso-Ramos, C.E.; Ravandi, F. Acute erythroleukemia: Diagnosis and management. *Expert Rev. Hematol.* **2010**, *3*, 705–718. [CrossRef] [PubMed]

2. Liu, W.; Hasserjian, R.P.; Hu, Y.; Zhang, L.; Miranda, R.N.; Medeiros, L.J.; Wang, S.A. Pure erythroid leukemia: A reassessment of the entity using the 2008 World Health Organization classification. *Mod. Pathol.* **2011**, *24*, 375–383. [CrossRef] [PubMed]

3. Zuo, Z.; Polski, J.M.; Kasyan, A.; Medeiros, L.J. Acute erythroid leukemia. *Arch. Pathol. Lab. Med.* **2010**, *134*, 1261–1270. [PubMed]

4. Kasyan, A.; Medeiros, L.J.; Zuo, Z.; Santos, F.P.; Ravandi-Kashani, F.; Miranda, R.; Vadhan-Raj, S.; Koeppen, H.; Bueso-Ramos, C.E. Acute erythroid leukemia as defined in the World Health Organization classification is a rare and pathogenetically heterogeneous disease. *Mod. Pathol.* **2010**, *23*, 1113–1126. [CrossRef] [PubMed]

5. Selby, D.M.; Valdez, R.; Schnitzer, B.; Ross, C.W.; Finn, W.G. Diagnostic criteria for acute erythroleukemia. *Blood* **2003**, *101*, 2895–2896. [CrossRef] [PubMed]

6. Arber, D.A.; Orazi, A.; Hasserjian, R.; Thiele, J.; Borowitz, M.J.; Le Beau, M.M.; Bloomfield, C.D.; Cazzola, M.; Vardiman, J.W. The 2016 revision to the World Health Organization classification of myeloid neoplasms and acute leukemia. *Blood* **2016**, *127*, 2391–2405. [CrossRef] [PubMed]

7. Arenillas, L.; Calvo, X.; Luno, E.; Senent, L.; Alonso, E.; Ramos, F.; Ardanaz, M.T.; Pedro, C.; Tormo, M.; Marco, V.; et al. Considering Bone Marrow Blasts From Nonerythroid Cellularity Improves the Prognostic Evaluation of Myelodysplastic Syndromes. *J. Clin. Oncol.* **2016**, *34*, 3284–3292. [CrossRef] [PubMed]

8. Calvo, X.; Arenillas, L.; Luno, E.; Senent, L.; Arnan, M.; Ramos, F.; Ardanaz, M.T.; Pedro, C.; Tormo, M.; Montoro, J.; et al. Erythroleukemia shares biological features and outcome with myelodysplastic syndromes with excess blasts: A rationale for its inclusion into future classifications of myelodysplastic syndromes. *Mod. Pathol.* **2016**, *29*, 1541–1551. [CrossRef] [PubMed]

9. Jogai, S.; Varma, N.; Garewal, G.; Das, R.; Varma, S. Acute erythroleukemia (AML-M6)—A study of clinicohematological, morphological and dysplastic features in 10 cases. *Indian J. Cancer* **2001**, *38*, 143–148. [PubMed]

10. Olopade, O.I.; Thangavelu, M.; Larson, R.A.; Mick, R.; Kowal-Vern, A.; Schumacher, H.R.; Le Beau, M.M.; Vardiman, J.W.; Rowley, J.D. Clinical, morphologic, and cytogenetic characteristics of 26 patients with acute erythroblastic leukemia. *Blood* **1992**, *80*, 2873–2882. [PubMed]

11. Domingo-Claros, A.; Larriba, I.; Rozman, M.; Irriguible, D.; Vallespi, T.; Aventin, A.; Ayats, R.; Milla, F.; Sole, F.; Florensa, L.; et al. Acute erythroid neoplastic proliferations. A biological study based on 62 patients. *Haematologica* **2002**, *87*, 148–153. [PubMed]

12. Lessard, M.; Struski, S.; Leymarie, V.; Flandrin, G.; Lafage-Pochitaloff, M.; Mozziconacci, M.J.; Talmant, P.; Bastard, C.; Charrin, C.; Baranger, L.; et al. Cytogenetic study of 75 erythroleukemias. *Cancer Genet. Cytogenet.* **2005**, *163*, 113–122. [CrossRef] [PubMed]

13. Mazzella, F.M.; Kowal-Vern, A.; Shrit, M.A.; Rector, J.T.; Cotelingam, J.D.; Schumacher, H.R. Effects of multidrug resistance gene expression in acute erythroleukemia. *Mod. Pathol.* **2000**, *13*, 407–413. [CrossRef] [PubMed]

14. Grossmann, V.; Bacher, U.; Haferlach, C.; Schnittger, S.; Potzinger, F.; Weissmann, S.; Roller, A.; Eder, C.; Fasan, A.; Zenger, M.; et al. Acute erythroid leukemia (AEL) can be separated into distinct prognostic subsets based on cytogenetic and molecular genetic characteristics. *Leukemia* **2013**, *27*, 1940–1943. [CrossRef] [PubMed]

15. Atkinson, J.; Hrisinko, M.A.; Weil, S.C. Erythroleukemia: A review of 15 cases meeting 1985 FAB criteria and survey of the literature. *Blood Rev.* **1992**, *6*, 204–214. [CrossRef]

16. Micci, F.; Thorsen, J.; Panagopoulos, I.; Nyquist, K.B.; Zeller, B.; Tierens, A.; Heim, S. High-throughput sequencing identifies an NFIA/CBFA2T3 fusion gene in acute erythroid leukemia with t(1;16)(p31;q24). *Leukemia* **2013**, *27*, 980–982. [CrossRef] [PubMed]

17. Srinivas, U.; Kumar, R.; Pati, H.P.; Saxena, R.; Tyagi, S. Sub classification and clinico-hematological correlation of 40 cases of acute erythroleukemia—Can proerythroblast/myeloblast and proerythroblast/total erythroid cell ratios help subclassify? *Hematology* **2007**, *12*, 381–385. [CrossRef] [PubMed]

18. Mazzella, F.M.; Kowal-Vern, A.; Shrit, M.A.; Wibowo, A.L.; Rector, J.T.; Cotelingam, J.D.; Collier, J.; Mikhael, A.; Cualing, H.; Schumacher, H.R. Acute erythroleukemia: Evaluation of 48 cases with reference to classification, cell proliferation, cytogenetics, and prognosis. *Am. J. Clin. Pathol.* **1998**, *110*, 590–598. [CrossRef] [PubMed]

19. Kowal-Vern, A.; Mazzella, F.M.; Cotelingam, J.D.; Shrit, M.A.; Rector, J.T.; Schumacher, H.R. Diagnosis and characterization of acute erythroleukemia subsets by determining the percentages of myeloblasts and proerythroblasts in 69 cases. *Am. J. Hematol.* **2000**, *65*, 5–13. [CrossRef]

20. Santos, F.P.; Faderl, S.; Garcia-Manero, G.; Koller, C.; Beran, M.; O'Brien, S.; Pierce, S.; Freireich, E.J.; Huang, X.; Borthakur, G.; et al. Adult acute erythroleukemia: An analysis of 91 patients treated at a single institution. *Leukemia* **2009**, *23*, 2275–2280. [CrossRef] [PubMed]

21. Pierdomenico, F.; Almeida, A. Treatment of acute erythroleukemia with Azacitidine: A case series. *Leuk. Res. Rep.* **2013**, *2*, 41–43. [CrossRef] [PubMed]

22. Colita, A.; Belhabri, A.; Chelghoum, Y.; Charrin, C.; Fiere, D.; Thomas, X. Prognostic factors and treatment effects on survival in acute myeloid leukemia of M6 subtype: A retrospective study of 54 cases. *Ann. Oncol.* **2001**, *12*, 451–455. [CrossRef] [PubMed]

23. Fenaux, P.; Mufti, G.J.; Hellstrom-Lindberg, E.; Santini, V.; Finelli, C.; Giagounidis, A.; Schoch, R.; Gattermann, N.; Sanz, G.; List, A.; et al. Efficacy of azacitidine compared with that of conventional care regimens in the treatment of higher-risk myelodysplastic syndromes: A randomised, open-label, phase III study. *Lancet Oncol.* **2009**, *10*, 223–232. [CrossRef]

24. Garcia-Manero, G.; Jabbour, E.; Borthakur, G.; Faderl, S.; Estrov, Z.; Yang, H.; Maddipoti, S.; Godley, L.A.; Gabrail, N.; Berdeja, J.G.; et al. Randomized open-label phase II study of decitabine in patients with low- or intermediate-risk myelodysplastic syndromes. *J. Clin. Oncol.* **2013**, *31*, 2548–2553. [CrossRef] [PubMed]

25. Pleyer, L.; Germing, U.; Sperr, W.R.; Linkesch, W.; Burgstaller, S.; Stauder, R.; Girschikofsky, M.; Schreder, M.; Pfeilstocker, M.; Lang, A.; et al. Azacitidine in CMML: Matched-pair analyses of daily-life patients reveal modest effects on clinical course and survival. *Leuk. Res.* **2014**, *38*, 475–483. [CrossRef] [PubMed]

26. Thorpe, M.; Montalvao, A.; Pierdomenico, F.; Moita, F.; Almeida, A. Treatment of chronic myelomonocytic leukemia with 5-Azacitidine: A case series and literature review. *Leuk. Res.* **2012**, *36*, 1071–1073. [CrossRef] [PubMed]

27. Fianchi, L.; Criscuolo, M.; Breccia, M.; Maurillo, L.; Salvi, F.; Musto, P.; Mansueto, G.; Gaidano, G.; Finelli, C.; Aloe-Spiriti, A.; et al. High rate of remissions in chronic myelomonocytic leukemia treated with 5-azacytidine: Results of an Italian retrospective study. *Leuk. Lymphoma* **2013**, *54*, 658–661. [CrossRef] [PubMed]

28. Fenaux, P.; Mufti, G.J.; Hellstrom-Lindberg, E.; Santini, V.; Gattermann, N.; Germing, U.; Sanz, G.; List, A.F.; Gore, S.; Seymour, J.F.; et al. Azacitidine prolongs overall survival compared with conventional care regimens in elderly patients with low bone marrow blast count acute myeloid leukemia. *J. Clin. Oncol.* **2010**, *28*, 562–569. [CrossRef] [PubMed]

29. Dombret, H.; Seymour, J.F.; Butrym, A.; Wierzbowska, A.; Selleslag, D.; Jang, J.H.; Kumar, R.; Cavenagh, J.; Schuh, A.C.; Candoni, A.; et al. International phase 3 study of azacitidine vs. conventional care regimens in older patients with newly diagnosed AML with >30% blasts. *Blood* **2015**, *126*, 291–299. [CrossRef] [PubMed]

30. Kantarjian, H.M.; Thomas, X.G.; Dmoszynska, A.; Wierzbowska, A.; Mazur, G.; Mayer, J.; Gau, J.P.; Chou, W.C.; Buckstein, R.; Cermak, J.; et al. Multicenter, randomized, open-label, phase III trial of decitabine versus patient choice, with physician advice, of either supportive care or low-dose cytarabine for the treatment of older patients with newly diagnosed acute myeloid leukemia. *J. Clin. Oncol.* **2012**, *30*, 2670–2677. [CrossRef] [PubMed]

31. Thepot, S.; Itzykson, R.; Seegers, V.; Recher, C.; Raffoux, E.; Quesnel, B.; Delaunay, J.; Cluzeau, T.; Marfaing Koka, A.; Stamatoullas, A.; et al. Azacitidine in untreated acute myeloid leukemia: A report on 149 patients. *Am. J. Hematol.* **2014**, *89*, 410–416. [CrossRef] [PubMed]

32. Pleyer, L.; Burgstaller, S.; Girschikofsky, M.; Linkesch, W.; Stauder, R.; Pfeilstocker, M.; Schreder, M.; Tinchon, C.; Sliwa, T.; Lang, A.; et al. Azacitidine in 302 patients with WHO-defined acute myeloid leukemia: Results from the Austrian Azacitidine Registry of the AGMT-Study Group. *Ann. Hematol.* **2014**, *93*, 1825–1838. [CrossRef] [PubMed]

33. Falantes, J.; Thepot, S.; Pleyer, L.; Maurillo, L.; Martínez-Robles, V.; Itzykson, R.; Bargay, J.; Stauder, R.; Venditti, A.; Martínez, M.P.; Seegers, V.; et al. Azacitidine in older patients with acute myeloid leukemia (AML). Results from the expanded international E-ALMA series (E-ALMA+) according to the MRC risk index score. *Blood* **2015**, *126*, 2554.

34. Pleyer, L.; Stauder, R.; Burgstaller, S.; Schreder, M.; Tinchon, C.; Pfeilstocker, M.; Steinkirchner, S.; Melchardt, T.; Mitrovic, M.; Girschikofsky, M.; et al. Azacitidine in patients with WHO-defined AML—Results of 155 patients from the Austrian Azacitidine Registry of the AGMT-Study Group. *J. Hematol. Oncol.* **2013**, *6*, 32. [CrossRef] [PubMed]

35. Maurillo, L.; Venditti, A.; Spagnoli, A.; Gaidano, G.; Ferrero, D.; Oliva, E.; Lunghi, M.; D'Arco, A.M.; Levis, A.; Pastore, D.; et al. Azacitidine for the treatment of patients with acute myeloid leukemia: Report of 82 patients enrolled in an Italian Compassionate Program. *Cancer* **2012**, *118*, 1014–1022. [CrossRef] [PubMed]

36. Pleyer, L.; Burgstaller, S.; Stauder, R.; Girschikofsky, M.; Sill, H.; Schlick, K.; Thaler, J.; Halter, B.; Machherndl-Spandl, S.; Zebisch, A.; et al. Azacitidine front-line in 339 patients with myelodysplastic syndromes and acute myeloid leukaemia: Comparison of French-American-British and World Health Organization classifications. *J. Hematol. Oncol.* **2016**, *9*, 39. [CrossRef] [PubMed]

37. Kadia, T.M.; Thomas, X.G.; Dmoszynska, A.; Wierzbowska, A.; Minden, M.; Arthur, C.; Delaunay, J.; Ravandi, F.; Kantarjian, H. Decitabine improves outcomes in older patients with acute myeloid leukemia and higher blast counts. *Am. J. Hematol.* **2015**, *90*, E139–E141. [CrossRef] [PubMed]

38. Mayer, J.; Arthur, C.; Delaunay, J.; Mazur, G.; Thomas, X.G.; Wierzbowska, A.; Ravandi, F.; Berrak, E.; Jones, M.; Li, Y.; et al. Multivariate and subgroup analyses of a randomized, multinational, phase 3 trial of decitabine vs. treatment choice of supportive care or cytarabine in older patients with newly diagnosed acute myeloid leukemia and poor- or intermediate-risk cytogenetics. *BMC Cancer* **2014**, *14*, 69. [CrossRef] [PubMed]

39. Gupta, N.; Miller, A.; Gandhi, S.; Ford, L.A.; Vigil, C.E.; Griffiths, E.A.; Thompson, J.E.; Wetzler, M.; Wang, E.S. Comparison of epigenetic versus standard induction chemotherapy for newly diagnosed acute myeloid leukemia patients ≥60 years old. *Am. J. Hematol.* **2015**, *90*, 639–646. [CrossRef] [PubMed]

40. Lao, Z.; Yiu, R.; Wong, G.C.; Ho, A. Treatment of elderly patients with acute myeloid leukemia with azacitidine results in fewer hospitalization days and infective complications but similar survival compared with intensive chemotherapy. *Asia Pac. J. Clin. Oncol.* **2015**, *11*, 54–61. [CrossRef] [PubMed]

41. Van der Helm, L.H.; Scheepers, E.R.; Veeger, N.J.; Daenen, S.M.; Mulder, A.B.; van den Berg, E.; Vellenga, E.; Huls, G. Azacitidine might be beneficial in a subgroup of older AML patients compared to intensive chemotherapy: A single centre retrospective study of 227 consecutive patients. *J. Hematol. Oncol.* **2013**, *6*, 29. [CrossRef] [PubMed]

42. Jabbour, E.; Mathisen, M.S.; Garcia-Manero, G.; Champlin, R.; Popat, U.; Khouri, I.; Giralt, S.; Kadia, T.; Chen, J.; Pierce, S.; et al. Allogeneic hematopoietic stem cell transplantation versus hypomethylating agents in patients with myelodysplastic syndrome: A retrospective case-control study. *Am. J. Hematol.* **2013**, *88*, 198–200. [CrossRef] [PubMed]

43. Ravandi, F.; Issa, J.P.; Garcia-Manero, G.; O'Brien, S.; Pierce, S.; Shan, J.; Borthakur, G.; Verstovsek, S.; Faderl, S.; Cortes, J.; et al. Superior outcome with hypomethylating therapy in patients with acute myeloid leukemia and high-risk myelodysplastic syndrome and chromosome 5 and 7 abnormalities. *Cancer* **2009**, *115*, 5746–5751. [CrossRef] [PubMed]

44. Almeida, A.; Ferreira, A.R.; Costa, M.J.; Silva, S.; Alnajjar, K.; Bogalho, I.; Pierdomenico, F.; Esteves, S.; Alpoim, M.; Braz, G.; et al. Clinical outcomes of AML patients treated with Azacitidine in Portugal: A retrospective multicenter study. *Leuk. Res. Rep.* **2017**, *7*, 6–10. [CrossRef] [PubMed]

45. Hansen, S.B.; Dufva, I.H.; Kjeldsen, L. Durable complete remission after azacitidine treatment in a patient with erythroleukaemia. *Eur. J. Haematol.* **2012**, *89*, 369–370. [CrossRef] [PubMed]

46. Hangai, S.; Nakamura, F.; Kamikubo, Y.; Honda, A.; Arai, S.; Nakagawa, M.; Ichikawa, M.; Kurokawa, M. Erythroleukemia showing early erythroid and cytogenetic responses to azacitidine therapy. *Ann. Hematol.* **2013**, *92*, 707–709. [CrossRef] [PubMed]

47. Vigil, C.E.; Cortes, J.; Kantarjian, H.; Garcia-Manero, G.; Lancet, J.; List, A. Hypomethylating Therapy for the Treatment of Acute Erythroleukemia Patients. *Blood* **2009**, *114*, 2069.

48. King, R.J.; Crouch, A.; Radojcic, V.; Marini, B.L.; Perissinotti, A.J.; Bixby, D. Therapeutic Outcomes of Patients with Acute Erythroid Leukemia Treated with Hypomethylating Agents. *Blood* **2016**, *128*, 5203.

49. Uchida, T.; Hagihara, M.; Hua, J.; Inoue, M. The effects of azacitidine on the response and prognosis of myelodysplastic syndrome and acute myeloid leukemia involving a bone marrow erythroblast frequency of >50. *Leuk. Res.* **2016**, *53*, 35–38. [CrossRef] [PubMed]

50. Steger, G.G.; Dittrich, C.; Chott, A.; Derfler, K.; Schwarzmeier, J.D. Long-term remission in a patient with erythroleukemia following interferon-α treatment. *J. Biol. Response Modif.* **1989**, *8*, 351–354.

51. Camera, A.; Volpicelli, M.; Villa, M.R.; Risitano, A.M.; Rossi, M.; Rotoli, B. Complete remission induced by high dose erythropoietin and granulocyte colony stimulating factor in acute erythroleukemia (AML-M6 with maturation). *Haematologica* **2002**, *87*, 1225–1227. [PubMed]

52. Creusot, F.; Acs, G.; Christman, J.K. Inhibition of DNA methyltransferase and induction of Friend erythroleukemia cell differentiation by 5-azacytidine and 5-aza-2'-deoxycytidine. *J. Biol. Chem.* **1982**, *257*, 2041–2048. [PubMed]

53. Gambari, R.; del Senno, L.; Barbieri, R.; Viola, L.; Tripodi, M.; Raschella, G.; Fantoni, A. Human leukemia K-562 cells: Induction of erythroid differentiation by 5-azacytidine. *Cell Differ.* **1984**, *14*, 87–97. [CrossRef]

54. Zucker, R.M.; Decal, D.L.; Whittington, K.B. 5-Azacytidine increases the synthesis of embryonic hemoglobin (E2) in murine erythroleukemic cells. *FEBS Lett.* **1983**, *162*, 436–441. [CrossRef]

55. Ando, T.; Nishimura, M.; Oka, Y. Decitabine (5-Aza-2'-deoxycytidine) decreased DNA methylation and expression of *MDR-1* gene in K562/ADM cells. *Leukemia* **2000**, *14*, 1915–1920. [CrossRef] [PubMed]

56. Efferth, T.; Futscher, B.W.; Osieka, R. 5-Azacytidine modulates the response of sensitive and multidrug-resistant K562 leukemic cells to cytostatic drugs. *Blood Cells Mol. Dis.* **2001**, *27*, 637–648. [CrossRef] [PubMed]

57. Pleyer, L.; Greil, R. Digging deep into "dirty" drugs—Modulation of the methylation machinery. *Drug Metab. Rev.* **2015**, *47*, 252–279. [CrossRef] [PubMed]

58. Damaj, G.; Duhamel, A.; Robin, M.; Beguin, Y.; Michallet, M.; Mohty, M.; Vigouroux, S.; Bories, P.; Garnier, A.; El Cheikh, J.; et al. Impact of azacitidine before allogeneic stem-cell transplantation for myelodysplastic syndromes: A study by the Societe Francaise de Greffe de Moelle et de Therapie-Cellulaire and the Groupe-Francophone des Myelodysplasies. *J. Clin. Oncol.* **2012**, *30*, 4533–4540. [CrossRef] [PubMed]

59. Figueroa, M.E.; Skrabanek, L.; Li, Y.; Jiemjit, A.; Fandy, T.E.; Paietta, E.; Fernandez, H.; Tallman, M.S.; Greally, J.M.; Carraway, H.; et al. MDS and secondary AML display unique patterns and abundance of aberrant DNA methylation. *Blood* **2009**, *114*, 3448–3458. [CrossRef]

60. Papaemmanuil, E.; Gerstung, M.; Bullinger, L.; Gaidzik, V.I.; Paschka, P.; Roberts, N.D.; Potter, N.E.; Heuser, M.; Thol, F.; Bolli, N.; et al. Genomic Classification and Prognosis in Acute Myeloid Leukemia. *N. Engl. J. Med.* **2016**, *374*, 2209–2221. [CrossRef] [PubMed]

61. Dohner, H.; Estey, E.H.; Amadori, S.; Appelbaum, F.R.; Buchner, T.; Burnett, A.K.; Dombret, H.; Fenaux, P.; Grimwade, D.; Larson, R.A.; et al. Diagnosis and management of acute myeloid leukemia in adults: Recommendations from an international expert panel, on behalf of the European LeukemiaNet. *Blood* **2010**, *115*, 453–474. [CrossRef] [PubMed]

62. Cheson, B.D.; Greenberg, P.L.; Bennett, J.M.; Lowenberg, B.; Wijermans, P.W.; Nimer, S.D.; Pinto, A.; Beran, M.; de Witte, T.M.; Stone, R.M.; et al. Clinical application and proposal for modification of the International Working Group (IWG) response criteria in myelodysplasia. *Blood* **2006**, *108*, 419–425. [CrossRef] [PubMed]

63. R: The R Project for Statistical Computing. Available online: https://www.r-project.org/ (accessed on 12 April 2017).

© 2017 by the authors. Licensee MDPI, Basel, Switzerland. This article is an open access article distributed under the terms and conditions of the Creative Commons Attribution (CC BY) license (http://creativecommons.org/licenses/by/4.0/).

MDPI AG

St. Alban-Anlage 66

4052 Basel, Switzerland

Tel. +41 61 683 77 34

Fax +41 61 302 89 18

http://www.mdpi.com

IJMS Editorial Office

E-mail: ijms@mdpi.com

http://www.mdpi.com/journal/ijms

www.ingramcontent.com/pod-product-compliance
Lightning Source LLC
Chambersburg PA
CBHW051852210326

41597CB00033B/5872